现代科学思维视域下的数理问题研究

《天水师范学院60周年校庆文库》编委会｜编

光明日报出版社

图书在版编目（CIP）数据

现代科学思维视域下的数理问题研究 /《天水师范学院 60 周年校庆文库》编委会编．--北京：光明日报出版社，2019.9

ISBN 978-7-5194-5511-8

Ⅰ.①现… Ⅱ.①天… Ⅲ.①数学—文集②物理学—文集 Ⅳ.①O1-53②O4-53

中国版本图书馆 CIP 数据核字（2019）第 189327 号

现代科学思维视域下的数理问题研究

XIANDAI KEXUE SIWEI SHIYU XIA DE SHULI WENTI YANJIU

编　　者：《天水师范学院 60 周年校庆文库》编委会

责任编辑：郭玫君　　　　责任校对：赵鸣鸣

封面设计：中联学林　　　　责任印制：曹　诤

出版发行：光明日报出版社

地　　址：北京市西城区永安路 106 号，100050

电　　话：010-67078251（咨询），63131930（邮购）

传　　真：010-67078227，67078255

网　　址：http：//book.gmw.cn

E - mail：guomeijun@gmw.cn

法律顾问：北京德恒律师事务所龚柳方律师

印　　刷：三河市华东印刷有限公司

装　　订：三河市华东印刷有限公司

本书如有破损、缺页、装订错误，请与本社联系调换，电话：010-67019571

开　　本：170mm×240mm

字　　数：231 千字　　　　印　　张：18.5

版　　次：2019 年 9 月第 1 版　　　　印　　次：2019 年 9 月第 1 次印刷

书　　号：ISBN 978-7-5194-5511-8

定　　价：89.00 元

《天水师范学院60周年校庆文库》

编委会

总　序

春秋代序，岁月倥偬，弦歌不断，薪火相传。不知不觉，天水师范学院就走过了它60年风雨发展的道路，迎来了它的甲子华诞。为了庆贺这一重要历史时刻的到来，学校以“守正·奋进”为主题，筹办了缤纷多样的庆祝活动，其中“学术华章”主题活动，就是希冀通过系列科研活动和学术成就的介绍，建构学校作为一个地方高校的公共学术形象，从一个特殊的渠道，对学校进行深层次也更具力度的宣传。

《天水师范学院60周年校庆文库》（以下简称《文库》）是“学术华章”主题活动的一个重要构成。《文库》共分9卷，分别为《现代性视域下的中国语言文学研究》《“一带一路”视域下的西北史地研究》《“一带一路”视域下的政治经济研究》《“一带一路”视域下的教师教育研究》《“一带一路”视域下的体育艺术研究》《生态文明视域下的生物学研究》《分子科学视域下的化学前沿问题研究》《现代科学思维视域下的数理问题研究》《新工科视域下的工程基础与应用研究》。每卷收录各自学科领域代表性科研骨干的代表性论文若干，集中体现了师院学术的传承和创新。编撰之目的，不仅在于生动展示每一学科60年来学术发展的历史和教学改革的面向，而且也在于具体梳理每一学科与时俱进的学脉传统和特色优势，从而体现传承学术传统，发扬学术精神，展示学科建设和科学研究的成就，砥砺后学奋进的良苦用心。

《文库》所选文章，自然不足以代表学校科研成绩的全部，近千名教职员工，60年孜孜以求，几代师院学人的学术心血，区区九卷书稿300多篇文章，个中内容，岂能一一尽显？但仅就目前所成文稿观视，师院数十年科研的旧貌新颜、变化特色，也大体有了一个较为清晰的眉目。

首先，《文库》真实凸显了几十年天水师范学院学术发展的历史痕迹，为人们全面了解学校的发展提供了一种直观的印象。师院的发展，根基于一些基础老学科的实力，如中文、历史、数学、物理、生物等，所以翻阅《文库》文稿，可以看到这些学科及其专业辉煌的历史成绩。张鸿勋、雒江生、杨儒成、张德华……，一个一个闪光的名字，他们的努力，成就了天水师范学院科研的初始高峰。但是随着时代的发展和社会需求的变化，新的学科和专业不断增生，新的学术成果也便不断涌现，教育、政法、资环等新学院的创建自是不用特别说明，单是工程学科方面出现的信息工程、光电子工程、机械工程、土木工程等新学科日新月异的发展，就足以说明学校从一个单一的传统师范教育为特色的学校向一个兼及师范教育但逐日向高水平应用型大学过渡的生动历史。

其次，《文库》具体显示了不同历史阶段不同师院学人不同的学术追求。张鸿勋、雒江生一代人对于敦煌俗文学、对于《诗经》《尚书》等大学术对象的文献考订和文化阐释，显见了他们扎实的文献、文字和学术史基本功以及贯通古今、熔冶正反的大视野、大胸襟，而雍际春、郭昭第、呼丽萍、刘雁翔、王弋博等中青年学者，则紧扣地方经济社会发展做文章，彰显地域性学术的应用价值，于他人用力薄弱或不及处，或成就了一家之言，或把论文写在陇原大地，结出了累累果实，发挥了地方高校科学研究服务区域经济社会发展的功能。

再次，《文库》直观说明了不同学科特别是不同学人治学的不同特点。张鸿勋、雒江生等前辈学者，其所做的更多是个人学术，其长处是几十年如一日，埋首苦干，皓首穷经，将治学和修身融贯于一体，在学术的拓展之中同时也提升了自己的做人境界。但其不足之处则在于厕身僻地小校之内，单兵作战，若非有超人之志，持之以恒，广为求索，自是难以取得理想之成果。即以张、雒诸师为例，以其用心用力，原本当有远愈于今日之成绩和声名，但其诸多未竟之研究，因一人之逝或衰，往往成为绝学，思之令人不能不扼腕以叹。所幸他们之遗憾，后为国家科研大势和学校科研政策所改变，经雍际春、呼丽萍等人之中介，至如今各学科纷纷之新锐，变单兵作战为团队攻坚，借助于梯队建设之良好机制运行，使一人之学成一众之学，前有所行，后有所随，断不因以人之故废以方向之学。

还有，《文库》形象展示了学校几十年科研变化和发展的趋势。从汉语到外语，变单兵作战为团队攻坚，在不断于学校内部挖掘潜力、建立梯队的同时，学校的一些科研骨干如邢永忠、王弋博、令维军、李艳红、陈于柱等，也融入了更大和更高一级的学科团队，从而不仅使个人的研究因之而不断升级，而且也带动学校的科研和国内甚至国际尖端研究初步接轨，让学校的声誉因之得以不断走向更远也更高更强的区域。

当然，前后贯通，整体比较，缺点和不足也是非常明显的，譬如科研实力的不均衡，个别学科长期的缺乏领军人物和突出的成绩；譬如和老一代学人相比，新一代学人人文情怀的式微等。本《文库》的编撰因此还有另外的一重意旨，那就是立此存照，在纵向和横向的多面比较之中，知古鉴今，知不足而后进，让更多的老师因之获得清晰的方向和内在的力量，通过自己积极而坚实的努力，为学校科研奉献更多的成果，在区域经济和周边社会的发展中提供更多的智慧，赢得更多的话语权和尊重。

六十年风云今复始，千万里长征又一步。谨祈《文库》的编撰和发行，能引起更多人对天水师范学院的关注和推助，让天水师范学院的发展能够不断取得新的辉煌。

是为序。

李正元　安涛

2019 年 8 月 26 日

目　录
CONTENTS

伏羲卦图中的布尔代数

侯维民*

论述了布尔代数和伏羲卦图的相关结构，指出伏羲卦图中蕴含的布尔代数的诸多模型，为中西方文化交流提供了新的例证。

1701年，德国哲学家莱布尼茨忽然宣称他与鲍威特(中国名白晋)用二进制序数破译了伏羲卦图(八卦图与六十四卦图)的数学秘密.莱布尼茨在该图中发现，用阴爻"--"与阳爻"—"可以表示万有的这一配列顺序，竟可以与他在数学上的新发明即以0与1表示一切数的二进制序数互相印证.由此他说：伏羲已先我得到二进制序数的关键.

然而，莱布尼茨这一震惊中外的发现仅是伏羲卦图蕴含的一少部分内容.笔者发现伏羲卦图无论在卦的符号上，还是在图的结构上都与1847年创立，目前仍在蓬勃发展的"布尔代数"一致.这一事实说明伏羲卦图不仅蕴含着二进制数原理，而且蕴含了"布尔代数"的基本模型.

本文首先简介了布尔代数的基础知识，接着论证了伏羲八卦、六十四卦分别与三维、六维布尔代数结构的一致性.

1、布尔代数简介

布尔代数是以英国数学家乔治·布尔(George·Boole，1813—1864)命名的一种代数系统.它由一个非空集合B和定义在B上的二元运算$+$与$\cdot$构成.它们满足下列条件：(1)运算$+$和$\cdot$都是交换的；(2)每一个运算对于另一个运算而言都是分配的；(3)对于运算$+$和$\cdot$，有互异的零元素0和单位元素1，使得对于所有$\alpha \in B$，恒有$\alpha+0=\alpha$，$\alpha \cdot 1=\alpha$；(4)对于每个$\alpha \in B$，都存在一个元素$\bar{\alpha} \in B$，这个元

---

* 作者简介：侯维民(1947—)，男，河南卫辉人，天水师范学院数学与统计学院教授、学士，主要从事代数学及数学教育研究。

素称为 α 的补元素，它满足 $\alpha+\bar{\alpha}=1,\alpha\cdot\bar{\alpha}=0$. 这个代数系统记作 $(B,\cdot,+,^{-},0,1)$.

利用布尔代数的 $+,\cdot$，可以诱导出 B 的一种偏序关系：

$$\alpha\leqslant\beta\Leftrightarrow\alpha+\beta=\beta(\text{或 }\alpha\cdot\beta=\alpha).$$

由于对于所有的 $\alpha\in B$，均有 $\alpha+0=\alpha,\alpha\cdot 1=\alpha$，可知 0 是 B 的最小元，1 是 B 的最大元.

由两个元素组成的集合 $B_2=\{0,1\}$ 上的布尔代数最常用到，B_2 上的 $+,\cdot,^{-}$ 的定义为：

$$\begin{gathered}0+0=0,0+1=1,1+0=1,1+1=1;\\0\cdot 0=0,0\cdot 1=0,1\cdot 0=0;1\cdot 1=1\\\bar{0}=1,\bar{1}=0.\end{gathered}\tag{1}$$

这个代数系统记作 $(B_2,\cdot,+,^{-},0,1)$.

利用 B_2，可以构造 B_2^n：

$$B_2^n=\{(a_1,a_2,\cdots,a_n)\mid a_i\in B_2,i=1,2,\cdots,n\}$$

B_2^n 的元素叫做 n 维布尔向量，a_i 叫它的第 i 个分量. B_2^n 的 $+,\cdot,^{-}$ 定义如下：

$$\begin{gathered}\text{设 }\alpha=(a_1,a_2,\cdots,a_n),\beta=(b_1,b_2,\cdots,b_n)\\\text{则 }\alpha+\beta=(a_1+b_1,a_2+b_2,\cdots,a_n+b_n)\\\alpha\cdot\beta=(a_1\cdot b_1,a_2\cdot b_2,\cdots,a_n\cdot b_n)\\\bar{\alpha}=(\bar{a}_1,\bar{a}_2,\cdots,\bar{a}_n)\end{gathered}\tag{2}$$

其中每一 $a_i+b_i,a_i\cdot b_i,\overline{ai}$ 的结果由(1)式确定. 这个代数系统记作 $(B_2^n,\cdot,+,^{-},0_n,1_n)$. 由简单的验证可知，$(B_2^n,\cdot,+,^{-},0_n,1_n)$ 也构成布尔代数，它叫做 n 维布尔代数.

与伏羲八卦相互印证的是 B_2^3，B_2^3 中的元素有 $2^3=8$ 个，它们分别是：

$$\begin{gathered}(1,1,1),(0,1,1),(1,0,1),(0,0,1),\\(1,1,0),(0,1,0),(1,0,0),(0,0,0).\end{gathered}\tag{3}$$

由(2)规定的运算易知 $0_3=(0,0,0)$ 是 B_2^3 零元，$1_3=(1,1,1)$ 为 B_2^3 单位元. 并且(1,1,1)与(0,0,0)，(0,1,1)与(1,0,0)，(1,0,1)与(0,1,0)，(0,0,1)与(1,1,0)分别互补.

容易验证，B_2^3 的部分非空子集关于上述规定的 $+,\cdot,-$ 也构成布尔代数，其中与伏羲八卦图密切相关的有：

$$C_1=\{(1,1,1),(0,1,1)\},\overline{C}_1=\{(0,0,0),(1,0,0)\}$$

$$C_2=\{(1,0,1),(0,0,1)\},\overline{C}_2=\{(0,1,0),(1,1,0)\}$$

$$C_3=\{(1,1,1),(0,1,1),(1,0,1),(0,0,1)\}$$

$$\overline{C}_3=\{(0,0,0),(1,0,0),(0,1,0),(1,1,0)\}$$

这里 $\overline{C}_i$ 是 $C_i(i=1,2,3)$ 中每一元素的补元组成的集合，称 C_i 与 $\overline{C}_i$ 是互补的布尔代数.

B_2^3 中的元素可由以下两种方式建立全序：

1.1　字典排序法：$\alpha=(a_1,a_2,a_3)$ 与 $\beta=(b_1,b_2,b_3)$ 中，若 $a_3>b_3$，则规定 $\alpha>\beta$；若 $a_3=b_3,a_2>b_2$，则规定 $\alpha>\beta$；若 $a_3=b_3,a_2=b_2,a_1>b_1$ 则规定 $\alpha>\beta$.

用这种排序法可得 B_2^3 中的全序序列为：

$$(1,1,1)>(0,1,1)>\cdots>(0,0,0) \tag{4}$$

1.2　对应数码法：即先令 B_2^3 中的每一向量 (a_1,a_2,a_3) 对应经换算式

$$a_1\cdot 2^0+a_2\cdot 2^1+a_3\cdot 2^2$$

算出的数码，再按数码从大到小的顺序排出 B_2^3 中的全部元素. 用这种方法，(3) 中的元素依次对 7、6、5、4、3、2、1、0，故同样可得全序序列(4).

2、伏羲卦图蕴含的布尔代数

《周易·系辞传》说："古者伏羲氏之王天下也，仰则观象于天，俯则观法于地，观鸟兽之文与地之宜，近取诸身，远取诸物，于是始作八卦."这就是说八卦是从仰观天文，俯察地理，中知人事，综合天地人三才之道才画出来的.《周易·说卦》又说："立天之道，曰阴与阳；立地之道，曰柔与刚；立人之道，曰仁与义."由此可知天地人中的每一才又可分阴(阴，柔，仁) 阳(阳，刚，义) 两种状态.

若用"--"表示阴，"—"表示阳，上述含义用数学中的向量方法描述，即下述八种状况：

(--，--，--)，(—，--，--)，(--，--，—)，(—，--，—)，

(--，—，--)，(—，—，--)，(--，—，—)，(—，—，—).

把它们竖向排列就成为八卦：

"☷"(坤)，"☶"(艮)，"☳"(震)，"☲"(离)

"☵"(坎)，"☴"(巽)，"☱"(兑)，"☰"(乾).

若用 0 表示"--"，1 表示"—"，八卦依次对应下述八个布尔向量：

$$(0,0,0),(1,0,0),(0,0,1),(1,0,1)$$

$$(0,1,0),(1,1,0),(0,1,1),(1,1,1).$$

它们恰是三维布尔向量 B_2^3 中的全部元素.

不仅如此，伏羲八卦图(见图一) 与三维布尔代数的结构也很一致.

$(B_2^3,\cdot,+,{}^{-},0_3,1_3)$ 中，0_3 与 1_3 两向量特别重要，0_3 既是零元，又是最小元；1_3 既是单位元，又是最大元. 在伏羲八卦中，$0_3=(0,0,0)$ 对应的坤卦 ☷ 及 $1_3=(1,1,1)$ 对应的乾卦 ☰ 特别重要.《周易·系辞传》说："乾坤成列，而易立乎其中矣."《周易·说卦》说："天地定位." 在伏羲八卦图中，乾卦位于最上方，坤卦位于最下方. 这些都证实了乾坤的重要.《周易·说卦》说："天地定位，山泽通气，雷风相薄，水火不相射，八卦相错." 这表明八卦中有乾坤，艮兑，震巽，坎离四对卦象相错. 与之对应，$(B_2^3,\cdot,+,{}^{-},0_3,1_3)$ 中有四对向量互补，并且互补的向量对应的卦象相错. 如(0,1,0) 与(1,0,1) 互补，它们对应的坎 ☵ 与离 ☲ 相错，这说明布尔代数中的互补与伏羲八卦中的相错是同一意思. 布尔代数把互补作为定义中的必备条件，伏羲八卦图(图一) 把相错的卦象排在关于圆心对称的位置上.

图一　伏羲八卦图

B_2^3 可以规定全序，伏羲八卦也有全序. 将本文第一部分中用字典排序法和对应数码法建立的全序序列(4) 用对应的八卦表示，即得

☰ > ☱ > ☲ > ☳ > ☴ > ☵ > ☶ > ☷.

这就是伏羲八卦衍生次序图规定的八卦顺序：乾一，兑二，离三，震四，巽五，坎六，艮七，坤八. 从乾卦开始沿反时针方向(周易称为"往") 依次排列乾，兑，离，震，再在它们关于圆心对称的位置上排上各自相错的卦象就得到了图一所示的伏羲八卦图.

B_2^3 中有三对互补的布尔代数 C_1 与 $\overline{C}_1$，C_2 与 $\overline{C}_2$，C_3 与 $\overline{C}_3$. 这个性质体现在伏羲八卦图(图一) 上，则是同一布尔代数的向量对应的卦象依次相邻，互补的布尔代数的向量对应的卦象关于圆心相对. 例如，C_3 中的向量对应的卦象依次是：乾 ☰，兑 ☱，离 ☲，震 ☳，他们在八卦图中依次位于图形的左上侧，周易把这四个卦象称做阳卦. $\overline{C}_3$ 中的向量对应的卦象依次是坤 ☷，艮 ☶，坎 ☵，巽 ☴，它们依次位于图形的右下侧，周易把这四个卦象称做阴卦.

将伏羲八卦中的任意两个上下相重，就得到伏羲六十四卦. 正如《周易·说卦传》所说："兼三才而两之，故易六画而成卦." 例如，下卦取震 ☳，上卦取坤 ☷，合起来就是六十四卦的复 ䷗. 伏羲六十四卦圆图与方图(见图二) 分别给出了六十四卦各卦的卦象及卦名. 仍用 0 表示阴爻 --，用 1 表示阳爻 —，并把爻的自上而下的竖排变为 0,1 的自左到右的横排,则图二中自坤䷁到乾䷀的各卦依次对应(0,0,0,0,0,0),(1,0,0,0,0,0),…,(1,1,1,1,1,1),它们正是六维布尔代数 B_2^6. 其零元 0_6 即坤䷁对应的(0,0,0,0,0,0),单位元 I_6 即乾䷀对应的(1,1,1,1,1,1),B_2^6 中有 32 对元素互补,六十四卦中有 32 对卦象相错,并且互补的向量对应的卦象相错. 例如,(0,0,1,1,1,0) 与(1,1,0,0,0,1) 互补,它们对应的恒卦䷟与益卦䷩相错. 为叙述简捷,以下将六十四卦的各卦象同它们对应的布尔向量等同对待,将互补与相错等同对待.

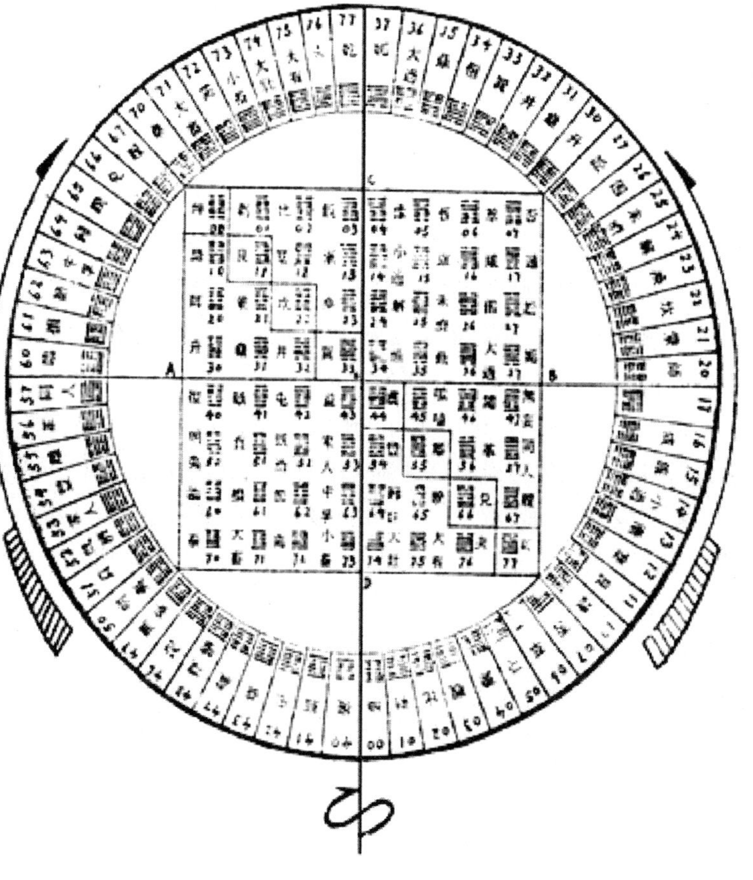

图二　六十四卦圆图与方图合图

伏羲六十四卦圆图与方图的构成,也可由对应数码法及字典排序法确定.

对应数码法是将各卦上卦对应的数码作为个位,下卦对应的数码作为“十”位,两个数码合起来作为该卦对应的八进制数码. 例如,䷖对应 02,䷋对应 07,则

䷟对应 72_8. 以对应数码的“十”位数确定各卦所在的行，个位数确定各卦所在的列，就可确定各卦在图二方图中的位置(注意首行(列)为第0行(列)). 从圆图下方的坤卦开始按反时针方向依次排列对应数码为00,01,…,37的各卦，再分别在它们关于圆心对称的位置上填上各自的相错卦，就可确定各卦在圆图中的位置. 六十四卦的字典排序法与八卦方法相仿.

在六十四卦方图中，相错的卦象位于与中心 o 对称的位置上. 各行的八个卦象都构成一个八元布尔代数. 从第0行开始，向下每两行，每四行分一组，各组的卦象分别构成一个十六元，三十二元布尔代数；并且其中互补的布尔代数所在的行与行中心线 AB 对称. 各行中，从位于第0列的卦象开始，向右每两个分一组，每四个分一组，各组的卦分别构成一个二元、四元布尔代数；并且互补的布尔代数位于与各行中点对称的位置上. 上述各元布尔代数均以其中对应数码最小的卦象为零元，最大的卦象为单位元. 列的情况亦然.

在六十四卦圆图中，相错的卦象位于与圆心对称的位置上. 自坤䷁起，沿反时针方向每两卦分一组，则各组都构成一个二元布尔代数；每四卦，每八卦，每十六卦，每三十二卦分一组，则各组分别构成一个四元、八元、十六元、三十二元布尔代数；并且同一布尔代数的卦象依次相邻，互补的布尔代数关于圆心相对. 上述各布尔代数均可像方图那样确定零元和单位元.

综上所述可知，伏羲八卦和六十四卦确实蕴含了布尔代数的基本模型.

伏羲卦图与布尔代数结构的一致性为中西方文化的交流，古代与现代文化的交流提供了新的线索. 例如：由现代人利用布尔代数研究逻辑推理，可以推想伏羲卦图很可能是我国先民研究逻辑规律的产物；而伏羲卦图的结构又为现代人研究布尔代数提供了各种各样的模型等.

(本文发表于1997年《周易研究》第3期总第三十三期)

置换群在多元多项式环因子分解中的应用

侯维民*

域上的多元多项式是单一分解环，但如何对其中的多项式因子分解却无一般方法可循.本文通过置换群对多元多项式的作用，给出了一类多元多项式的因子分解的一种方法.

1 引言

域上的多项式环是单一分解环，但如何对其中的多项式因子分解却无一般方法可循，文献[1～3]已对其中几类多项式的因子分解展开讨论，本文通过置换群对多元多项式的作用，给出了另一类多元多项式的因子分解的一种方法.

2 多项式的置换群

设 $X=(x_1,x_2,\cdots,x_n)$，则 X 的所有置换作成置换群

$$E(X)=\{\tau_1,\tau_2,\cdots,\tau_n\}$$

对于域 F 上的多项式环 $F(X)$ 上的多项式 $f(X)$，所有使 $f(X)=f(\tau_i(X))$ 的置换 τ_i 组成集合 H_f，不难证明 H_f 作成 $E(X)$ 的子群.把 H_f 叫做 $f(X)$ 的置换群，或者说 $f(X)$ 在 H_f 下不变.

例 1　设 $f(X)=x_1^4+x_1^2x_2^2+x_2^4+x_3^2$，则 $H_f=\{\tau_0=(x_1),\tau_1=(x_1,x_2)\}$.

式中：(x_1) 表示恒等置换，(x_1,x_2) 表示 x_1 与 x_2 的对换.

$f(X)$ 的置换群中有一类最为常见，它的元素使 X 中的未定元 $x_{i_1},x_{i_2},\cdots,x_{i_k}(1\leqslant k\leqslant n)$ 任意置换而不改变其它未定元.这类群显然与 k 次对称群 S_k 同构.以下称这样的群为关于 $x_{i_1},x_{i_2},\cdots,x_{i_k}$ 的对称群 S_k.

---

* 作者简介：侯维民(1947—)，男，河南卫辉人，天水师范学院数学与统计学院教授、学士，主要从事代数学及数学教育研究。

3 置换多项式和置换因式积

定义 1 设 $g(X)$ 是 $F(X)$ 中的次数大于零的多项式，H 是 $E(X)$ 的子群，把对 X 经过 H 中的每一置换 τ_i 作用得到的多项式 $g(\tau_i(X))$ 都叫 $g(X)$ 关于 H 的置换多项式.

例 2 令 $g(X)=x_1+x_2^2+3x_3^2$,

$$H=\{\tau_0=(x_1),\tau_1=(x_1,x_2)\},$$

则 $g(\tau_0(X))=x_1+x_2^2+3x_3^2, g(\tau_1(X))=x_2+x_1^2+3x_3^2$.

任一多项式关于群 H 的置换多项式具有如下性质.

性质 1 同一多项式的各置换多项式有相同的次数.

证明 设 $g(X)=\sum a_{k_1k_2\cdots k_n}x_1^{k_1}x_2^{k_2}\cdots x_n^{k_n}$，用 H 中的置换 τ 对 X 作用只改变各项未定元的下标，并不改变各项的次数，所以 $g(X)$ 和 $g(\tau(X))$ 有相同的次数.

性质 2 同一多项式的各置换多项式同时可约，或同时不可约.

证明 设 $\tau_i,\tau_j\in H$，$g(X)$ 的置换多项式 $g(\tau_i(X))$ 不可约，而 $g(\tau_j(X))$ 可约，则存在 $p(X),q(X)(\partial(p(X))>0,\partial(q(X))>0)$，使

$$g(\tau_j(X))=p(X)q(X)$$

由于 H 是有限群，必存在 τ_k，使 $\tau_k\tau_j=\tau_i$，从而

$$g(\tau_k\tau_j(X))=p(\tau_k(X))q(\tau_k(X)),$$

即 $g(\tau_i(X))=p(\tau_k(X))q(\tau_k(X))$.

又由性质 1 知 $\partial(p(\tau_k(X)))>0,\partial(q(\tau_k(X)))>0$，这与 $g(\tau_i(X))$ 不可约矛盾.

性质 3 同一不可约多项式的各置换多项式要么互不整除，要么相伴.

证明 设 $g(\tau_i(X))$ 与 $g(\tau_j(X))$ 都是不可约多项式 $g(X)$ 的置换多项式，若 $g(\tau_i(X))\mid g(\tau_j(X))$，由于它们有相同的次数，只能 $g(\tau_i(X))=cg(\tau_j(X))$，即二者相伴.

记 W 是不可约多项式 $g(X)$ 关于群 H 的所有置换多项式的集合，置换多项式的相伴显然是 W 元素间的一种等价关系. 这个等价关系决定 W 的一个分类，使得同一类内的多项式相伴，不同类间的多项式互不整除.

为证明性质 4，先给出

引理 1 设 $\tau=(x_i,x_j)$ 是 X 的任一对换，如果 $f(\tau(X))=-f(X)$，则 $(x_i-x_j)\mid f(X)$.

证明 见参考文献[4].

性质4　设 S_k 是关于 $x_{i_1},\cdots,x_{i_l},\cdots,x_{i_t},\cdots,x_{i_k}$ 的对称群，不可约多项式 $g(X)\neq a(x_{i_l}-x_{i_t})$，$(x_{i_l},x_{i_t})\in S_k$，则 $g(X)$ 关于 S_k 的所有互不相伴的置换多项式的积在 S_k 下不变.

证明　从 $g(X)$ 关于 S_k 的置换多项式集合 W 的各类中分别选取一个代表元

$$g(\tau_0(X)),g(\tau_1(X)),\cdots,g(\tau_s(X)),$$

其中：τ_0 是 S_k 的恒等置换.

$$令\ f(X)=g(\tau_0(X))g(\tau_1(X))\cdots g(\tau_s(X)),\tag{1}$$

对于 S_k 的任一对换 $\tau=(x_{i_l},x_{i_t})$，有

$$f(\tau(X))=g(\tau\tau_0(X))g(\tau\tau_1(X))\cdots g(\tau\tau_s(X)).\tag{2}$$

(2) 式中各因式仍分别属于 W 的各类，否则若有

$$g(\tau\tau_{j_0}(X))=cg(\tau\tau_{i_0}(X))$$

由 $g(\tau^{-1}\tau\tau_{j_0}(X))=cg(\tau^{-1}\tau\tau_{i_0}(X))$ 得

$$g(\tau_{j_0}(X))=cg(\tau_{i_0}(X))$$

即 $g(\tau_{j_0}(X))$ 与 $g(\tau_{i_0}(X))$ 位于 W 的同一类，这与假设矛盾，于是必有

$$g(\tau\tau_m(X))=c_pg(\tau_p(X))(0\leqslant m,p\leqslant s)\tag{3}$$

将(3) 的各式代入(2)，并且令 $c=c_0c_1\cdots c_s$，得

$$f(\tau(X))=cf(X).\tag{4}$$

又由 $f(\tau^{-1}\tau(X))=cf(\tau^{-1}(X))$ 及 $\tau^{-1}=\tau$，得

$$f(X)=cf(\tau(X)).\tag{5}$$

将(5) 代入(4) 得 $c^2=1$，即 $c=\pm 1$.

当 $c=-1$ 时，(4) 式变为 $f(\tau(X))=-f(X)$. 引理1，$f(X)$ 必有因式(x_{i_l},x_{i_t})，这与 $g(X)\neq a(x_{i_l},x_{i_t})$ 矛盾，所以 $c=1$，即 $f(\tau(X))=f(X)$

由于 τ 是 S_k 中的任意对换，而 S_k 中的置换又都可以表成对换的乘积，所以 $f(X)$ 在 S_k 下不变.

性质5　若 $g(X)=a(x_{i_l},x_{i_t})$，$(x_{i_l},x_{i_t})\in S_k$，则 $g(X)$ 关于 S_k 的所有置换多项式的积 $\prod\limits_{1\leqslant l\neq t\leqslant k}a(x_{i_l}-x_{i_t})$ 在 S_k 下不变.

证明　令 $f(x)=\prod\limits_{1\leqslant l\neq t\leqslant k}a(x_{i_l}-x_{i_t})$，取 $\tau(x_{i_c}-x_{i_d})\in S_k$，将 $f(X)$ 改写成

$$f(X)=a^2(x_{i_c}-x_{i_d})(x_{i_d}-x_{i_c})\prod_{\substack{1\leqslant b\leqslant k\\ b\neq c,d}}a^4(x_{i_c}-x_{i_b})(x_{i_b}-x_{i_c})$$

$$\cdot(x_{i_d}-x_{i_b})(x_{i_b}-x_{i_d})\cdot\prod_{\substack{1\leq l\neq t\leq k\\ l,t=c,d}}a(x_{i_l}-x_{i_t})\tag{6}$$

则由于

$$f(\tau(X)) = a^2(x_{i_d} - x_{i_c})(x_{i_c} - x_{i_d}) \prod_{\substack{1 \leqslant b \leqslant k \\ b \neq c,d}} (x_{i_d} - x_{i_b})(x_{i_b} - x_{i_d})$$
$$\cdot (x_{i_c} - x_{i_b})(x_{i_b} - x_{i_c}) \cdot \prod_{\substack{1 \leq l \neq t \leq k \\ l,t = c,d}} a(x_{i_l} - x_{i_t}) \tag{7}$$

对照(6)式与(7)式,可得 $f(\tau(X)) = f(X)$. 由 S_k 中对换 τ 的任意性,性质得证.

定义 2　设不可约多项式 $g(X) \neq a(x_{i_l} - x_{i_t})$,把 $g(X)$ 关于 S_k 的所有互不相伴的置换多项式的积 $\prod\limits_{S_k} g(X)$ 叫它关于 S_k 的不相伴置换因式积,而把 $a(x_{i_l} - x_{i_t})$ 关于 S_k 的全部置换多项式的积 $\prod\limits_{1 \leqslant l \neq t \leqslant k} a(x_{i_l} - x_{i_t})$ 叫它关于 S_k 的完全置换因式积.

4　多元多项式因子分解定理

引理 2　设 $f(X)$,$g(X)$ 都在 S_k 下不变,若 $f(X) = g(X)h(X)$,则 $h(X)$ 也在 S_k 下不变.

证明　由

$$f(X) = g(X)h(X), \tag{8}$$

任取 $\tau \in S_k$,得

$$f(\tau(X)) = g(\tau(X))h(\tau(X)). \tag{9}$$

但已知 $f(\tau(X)) = f(X)$,$g(\tau(X)) = g(X)$,对照(8)式与(9)式得 $h(X) = h(\tau(X))$. 即 $h(X)$ 也在 S_k 下不变.

定理　设在关于 $x_{i_1},\cdots,x_{i_l},\cdots,x_{i_t},\cdots,x_{i_k}$ 的置换群 S_k 下不变的多项式 $f(X)$ 可约,其不可约分解式中只可能含有关于 S_k 的不相伴置换因式积 $\prod\limits_{S_k} g(X)$($g(X) \neq a(x_{i_l} - x_{i_t})$),及关于 S_k 的完全置换因式积 $\prod\limits_{1 \leqslant l \neq t \leqslant k} a(x_{i_l} - x_{i_t})$.

证明　设 $f(X)$ 可约,则存在不可约多项式 $g(X)$,使

$$f(X) = g(X)h(X)(\partial(g(X)) > 0, \partial(h(x)) > 0). \tag{10}$$

(Ⅰ)当 $g(X) \neq a(x_{i_l} - x_{i_t})$ 时,设 $g(X)$. 关于 S_k 的所有互不相伴的置换多项式分别为 $g(X)$,$g(\tau_1(X))$,$\cdots$,$g(\tau_s(X))$,由置换多项式的性质 3,它们也是互不整除的. 由(10)式得

$$f(\tau_1(X)) = g(\tau_1(X))h(\tau_1(x)), \tag{11}$$

又已知 $f(\tau_1(X)) = f(X)$,由(10),(11)两式得

$$g(\tau_1(X))h(\tau_1(X)) = g(X)h(X),$$

但 $g(\tau_1(X)) \nmid g(X)$,只能 $g(\tau_1(X)) \mid h(X)$. 令

$$h(X)=g(\tau_1(X))h_1(X), \tag{12}$$

将(12)代入(10),得

$$f(X)=g(X)g(\tau_1(X))h_1(X).$$

同理,由 $f(\tau_2(X))=g(\tau_2(X))h_2(\tau_2(X))$ 及 $f(\tau_2(X))=f(X)$,得

$$f(X)=g(X)g(\tau_1(X))g(\tau_2(X))h_2(X).$$

这样继续下去,可得

$$f(X)=g(X)g(\tau_1(X))\cdots g(\tau_S(x))U(x)=\prod_{S_k}g(X)u(X). \tag{13}$$

由置换多项式性质 4,$\prod\limits_{S_k}g(X)$ 在 S_k 下不变,又由引理 2,$u(X)$ 也在 S_k 下不变.

(Ⅱ)当 $g(X)=a(x_{i_l}-x_{i_t})$ 时,设 $g(X)$ 关于 S_k 的所有互不相伴的置换多项式分别是 $a(x_{i_1}-x_{i_2}),a(x_{i_1}-x_{i_3}),\cdots,a(x_{i_{r-1}}-x_{i_r})$,由(Ⅰ)的证明可知,$\prod\limits_{1\leqslant l<t\leqslant k}a(x_{i_l}-x_{i_t})$ 是 $f(X)$ 的因式. 以下证明若 $a(x_{i_l}-x_{i_t})$ 是 $f(X)$ 的因式,$a^2(x_{i_l}-x_{i_t})(x_{i_t}-x_{i_l})$ 也是 $f(X)$ 的因式.

设 $f(X)=a(x_{i_l}-x_{i_t})h(X)$,取 $\tau=(x_{i_l},x_{i_t})$,
则 $f(\tau(X))=a(x_{i_t}-x_{i_l})h(\tau(X))=-a(x_{i_l}-x_{i_t})h(\tau(X))$.
由 $f(\tau(X))=f(X)$ 得 $h(\tau(X))=-h(X)$,从而 $a(x_{i_l}-x_{i_t})$ 是 $h(X)$ 的因式,$a^2(x_{i_l}-x_{i_t})(x_{i_t}-x_{i_l})$ 是 $f(X)$ 的因式. 因此可设

$$f(X)=\prod_{1\leqslant l\neq t\leqslant k}a(x_{i_l}-x_{i_t})v(X)$$

由置换多项式性质 5,可知 $\prod\limits_{1\leqslant l\neq t\leqslant k}a(x_{i_l}-x_{i_t})$ 在 S_k 下不变,再由引理 2,可知 $v(X)$ 也在 S_k 下不变.

在以上两种情况的讨论中,若(Ⅰ)中的 $u(X)$,(Ⅱ)中的 $v(X)$ 是不可约多项式,此时 $u(X),v(X)$ 关于 S_k 的各置换多项式相同,定理成立. 否则按以上两种情况继续讨论. 最后可得 $f(X)$ 的典型分解式

$$f(X)=\Big(\prod_{1\leqslant l\neq t\leqslant k}a(x_{i_l}-x_{i_t})\Big)^m\Big(\prod_{S_k}g_1(X)\Big)^{\alpha_1}\cdots\Big(\prod_{S_k}g_i(X)\Big)^{\alpha_i}, \tag{14}$$

于是定理得证.

例 3　在复数域上分解

$$x_2^2x_3+x_2x_3^2+x_2^2x_4+x_2x_4^2+x_3^2x_4+x_3x_4^2+2x_2x_3x_4$$
$$-x_1(3x_2x_3+3x_2x_4+2x_3x_4-x_2^2-x_3^2-x_4^2)+2x_1^2(x_2+x_3+x_4). \tag{15}$$

解　原式的置换群是关于 x_2,x_3,x_4 的对称群 S_3. 先令 $x_1=0$,原式变为

$$x_2^2x_3+x_2x_3^2+x_2^2x_4+x_2x_4^2+x_3^2x_4+x_3x_4^2+2x_2x_3x_4 \tag{16}$$

令 $x_2=-x_3$，(16) 式为 0，从而 (x_2+x_3) 是(16) 式的因式. 考虑到(15) 式是 x_1,x_2,x_3,x_4 的三次齐次，$(x_1+x_2+x_3)$ 有可能为(15) 式的因式，并且若它是(15) 式的因式，由本定理关于 S_3 的不相伴置换因式积 $\prod_{S_3}(x_1+x_2+x_3)$ 也一定是(15) 式的因式.

$$\text{令原式} = l(x_1+x_2+x_3)(x_1+x_2+x_4)(x_1+x_3+x_4),$$

比较两边系数得 $l=1$，分解完毕.

例 4　求证 $x_1^2+x_2^2+x_3^2+x_1x_2+x_1x_3+x_2x_3$ 在复数域上不可约.

证明　原式的置换群是关于 x_1,x_2,x_3 的对称群 S_3. 注意到齐次多项式的特点，原式的不可约分解式中只可能含有关于 S_3 的不相伴置换因式积 $\prod_{S_3}(x_1+x_2+x_3)$，$\prod_{S_3}(x_1+x_2)$，以及关于 S_3 的完全置换因式积 $\prod_{1\leqslant i\neq j\leqslant 3}(x_i-x_j)$. 但 $\prod_{1\leqslant i\neq j\leqslant 3}(x_i-x_j)$ 的次数为 $A_3^2=6$，$\prod_{1\leqslant i\neq j\leqslant 3}(x_1+x_2)$ 的次数为 $C_3^2=3$，故 $\prod_{1\leqslant i\neq j\leqslant 3}(x_i-x_j)$，$\prod_{S_3}(x_1+x_2)$ 均不可能是原式的因式.

将 $x_1=-x_2-x_3$ 代入原式，原式不为 0，故 $x_1+x_2+x_3$ 也不是原式的因式. 所以原式在复数域上不可约.

参考文献

[1]王瑞. 惟一分解整环上不可约多项式的若干结构[J]. 数学研究与评论，1999，19(2)：367－373.

[2]张传林. 有理数域上多项式因子分解的一个注记[J]. 杭州大学学报(自然科学版)，1997，24(4)：277－280.

[3]姜同松，张春国. R 上含有一个参数的 n 元二次多项式因式分解[J]. 淮北煤师院学报，1997，18(1)：8－10.

[4]程汉普. 数与式[M]. 成都：四川人民出版社，1984：194.

(本文发表于 2000 年《兰州大学学报(自然科学版版)》第 36 卷第 6 期)

返回式框图学习法

潘书林*

返回式框图学习法是在每学到一个重要概念后，依托框图，返回思辨它与以前学过的重要概念间的关系，在辩证唯物主义的基本观点、脑科学的研究成果的启示下，尝试用返回式框图学习法学习数学，从发展中看联系，在联系的基础上求发展，利用该方法可以让学生体味学习的成功感，进而增强学习兴趣，达到减负增效、学会学习的目的。

随着科技的迅猛发展，综合国力的竞争日趋激烈。国运兴衰，系于教育；教育振兴，全民有责。21 世纪要求每一个人都要学会学习，指导每一个学生学会学习，是摆在每位教师面前的严峻问题。

返回式框图学习法，是在每学到一个重要概念后，依托框图，返回思辨它与以前学过的重要概念特别是作为本课程研究的主要对象的概念的关系，从中获得启发以至创造。

世界是普遍联系和永恒发展的，它是由纵横交错的联系和发展构成的过程的集合体。联系、发展、过程，是客观世界的基本特征，也是唯物辩证法的基本思想。而科学地论证世界的普遍联系，则是唯物辩证法的逻辑起点[1]。

主要矛盾是在一个矛盾体系中居于支配地位、对事物的发展过程起决定作用的矛盾。抓住主要矛盾，有利于理清事物发展过程的主要线索。

脑科学的研究证明演绎推理的心理模型的存在，理模型理论对教育的启示是：改善我们现有课程知识的呈现方式，课程知识不仅以语言和符号的形式呈现，而且以图形的方式呈现，可能既激发 2 种工作记忆的并行编码，又帮助演绎推理的进行[2]。

---

* 作者简介：潘书林（1946—），男，甘肃天水人，天水师范学院数学与统计学院教授、学士，主要从事师范院校学习论的研究与实践。

兴趣是最好的老师,但不是每个学生对学习数学都感兴趣。兴趣需要培养,使原来没有兴趣的人有兴趣,使原来有兴趣的人的兴趣更加稳固和持久,以至升华为对数学的热爱。学生通过运用新知识解决原来不能解决的问题或解决问题的方法比原来更有效,从而体味到成功的乐趣,是培养兴趣进而升华为对数学的热爱的一个有效方法。只有热爱数学才能在学习数学中不屈不挠,百折不回。在辩证唯物主义的基本观点、脑科学的研究成果的启示下,尝试用返回式框图学习法学习数学,从发展中看联系,在联系的基础上求发展,抓住了主要矛盾,实现“从薄到厚”又“从厚到薄”的辩证的读书过程,使学生及时体味学习的成功感,完成了从被动接受到主动探索的转化,达到“增效减负”、学会学习的目的。

下面以一元微积分的学习为例,说明返回式框图学习法的应用。如图 1:

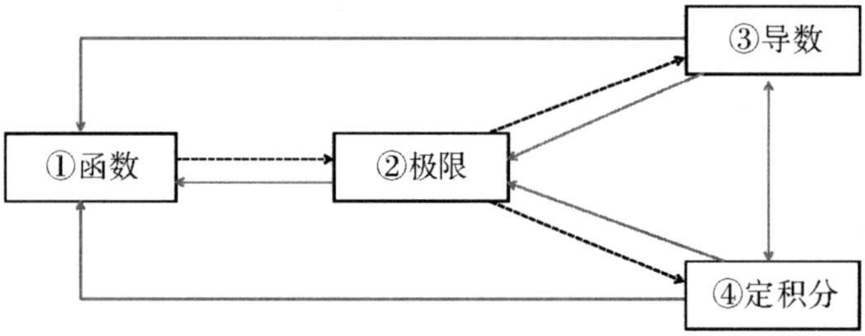

图 1　返回式框图学习法的应用

首先学习函数的概念①f:D→M,即函数是从实数集 D 到实数集 M 的映射,它是本课程研究的对象。

从①出发,观察函数当自变量在一个无限变化过程中,因变量变化的趋势,就产生了极限的概念②。

②→①,用极限返回研究函数,若极限值等于函数值,则函数在该点连续。进而把函数分成连续函数与非连续函数。有时判断一个函数是连续函数十分容易,而连续函数的极限是很好求的。另外,若函数的图像有渐近线,可以用相应函数的极限求得。有了渐近线,函数图像无限伸展的部分将了如指掌。

从②出发,考察函数增量比的极限得到导数的概念③,应用导数可以方便地求得曲线的切线和法线。

③→①,利用导数返回研究函数,考察函数增量与导数的关系便产生了微分的概念。因为求导是一种运算,研究其逆运算,就得到不定积分的概念。特别借助微分中值定理对函数性态的研究空前深入,可导函数或除有限个点以外可导的函数的作图问题彻底解决,函数的两点一线(极值点、拐点、渐近线)尽收眼底,函数的变化趋势一目了然。

③→②，利用导数返回研究函数的极限，方法又有新的突破。罗比塔法则求不定式的极限势如破竹，泰勒公式又是求极限的一把利剑。

再从②出发，考虑函数积分和的极限，得到定积分的概念④。它在几何与物理学上有许多应用。

④→①，利用定积分返回研究函数，讨论函数的可积性，可积的必要条件，充要条件，充分条件。特别变上限的定积分是上限的函数，它是函数家族中的重要成员，使我们对函数的认识又进了一步。

④→②，利用定积分研究函数的极限，产生了用定积分的定义求数列极限的方法。

④→③，当被积函数连续时，借助积分中值定理，推得变上限的定积分所确定的函数可导，这就是著名的微积分学基本定理，它能方便地推出牛顿—莱布尼兹公式。使得求定积分的问题变成求被积函数的一个原函数在积分区间的增量问题，其简便程度与用定积分定义求定积分不可同日而语。

由以上返回式框图学习法看出，一元函数的微积分学包括 7 个基本概念，即函数、极限、连续、导数、微分、不定积分、定积分。而连续是极限派生的概念，微分、不定积分又是导数派生的概念，于是函数、极限、导数、定积分这 4 个概念的重要性就凸显出来。又因为导数是函数增量比极限，定积分是函数积分和极限，所以函数、极限就成了最重要的概念，前者是我们研究的对象，后者是我们研究的工具，一句话，一元微积分学是利用极限研究一元函数的学问。真正理解这一句话可谓把书读薄了。

参考文献

[1]国家教委社科司.马克思主义原理[M].北京：高等教育出版社，1993:85.
[2]吴刚.从脑科学看素质教育[N].北京：中国教育报，2000－02－14(4).

（本文发表于 2001 年《数学教育学报》第 10 卷第 2 期）

关于可估函数 LS 估计相合条件的一个问题

杨复兴*

设有线性回归 $Y_i = x_i'\beta_0 + e_i, 1 \le i \le n, n \ge 1$. 本文在 $S_n = \sum_{i=1}^{n} x_i x_i'$ 的逆可以不存在的情况下，对任一特定的可估函数 $c'\beta_0$，给出了其 LS 估计为相合的充要条件.

1. 引理及主要结果

设有线性回归模型：

$$Y_i = x_i'\beta_0 + e_i, 1 \le i \le n, n \ge 1 \tag{1}$$

其中 $x_1, x_2 \cdots x_n$ 是已知的 P 维向量 $\beta_0 = (\beta_{01}, \beta_{02} \cdots \beta_{0p})'$ 是未知的 p 维回归系数向量，$e_1, e_2 \cdots e_n$ 是随机误差，满足 $Ee_i = 0, i \ge 1$. β_0 的 Ls 估计记为 $\hat{\beta}_n$. 对已知的 P 维常向量 c，$c'\beta_0$ 的 Ls 估计就是 $c'\hat{\beta}_n$. 以 $\overline{\lambda_n}$ 和 $\underline{\lambda_n}$ 记 $Cov(e_1, e_2 \cdots e_n)$ 的最大和最小特征根，当 $\overline{\lambda_n} = \underline{\lambda_n}$ 时，$Cov(e_1, e_2 \cdots e_n) = \sigma^2 I_n$，这时称 $e_1, e_2 \cdots e_n$ 满足 Gauss—Markov(GM)条件.

Drygas 在[1]中证明了：若 $0 < \inf_n \underline{\lambda_n} < \sup_n \overline{\lambda_n} < \infty$，以 ρ_n 记 $S_n = \sum_{i=1}^{n} x_i x_i'$ 的最小非 0 特征根，则“对一切 P 维向量 c，当 $c'\beta_0$ 可估时，$c'\hat{\beta}_n$ 为 $c'\beta_0$ 的弱相合估计”的充要条件是 $\rho_n \to \infty$，当 $n \to \infty$ 时。陈希孺等在[2]中，用另一种方法证明：在 GM 条件下，当 S_n^{-1} 存在（对充分大的 n）时，对任一特定的 P 维向量 c，$c'\hat{\beta}_n$ 为 $c'\beta_0$ 的弱相合估计的充要条件是

$$\lim_{n \to \infty} c' S_n^{-1} c = 0 \tag{2}$$

---

* 作者简介：杨复兴(1948—)，男，河南偃师人，天水师范学院数学与统计学院教授、硕士，主要从事概率统计方面的研究。

其证明对 $0 < \inf_n \underline{\lambda_n} < \sup_n \overline{\lambda_n} < \infty$ 的更一般的情况也有效. 又在上述两个结果中,

均方相合与弱相合等价,即所述条件也是均方相合的充要条件。

[1]中的结果是针对全体的可估函数. 对一个特定的 c, $c'\beta_0$ 可估,则 $\rho_n \to \infty$ 只是 $c'\hat{\beta}_n$ 弱相合的充分而非必要之条件. [2]中的结果虽是针对特定的 c,但其中假定了 S_n^{-1} 存在. 而如所周知,即使对任何 n, S_n^{-1} 不存在,并不妨碍对某些特定的可估函数 $c'\beta_0$ 的 Ls 估计 $c'\hat{\beta}_n$ 为相合. 所以,以上两个结果都缺失了一点:如不假定 S_n^{-1} 存在,对一个特定的 c, $c'\beta_0$ 可估,则 $c'\hat{\beta}_n$ 相合的条件如何. 本文回答了这个问题,结果是:

定理 1.1 沿用前面的记号,并假定 $0 < \inf_n \underline{\lambda_n} < \sup_n \overline{\lambda_n} < \infty$,设 $c'\beta_0$ 可估,则 $c'\hat{\beta}_n$ 为 $c'\beta_0$ 的(弱和均方)相合估计的充要条件是

$$\sum_{i=1}^{\infty} (x_i'\alpha)^2 = \infty, \text{对任何满足条件 } c'\alpha \neq 0 \text{ 的 } \alpha. \tag{3}$$

2. 一个引理

定理 1.1 的证明依赖下述具有独立意义的引理.

引理 2.1 设 $\{f_n(\beta), n = 1,2,\cdots\}$ 是一串定义在 R^p 上的连续函数,满足条件:

(1)非降性: $f_1(\beta) \leq f_2(\beta) \leq \cdots$, 对一切 $\beta \in R^p$;

(2)对某个 n_0, 有 $f_{n_0}(\beta) \to \infty$, 当 $\|\beta\| \to \infty$;

(3) $\lim_{n\to\infty} f_n(\beta) = \infty$, 对任何 $\beta \in R^p$,

则当 $n \to \infty$ 时,对 $\beta \in R^p$ 一致地有 $\lim_{n\to\infty} f_n(\beta) = \infty$.

证明:任给 $M > 0$, 据(1),(2)知:当 $n \geq n_0$ 时,有 $f_n(\beta) \to \infty$, 当 $\|\beta\| \to \infty$. 定义

$$A_n = \{\beta: f_n(\beta) \leq M\}, n \geq 1$$

由(1)知 $A_1 \supset A_2 \supset \cdots$. 由于当 $n \geq n_0$ 时, $f_n(\beta) \to \infty$, 当 $\|\beta\| \to \infty$, 知 A_n 当 $n \geq n_0$ 时为有界集,又因 f_n 连续,知 A_n 为闭集. 因此,只有两种可能的情况

a. 对一切 n, $A_n \neq \varphi$, 这时 $\bigcap_{n=1}^{\infty} A_n \equiv A \neq \varphi$;

b. 对某个 n, 有 $A_n = \varphi$, 这时 $A_m = \varphi$, 当 $m \geq n$ 时(n 可与 M 有关).

情况 a 不可能,因若不然,取 $a \in A$, 则将有 $f_n(a) \leq M$ 对一切 n, 与(3)矛盾. 故有 $A_m = \varphi$, 当 $m \geq n$, 即 $f_m(\beta) \geq M$ 对一切 $m \geq n$ 及 $\beta \in R^p$, 由 M 的任意性,

证明了引理的结论。

3. 定理 1.1 的证明

先考虑 $c=(1,0.\cdots 0)'$ 的特例,即 $c'\beta_0=\beta_{01}$,而 $c'\alpha\neq 0$,$\alpha=(\alpha_1,\alpha_2\cdots\alpha_p)'$,归结为 $\alpha_1\neq 0$,故不失普遍性,取 $\alpha_1=1$,记

$$x_i=(x_{1i},x_{2i}\cdots x_{pi})',\ i\geq 1;$$

$$\gamma_{jn}=(x_{j1},x_{j2}\cdots x_{jn})',\ j=1,2,\cdots p;$$

$$\delta=(\alpha_1,\alpha_2\cdots\alpha_p)';$$

$$\Gamma_n=(\gamma_{2n},\cdots\gamma_{pn}),$$

$$M_n=\gamma_{2n},\cdots\gamma_{pn}\ \text{生成的线性子空间}.$$

在这些记号下,条件(3)转化为

$$f_n(\delta)\equiv\|\gamma_{1n}-\Gamma_n\delta\|^2\rightarrow\infty,\ \text{对任何}\ \delta\in R^{p-1}\ . \tag{4}$$

以 ξ_n 记 γ_{1n} 在 M_n 上的投影,而 $\eta_n=\gamma_{1n}-\xi_n\equiv\{h_{n1},\cdots h_{nn}\}'$,在[2](见[2] $p_{37}-p_{39}$)中证明了

$$\hat{\beta}_{01}-\beta_{01}=\sum_{i=1}^{n}h_{ni}e_i/h_n,\ h_n=\sum_{i=1}^{n}h_{ni}^2\ . \tag{5}$$

[2]在证明(5)式时附加了条件 S_n^{-1} 存在.但容易看出,(5)式之成立不依赖这一条件.因为,若 Γ_n 之秩为 $r-1$,则在 $\gamma_{2n},\cdots\gamma_{pn}$ 中选 $r-1$ 个线性无关的向量,不妨假定就是 $\gamma_{2n},\cdots\gamma_{rn}$,令 $X_n=\{\gamma_{1n},\gamma_{2n},\cdots\gamma_{rn}\}$,可将模型(1)改写为

$$Y_{(n)}=X_n\beta_0^*+e_{(n)},\ Y_{(n)}=(Y_1,Y_2,\cdots Y_n)',\ e_{(n)}=(e_1,e_2,\cdots e_n)' \tag{6}$$

而 β_0^* 为 r 维向量,其第一分量 β_{01}^* 就是 β_{01},(这用到 β_{01} 可估,因而 $\gamma_{1n}\notin M_n$),因为 $\gamma_{1n},\gamma_{2n},\cdots\gamma_{rn}$ 线性无关,知 $(X_n'X_n)^{-1}$ 存在.故对模型(6),β_{01}^* 的 LS 估计,也即 β_{01} 的 LS 估计,适用(5)式.但现在 $(h_{n1},\cdots h_{nn})'$ 要理解为 $\gamma_{1n}-\xi_n^*$,其中 ξ_n^* 为 γ_{1n} 在 $M_n^*=\gamma_{2n}\cdots\gamma_{rn}$ 生成的线性子空间上的投影.由 $M_n^*=M_n$ 及投影的唯一性,知在新意义下 $\{h_{n1},\cdots h_{nn}\}'\equiv\gamma_{1n}-\xi_n$,即与原来的 $\{h_{n1},\cdots h_{nn}\}'$ 一致,这就证明了(5)式。

在 $0<\inf\limits_n\underline{\lambda_n}<\sup\limits_n\overline{\lambda_n}<\infty$ 的条件下,按]2] $p_{41}-p_{44}$ 的证法,易见 β_{01}^* 为弱相合及均方相合的充要条件都是

$$\lim_{n\to\infty}h_n=\infty \tag{7}$$

而据 h_n 及(4)式的 f_n 的定义,(7)式的意义是

$$\inf_{\delta\in R^{p-1}}f_n(\delta)\rightarrow\infty,\ \text{当}\ n\rightarrow\infty \tag{8}$$

据引理 1.1,(8)式等价于(4)式,这就在 $c=(1,0.\cdots 0)'$ 的特例下证明了定理

1 的结论.

现在考虑一般情况,设 $c'\beta_0$ 可估.若 $c=0$,定理不证自明.故设 $c\neq 0$,作非异方阵 D,其第一行为 c',令 $\theta_0=(\theta_{01},\cdots\theta_{0p})'=D\beta_0$,$\tilde{x}_i=(D^{-1})'x_i$,而将模型(1)写为

$$Y_i=\tilde{x}_i'\theta_0+e_i,1\leq i\leq n,n\geq 1$$

因 $\theta_{01}=c'\beta_0$,按已证部分,记 θ_0 的 LS 估计为 $\hat{\theta}_n=(\hat{\theta}_{n1},\cdots\hat{\theta}_{np})'$,有

$c'\hat{\beta}_n$ 相合 $\Leftrightarrow \hat{\theta}_{n1}$ 相合 $\Leftrightarrow \sum_{i=1}^{\infty}\|\tilde{x}_i'\alpha\|^2=\infty$,当 α 的第一分量 $\neq 0$

$\Leftrightarrow \sum_{i=1}^{\infty}\|x_i'D^{-1}\alpha\|^2=\infty$,当 α 的第一分量 $\neq 0$

$\Leftrightarrow \sum_{i=1}^{\infty}\|x_i'\beta\|^2=\infty$,当 β 的第一分量 $\neq 0$

$\Leftrightarrow \sum_{i=1}^{\infty}\|x_i'\beta\|^2=\infty$,当 $c'\beta\neq 0$

最后一步用到 D 的第一行为 c'.定理 1 证毕.

例:$p=3$,$x_{2i+1}=(1,1,2)'$,$x_{2i+2}=(1,-1,0)'$,$i=0,1.\cdots$,问 $c'\beta=c_1\beta_1+c_2\beta_2+c_3\beta_3$ 的 LS 估计何时为相合? 按定理 1,施加在 c_1,c_2,c_3 上的(充要)条件为

$$c_1\alpha_1+c_2\alpha_2+c_3\alpha_3\neq 0\Rightarrow|\alpha_1-\alpha_2|+|\alpha_1+\alpha_2+2\alpha_3|\neq 0,$$

容易看出,此条件等价于 $c_1+c_2=c_3$.

注:不难验证,在 S_n^{-1} 存在时,条件(3)与(2)等价.事实上,对 $c=(1,0,\cdots 0)'$,(2)式就是 S_n^{-1} 的 (1,1) 元 $u_n\to 0$,而如所周知 $u_n=1/h_n$,h_n 的定义见(5),故(2)归结为 $h_n\to 0$,即(8)式,而据引理 1.1,此等价于(4)即条件(3).对一般的 c,不妨设 $\|c\|=1$,作正交阵 $D=(d_{jk})$,其第一行为 c,然后施加定理 1.1 证明中的变换,把 x_i 变为 $\tilde{x}_i=Dx_i$,$i=1,2,\cdots n$,用由 $x_1,x_2,\cdots x_n$ 做成 γ_{jn},$j=1,\cdots p$ 的方法,由 $\tilde{x}_1,\cdots\tilde{x}_n$ 做成向量 $\tilde{\gamma}_{jn}$ $j=1,\cdots p$,则

$$\tilde{\gamma}_{jn}=\sum_{k=1}^{p}d_{jk}\gamma_{kn},j=1,\cdots p \tag{9}$$

由 $\tilde{\gamma}_{jn}$ $j=1,\cdots p$,仿照由 γ_{jn}($j=1,\cdots p$)算出 η_n 的方法算出 $\tilde{\eta}_n$.由(9)及变换 D 的正交性,知 $\|\tilde{\eta}_n\|=\|\eta_n\|$,于是,归结到已证的 $c=(1,0.\cdots 0)'$ 的特例。

参考文献

[1] H. Drygas, weak and strongconsistency of the leastsquaresestimateinregres-

sionmodelz. Wahrsch. Verw. Gebiete,34(1976),119－127.

[2] 陈希孺.线性模型参数的估计理论[M].北京:科学出版社,1985.

(本文发表于2004年《应用概率统计》12卷第1期)

独立误差下线性回归最小二乘估计相合性的必要条件

杨复兴*

设线性回归 $Y_i = x_i'\beta + e_i, i = 1, \cdots n \cdots$，其中 $x_1, x_2 \cdots$ 为已知的 p 维向量，e_1，$e_2 \cdots$ 为随机误差，本文证明了：如果 $e_1, e_2 \cdots$ 独立，每一个非退化，则 $S_n^{-1} = (\sum_{i=1}^{n} x_i x_i')^{-1} \to 0$ 是 β 的最小二乘估计相合的必要条件，注意此处对 e_i 的期望和方差没有施加任何条件。

1. 问题的提出

考虑线性回归模型：

$$Y_i = x_i'\beta + e_i, 1 \le i \le n, n \geq 1 \tag{1.1}$$

这里 $x_1, x_2 \cdots x_n$ 是已知的 p 维(列)向量，x_i' 为 x_i 的转置，(一下凡不加“ ′ ”的向量全为列向量)，β 为未知的 P 维回归系数向量，$e_1, e_2 \cdots$ 是随机误差，Y_1，$Y_2 \cdots$ 为因变量的已知观测值，记

$$S_n = x_1 x_1' + \cdots x_n x_n' \tag{1.2}$$

满它 p 阶方阵，以下恒假定当 n 充分大时，S_n 为满秩，这时 β 的最小二乘(LS) 估计为：

$$\hat{\beta}_n = (\hat{\beta}_{1n}, \hat{\beta}_{2n} \cdots \hat{\beta}_{pn})' = S_n^{-1} \sum_{i=1}^{n} x_i y_i \tag{1.3}$$

一般地，对一个给定的 p 维常向量 c，$c'\beta$ 的 LS 估计就定义为 $c'\hat{\beta}_n$，特别地，β 的第一分量 β_1 的 LS 估计就是 $\hat{\beta}_{1n}$.

当 $n \to \infty$ 时，$\hat{\beta}_n \to \beta, pr$. 则称 $\hat{\beta}_n$ 为(弱)相合的，这时必有 $c'\hat{\beta}_n \to c'\beta, pr$. 即对

---

* 作者简介：杨复兴(1948—)，男，河南偃师人，天水师范学院数学与统计学院教授、硕士，主要从事概率统计方面的研究。

一切常向量 c，$c'\hat{\beta}_n$ 为 $c'\beta$ 的弱相合估计，反过来也成立。因此，关于 $\hat{\beta}_n$ 的相合性可归结为其分量的相合性。

LS 估计的相合性取决于两方面的条件：一是随机误差列 $\{e_i\}$ 所满足的条件，二是 自变量取值的序列 $\{x_i\}$，通常在研究这个问题时都是采取下面的方式：给定施加于 $\{e_i\}$ 上的一组条件，在此前提下去研究，为使 $\hat{\beta}_n$ 相合，序列 $\{x_i\}$ 所应满足的条件. 例如 Gauss－Markov 条件是指 $Ee_i = 0$，$Ee_ie_j = \sigma^2\delta_{ij}$，$i,j = 1,2,\cdots$，$0 < \sigma^2 < \infty$，$\sigma^2$ 未知，其中 $\delta_{ij} = 1$ 或 0，视 $i = j$ 或否而定.

在 G－M 条件下，文[1]证明了 $\hat{\beta}_n$ 弱相合的充分条件是

$$S_n^{-1} \to 0, n \to \infty \tag{1.4}$$

或更一般地，$c'\hat{\beta}_n$ 相合的充分条件是

$$c'S_n^{-1}c \to 0, n \to \infty \tag{1.5}$$

这是一个初浅的结果，因为在 G－M 条件下，$c'\hat{\beta}_n$ 为 $c'\beta$ 的无偏估计，且 $Var(c'\hat{\beta}_n) = c'S_n^{-1}c$，因此，当(1.5)成立时，$c'\hat{\beta}_n$ 为 $c'\beta$ 的二阶矩相合（均方相合）估计，当然是其弱相合估计. 困难的是其反面部分：对 $\hat{\beta}_n$ 的相合来说，条件(1.4)是否必要？1976 年，Drygas[2]解决了这个问题，他证明了在满足 G－M 条件下，条件(1.4)的必要性，综合 Eicke 和 Drygas 的结果，有下面的定理：

定理 1.1　设在模型(1.2)中误差序列 $\{e_i\}$ 满足 G－M 条件，则 $\hat{\beta}_n$（弱）相合的充要条件是(1.4)，或更一般地，$c'\hat{\beta}_n$ 相合的充要条件是(1.5).

这是关于 $\hat{\beta}_n$ 相合性问题的一个基本结果，它表明 LS 估计的一个特异性质：在 G－M 条件下，LS 估计的弱相合性与均方相合性等价.

根据这个结果，自然产生下面的问题：如果把对误差序列 $\{e_i\}$ 的条件有所改变，则 $S_n^{-1} \to 0$ 这个条件对 LS 估计 $\hat{\beta}_n$ 的相合性是否仍为充分或必要？在充分性问题方面，陈希孺[3]曾证明以下的结果：若假定 $e_1, e_2\cdots$ 为 $i.i.d$ ' $Ee_1 = 0$，$0 < E|e_1|^r < \infty$，对某个 r，$1 \leq r < 2$，则 $S_n^{-1} \to 0$ 这个条件仍为必要但不再为充分，他并且在这个情况下给出了某种意义下的充要条件. 另外，也很容易举例证明：即使在对误差方差的条件大大弱化的情况下，仍有可能 $S_n^{-1} \to 0$ 是 $\hat{\beta}_n$ 相合的充分条件.

例 设在模型(1.1)中 $p = 1$，误差 $e_1, e_2\cdots$ 独 立，e_n 的分布为

$$\begin{cases} P(e_n = 0) = 1 - n^{-2} \\ e_n \text{ 在 } R^1 - \{0\} \text{ 有概率密度} \dfrac{1}{\pi^2}(^1 + x^2) - 1, n = 1,2,\cdots \end{cases} \tag{1.6}$$

则 e_n 的期望不存在，更谈不上方差，但 β 的 LS 估计为

$$\hat{\beta}_n = \sum_{i=1}^{n} x_i y_i / S_n = \beta + \sum_{i=1}^{n} x_i e_i / S_n , S_n = \sum_{i=1}^{n} x_i^2 ,$$

设 $\sum_{i=1}^{n} x_i^2 = \infty$，因而 $S_n^{-1} \to 0$，为证 $\hat{\beta}_n$ 相合，只须证

$$\lim_{n\to\infty} x_i e_i / S_n = 0 , a.s \tag{1.7}$$

但由(1.6)有

$$\sum_{n=1}^{\infty} P(e_n \neq 0) < \infty$$

按 Borel－Cantel 引理，知 $P(e_n \neq 0, i.o.) = 0$，即以概率 1 成立：当 n 充分大时，有 $e_n = 0$，再加上 $S_n \to \infty$，即得(1.7)，于是证明了 $\hat{\beta}_n$ 的（强）相合性.

关于必要性方面，所知还很少，直观上看有理由相信：这个条件，在对误差 $\{e_n\}$ 很一般的假定下，应为必要的. 本文结果在一个重要特例上印证了这一直观看法，结果表明：只要误差 $e_1, e_2 \cdots$ 独立（不必同分布），则为了 LS 估计 $\hat{\beta}_n$ 应为相合，条件 $S_n^{-1} \to 0$ 是必要的.

2. 误差独立的情况

定理 2.1　若在模型(1.1)中随机误差相互独立，且都不退化，则为 LS 估计 $\hat{\beta}_n$ 相合，条件 $S_n^{-1} \to 0$　是必要的.

"误差 e_i 不退化"是指不存在常数 a，使得 $P(e_i \neq a) = 1$.

证明　记 $\beta = (\beta_1, \cdots \beta_p)'$，$\hat{\beta}_n = (\hat{\beta}_{1n}, \hat{\beta}_{2n}, \cdots \hat{\beta}_{pn})'$，取 β 的一个分量，例如 β_1 来讨论，往证如果 $\hat{\beta}_{1n}$ 是 β_1 的相合估计，则必有

$$\lim_{n\to\infty} u_n = 0 \tag{2.1}$$

其中 u_n 是 S_n^{-1} 的 (1.1) 元。

不妨设 $p > 1$（p 是 β 的维数），$p = 1$ 的情况简单，其证明不难从 $p > 1$ 的证法看出.

把 x_i 记为 $(x_{i1}, \cdots x_{ip})'$，令 $T_i = (x_{i2}, \cdots x_{ip})'$，$i = 1, 2, \cdots$，则

$$S_n = \begin{bmatrix} \sum_{i=1}^{n} x_{i1}^2 & K_n' \\ K_n & H_n \end{bmatrix}$$

其中 $K_n = \sum_{i=1}^{n} x_{i1} T_i$，$H_n = \sum_{i=1}^{n} T_i T_i'$. 定义

$$h_{in} = x_{i1} - K_n' H_n^{-1} T_i, \quad i = 1,2,\cdots \tag{2.2}$$

$$\hat{\beta}_{1n} = \beta_1 + \sum_{i=1}^{n} h_{ni} e_i / \sum_{i=1}^{n} h_{ni}^2 \tag{2.3}$$

且

$$\sum_{i=1}^{n} h_{ni}^2 = u_n^{-1} \tag{2.4}$$

又当 $n > k$ 时,必有

$$\sum_{i=1}^{k} h_{ni}^2 \geq \sum_{i=1}^{k} h_{ki}^2 \tag{2.5}$$

这里及以下总假定 n,k 足够大,使得 S_n^{-1}, S_k^{-1} 存在.

先证明(2.5),考虑最小二乘问题

$$\min_q \sum_{i=1}^{k} (x_{i1} - T_i' q)^2$$

即要找 $p-1$ 维向量 q,使 $\sum_{i=1}^{k} (x_{i1} - T_i' q)^2$ 达到最小,按公式(1.3),其解是

$$q = \sum_{i=1}^{k} (T_i T_i')^{-1} \sum_{i=1}^{k} x_{i1} T_i = H_k^{-1} K_k$$

因此,对 $p-1$ 维向量 $H_n^{-1} K_n$,有

$$\sum_{i=1}^{k} (x_{i1} - T_i' H_n^{-1} K_n)^2 \geq \sum_{i=1}^{k} (x_{i1} - T_i' H_k^{-1} K_k)^2$$

按 h_{ni} 的定义(2.2),此式即(2.5).

记 $a = \sum_{i=1}^{n} x_{i1}^2$,按四块求逆公式(例如,见[4] P_{33}),有

$$S_n^{-1} = \begin{bmatrix} a & K_n' \\ K_n & H_n \end{bmatrix}^{-1} = \begin{bmatrix} (a - K_n' H_n^{-1} K_n)^{-1} & -a^{-1} K_n' (H_n - K_n K_n')^{-1} \\ * & * \end{bmatrix} \tag{2.6}$$

此式与(1.3)结合,得到

$$\hat{\beta}_{1n} - \beta_1 = \sum_{i=1}^{n} \left(\frac{x_{i1}}{a - K_n' H_n^{-1} K_n} - \frac{K_n' (H_n - K_n K_n')^{-1} T_i}{a} \right) e_i$$

注意到

$$\begin{aligned} & \frac{K_n' (H_n - K_n K_n')^{-1}}{a} \\ = & \frac{(1 - a^{-1} K_n' H_n^{-1} K_n) K_n' (H_n - a^{-1} K_n K_n')^{-1})}{a - K_n' H_n^{-1} K_n} \\ = & \frac{K_n' (1 - a^{-1} H_n^{-1} K_n K_n') \left[H_n (1 - a^{-1} H_n^{-1} K_n K_n') \right]^{-1}}{a - K_n' H_n^{-1} K_n} \\ = & K_n' H_n^{-1} / a - K_n' H_n^{-1} K_n \end{aligned}$$

此与上式结合,并注意 h_{ni} 的定义(2.2),可得

$$\hat{\beta}_{1n}-\beta_1=\sum_{i=1}^{n}(x_{i1}-K_n'H_n^{-1}\Gamma_i)e_i/(a-K_n'H_n^{-1}K_n)$$

$$=\sum_{i=1}^{n}h_{ni}e_i/(a-K_n'H_n^{-1}K_n)$$

因为

$$\sum_{i=1}^{n}h_{ni}^2=\sum_{i=1}^{n}(x_{i1}-K_n'H_n^{-1}\mathrm{T}_i)^2$$

$$=\sum_{i=1}^{n}x_{i1}^2-2K_n'H_n^{-1}\sum_{i=1}^{n}x_i\mathrm{T}_i+K_n'H_n^{-1}(\sum_{i=1}^{n}\mathrm{T}_i\mathrm{T}_i')H_n^{-1}K_n \tag{2.7}$$

$$=a-2K_n'H_n^{-1}K_n+K_n'H_n^{-1}K_n$$

$$=a-K_n'H_n^{-1}K_n$$

此与上式结合,得到

$$\hat{\beta}_{1n}-\beta_1=\sum_{i=1}^{n}h_{ni}e_i/\sum_{i=1}^{n}h_{ni}^2$$

这证明了(2.3),最后(2.4)由(2.6)与(2.7)得出.

有了以上的准备,我们可以着手证明定理 2.1 中对 $\{e_i\}$ 所施加的条件的充分性.这只需证明:设 $\{e_i\}$ 满足定理 2.1 中的条件而(1.4)不成立,则 $\hat{\beta}_{1n}$ 不是 β_1 的相合估计.

由 S_n 的定义,并注意到 $x_ix_i'\geq 0, i\geq 1$,可知,当 $n>m$ 时,$S_n\geq S_m$,因而 $S_n^{-1}\leq S_m^{-1}$,由此可知,$u_n\leq u_m$,当 $n>m$ 时,再由(2.1)不成立可知,存在常数 $d^{-1}>0$,使 u_n 当 n 上升时非增并当 $n\to\infty$ 时,收敛于 d^{-1},因而,当 n 上升时,$\sum_{i=1}^{n}h_{ni}^2\uparrow d<\infty$,此与(2.3)结合,可知,为了 $\hat{\beta}_{1n}$ 为 β_1 的相合估计,必须有

$$\sum_{i=1}^{n}h_{ni}e_i\to 0, pr.\ n\to\infty \tag{2.8}$$

固定 k,记 $kg^2=\sum_{i=1}^{k}h_{ki}^2$,有 $g>0$,注意到(2.5),可知对任何自然数 $n>k$,必存在与 n 有关的整数 j,使得 $|h_{nj}|\geq g$,必要时取自然数的子序列,不失普遍性可设 j 与 n 无关,又不失普遍性不妨设 $j=1$,即 $|h_{n1}|\geq g$ 对一切充分大的自然数 n 都成立.

以 φ_n,ψ,δ_n 分别记变量 $\sum_{i=1}^{n}h_{ni}e_i$,e_1,$\sum_{i=2}^{n}h_{ni}e_i$ 的特征函数,由独立性假定,有

$\varphi_n(t)=\psi(h_{n1}t)\delta_n(t)$，又因一切 e_i 均非退化，故 e_1 非退化，因而存在 $l\neq 0$，使得 $|\psi(l)|<1$，因此由 $\varphi_n(l/h_{n1})=\psi(l)\delta_n(l/h_{n1})$，$|\psi(l)|<1$ 以及 $|\delta_n(l/h_{n1}|\leq 1$，可知，当 $n\to\infty$ 时，

$$\limsup|\varphi_n(l/h_{n1})|\leq|\psi(l)|<1$$

再由 $|h_{n1}|\geq g>0$，可知 $\{l/h_{n1}\}$ 为有界序列. 这证明了：当 $n\to\infty$ 时，φ_n 在有界区间上不能一致收敛于 1，因而(2.8)不能成立. 如前所述，这证明了 $\hat{\beta}_{1n}$ 不是 β_1 的相合估计. 定理 2.1 证毕.

3. 误差有退化的情况

如果误差 $e_1,e_2\cdots$ 中有一个或 n 个为退化，那么 S_n^{-1}，是否仍为必要 ？本节分几种情形来考察这个问题.

(a)中存在两个或更多的退化者.

这时容易证明：为了 $\hat{\beta}_n$ 相合，条件(1.4)可以是不必要的.

例 3.1　设 $P(e_1=c_1)=P(e_2=c_2)=1$. 考察线性模型

$$Y_i=x_i\beta+e_i,1\leq i\leq n,n\geq 1 \tag{3.1}$$

β 为一维，令 $x_1=c_2$，$x_2=-c_1$，$x_3=x_4=\cdots=0$，条件(1.4)显然不满足，但 $\hat{\beta}=\beta+(x_1c_1+x_2c_2)/(c_1^2+c_2^2)=\beta$，它当然是相合的.

$e_1,e_2\cdots$ 中有一个退化为 0.

这时也很显然，为了 $\hat{\beta}_n$ 相合，条件(1.4) 并非必要，作为例子，不妨设 $P(e_1=0)=1$，考察线性模型(3.1)，令 $x_1=1$，$x_2=x_3=\cdots=0$，对任何 $n\geq 1$，总有 $\hat{\beta}=Y_1=\beta$.

在上面的定理 2.1 中，我们并未假定误差 $e_1,e_2\cdots$ 的期望为 0，从实用的角度看这个条件是必要的，因为不然的话，不仅 e_i 作为“误差”的意义不明（除非另附条件，如 $med(e_i)=0$ 之类），且更重要的是，LS 估计也失掉了根据. 如果把条件 $E(e_i)=0,i\geq 1$ 附加在定理 2.1 对 $\{e_i\}$ 的条件中，则可推出，若某个 e_i 退化，它只能退化于 0，再结合刚才讨论的情况(b)，就可以知道，若 $\{e_i\}$ 中有退化的，则定理 2.1 必然不真，这就是说，在这个补充假定下，“一切 e_i 均非退化”是定理 2.1 正确的充要条件.

作为纯粹理论上的兴趣，我们不妨考察一下还剩下的一种情况

(c) $\{e_i\}$ 中恰有一个退化，且不退化到 0.

在这种情况下，定理 2.1 是否正确，取决于 $\{e_i\}$ 的具体分布，两种可能性都

有,定理 2.1 正确的例子很容易举出,例如 $P(e_1=1)=1$, $e_i \sim N(0,1)$, $i>1$,这很容易验证,细节从略.反面的例子(使定理 2.1 不真的例子)较复杂一些.

例 3.2 考虑线性模型(1.1), $p=2$, $e_1, e_2 \cdots$ 分布为 $P(e_1=1)=1$,而

$$e_i \sim R(1-1/i, 1+1/i), i \geq 2 \tag{3.2}$$

又令 $x_1=\begin{bmatrix}1\\1\end{bmatrix}$, $x_2=x_3=\cdots=\begin{bmatrix}0\\1\end{bmatrix}$,对这个模型,有

$$S_n^{-1}=\begin{bmatrix}\frac{n}{n-1} & -\frac{1}{n-1}\\ -\frac{1}{n-1} & \frac{1}{n-1}\end{bmatrix}$$

于是,对于 u_n,即 S_n^{-1} 的(1.1)元,不随 $n\to\infty$ 而趋于 0.另一方面,β_1 的 LS 估计为 $\hat{\beta}_{1n}=Y_1-\frac{1}{n-1}(Y_2+\cdots+Y_n)$,因为 $Y_1=\beta_1+\beta_2+1$, $Y_I=\beta_2+e_i$, $i\geq 2$,有

$$\hat{\beta}_{1n}=\beta_1+1-\frac{1}{n-1}(e_2+\cdots+e_n) \tag{3.3}$$

按 e_i 的分布(3.2),有

$$\left|1-\frac{1}{n-1}(e_2+\cdots+e_n)\right| \leq \frac{1}{n-1}\left(\sum_{i=2}^{n}\frac{1}{i}\right)\to 0, n\to\infty$$

于是,由(3.3)得 $\hat{\beta}_{1n}\to\beta_1$,当 $n\to\infty$ 时,即 $\hat{\beta}_{1n}$ 为 β_1 的相合估计,这表明在本模型中,定理 2.1 不成立.

参考文献

[1] Eicker, F. Asymptotic normality and consistency of the least squares estimates for families of linear regression [J], Ann. Math, Statist., 34(1963)447-463.

[2] H. Drygas, weak and strongconsistency of the leastsquaresestimateinregressionmodelz. Wahrsch. Verw. Gebiete, 34(1976), 119-127.

[3]陈希孺,等.低阶矩条件下线性回归最小二乘估计弱相合的充要条件[J].中国科学 A 辑,1995,25(4):349-358.

[4]Rao. C. R., Linear Statistical inference and Its Application [M], John Wiley, New York, 1973.

(本文发表于 2004 年《数学年刊》第 25 卷 A 辑第 2 期)

对推广 Raabe 判别法的再讨论

杨钟玄*

推广 Raabe 判别法是新近提出的关于正项级数敛散性问题一种普遍性方法. 通过对它的进一步探讨，推出了几种常用判别法，同时得到了推广 Raabe 判别法与经典的 Kummer 判别法的关系.

《大学数学》2005 年第 2 期刊登了唐翠娥老师的文 [1]，提出了关于正项级数敛散问题的一种普遍性方法，可称为推广 Raabe 判别法. 笔者经过进一步探讨，得出了一些有趣的结果. 为论述方便，先叙述 Raabe 判别法、Gauss 判别法与推广 Raabe 判别法如下：

命题 1 （Raabe 判别法） 设 $\sum u_n$ 是正项级数，如果

$$\lim_{n\to\infty} n\left(\frac{u_n}{u_{n+1}}-1\right)=r \tag{1}$$

则 $r>1$ 时级数收敛；$r<1$ 时级数发散.

命题 2 （Gauss 判别法） 设 $\sum u_n$ 都是正项级数，如果

$$\lim_{n\to\infty}\ln n\left[n\left(\frac{u_n}{u_{n+1}}-1\right)-1\right]=r, \tag{2}$$

则 $r>1$ 时级数收敛；$r<1$ 时级数发散.

命题 3 （推广 Raabe 判别法） 设 $\sum u_n$ 与 $\sum a_n$ 是正项级数，并且 $\sum a_n$ 发散，如果

$$\lim_{n\to\infty}\frac{a_1+a_2+\cdots+a_n}{a_n}\left(\frac{u_n}{u_{n+1}}-\frac{a_n}{a_{n+1}}\right)=R. \tag{3}$$

则 $R>1$ 时级数 $\sum u_n$ 收敛；$R<1$ 时级数 $\sum u_n$ 发散.

---

* 作者简介：杨钟玄（1946—），男，甘肃天水人，天水师范学院数学与统计学院教授、学士，主要从事数学分析一研究。

文[1]通过命题3给出了当 $a_n\equiv 1$ 时的特例情形，即通常的 *Raabe* 判别法，而未能考虑其他情形和得出相应的具体判别法．为了说明推广 *Raabe* 判别法的具体作用以及与其他判别法的联系，下面进行一些补充讨论．

1．取 $a_n=n!$，它显然满足命题 3 中条件，这时

$$\frac{a_1+a_2+\cdots+a_n}{a_n}\left(\frac{u_n}{u_{n+1}}-\frac{a_n}{a_{n+1}}\right)=\frac{\sum\limits_{k=1}^{n}k!}{n}\left(\frac{u_n}{u_{n+1}}-\frac{1}{n+1}\right).$$

因为当 $n>2$ 时，

$$n!<\sum_{k=1}^{n}k!<(n-2)(n-2)!+(n-1)!+n!<2(n-1)!+n!$$

所以，$n>2$ 时，有 $1<\frac{1}{n!}\sum\limits_{k=1}^{n}k!<\frac{2}{n}+1\to 1(n\to\infty)$．所以

$$\lim_{n\to\infty}\left(\frac{1}{n!}\sum_{k=1}^{n}k!\right)\left(\frac{u_n}{u_{n+1}}-\frac{a_n}{a_{n+1}}\right)=\lim_{n\to\infty}\frac{u_n}{u_{n+1}}.$$

当 $\lim\limits_{n\to\infty}\frac{u_n}{u_{n+1}}>1$（或 <1），有 $\frac{a_1+a_2+\cdots+a_n}{a_n}\left(\frac{u_n}{u_{n+1}}-\frac{a_n}{a_{n+1}}\right)>1$（或 <1）．由命题 3 知 $\sum u_n$ 收敛（或发散）．于是我们推出了熟知的 *D' Alembert* 判别法．

2．取 $a_n=n$，它也满足命题 3 中条件，这时

$$\lim_{n\to\infty}\frac{a_1+a_2+\cdots+a_n}{a_n}\left(\frac{u_n}{u_{n+1}}-\frac{a_n}{a_{n+1}}\right)=\lim_{n\to\infty}\frac{n+1}{2}\left(\frac{u_n}{u_{n+1}}-\frac{n}{n+1}\right)$$

$$=\lim_{n\to\infty}\frac{n}{2}\left(\frac{u_n}{u_{n+1}}-\frac{n}{n+1}\right).$$

由此利用命题3可得出一种具体的新判别法，即

命题 4　设 $\sum u_n$ 是正项级数，如果

$$\lim_{n\to\infty}\frac{n}{2}\left(\frac{u_n}{u_{n+1}}-\frac{n}{n+1}\right)=l \tag{4}$$

存在，则 $l>1$ 时级数收敛；$l<1$ 时级数发散．

易知此判别法强于 *D' Alembert* 判别法，因为当 $\lim\limits_{n\to\infty}\frac{u_{n+1}}{u_n}=q<1$（或 >1）时，(4) 式中极限 $l=+\infty$（或 $-\infty$）．

下面给出命题 4 与 *Raabe* 判别法（命题 1）的关系．

命题 5　命题 4 中的判别法与 *Raabe* 判别法等价．

证　事实上，由于

$$\frac{n}{2}\left(\frac{u_n}{u_{n+1}}-\frac{n}{n+1}\right)=\frac{n}{2}\left(\frac{u_n}{u_{n+1}}-1+1-\frac{n}{n+1}\right)=\frac{n}{2}\left(\frac{u_n}{u_{n+1}}-1\right)+\frac{1}{2}\frac{n}{n+1},$$

$$n(\frac{u_n}{u_{n+1}}-1)=2\cdot\frac{n}{2}(\frac{u_n}{u_{n+1}}-\frac{n}{n+1}+\frac{n}{n+1}-1)$$
$$=2\cdot\frac{n}{2}(\frac{u_n}{u_{n+1}}-\frac{n}{n+1})-\frac{n}{n+1},$$

从而(1)式与(4)式中的极限存在性相同,并且极限值同时大,1或小于1.于是命题5成立.

3.取 $a_n=\frac{1}{n}$,则它也满足命题3中条件.由于

$$\sum_{k=1}^{n}a_k=\ln n+C+\varepsilon_n$$

其中 C 为 *Euler* 常数,$\lim\varepsilon_n=0$(参见文[2]),因此

$$\lim_{n\to\infty}\frac{a_1+a_2+\cdots+a_n}{a_n}(\frac{u_n}{u_{n+1}}-\frac{a_n}{a_{n+1}})=\lim_{n\to\infty}n(\ln n+C+\varepsilon_n)(\frac{u_n}{u_{n+1}}-\frac{n+1}{n})$$
$$=\lim_{n\to\infty}\frac{\ln n+C+\varepsilon_n}{\ln n}\cdot\ln n(n\frac{u_n}{u_{n+1}}-n-1)$$
$$=\lim_{n\to\infty}\ln n[n(\frac{u_n}{u_{n+1}}-1)-1]$$

所以当 $\lim\limits_{n\to\infty}\ln n[n(\frac{u_n}{u_{n+1}}-1)-1]>1$(或 <1),由上式及命题3,就知道 $\sum u_n$ 收敛(或发散).于是我们又推出了 *Gauss* 判别法.

4.推广 *Raabe* 判别法与 *Kummer* 判别法的关系.

命题6 设正项级数 $\sum a_n$ 发散,则级数 $\sum\frac{a_n}{S_n^p}$,当 $p>1$ 时收敛,当 $p\leq 1$ 时发散,其中 $S_n=\sum\limits_{k=1}^{n}a_k$.

此命题的证明可参见文[1]或文[2].

命题7 (*Kummer* 判别法)设 $\sum u_n$ 是正项级数,$\{c_n\}_{n=1}^{\infty}$ 是使级数 $\sum\frac{1}{c_n}$ 发散的一个正数数列.如果极限

$$\lim_{n\to\infty}(c_n\frac{u_n}{u_{n+1}}-c_{n+1})=l \tag{5}$$

存在(有穷或无穷的),则当 $l>0$ 时 $\sum u_n$ 收敛;当 $l<0$ 时 $\sum u_n$ 发散.

此命题的证明可参见文[2].

Kummer 判别法也是一种非常普遍的判别法,D'Alembert 判别法、Raabe 判别法、Gauss 判别法都是它的特例.(分别取 $c_n=1$,n 与 $n\ln n$).我们有理由猜测,推

广 Raabe 判别法与 Kummer 判别法之间应该有密切的内在联系.

命题 8 推广 Raabe 判别法与 Kummer 判别法等价.

证(i)由推广 Raabe 判别法推证 Kummer 判别法.

设命题 7 中(5) 式右边极限值 $l > 0$,令 $a_n = \frac{1}{c_n}$,则有假设知 $\sum a_n$ 发散,且 $\sum a_n = \infty$. 因此

$$\lim_{n\to\infty} \frac{\sum_{k=1}^{n} a_k}{a_n}\left(\frac{u_n}{u_{n+1}} - \frac{a_n}{a_{n+1}}\right) = \lim_{n\to\infty}\sum_{k=1}^{n} a_k \cdot \lim_{n\to\infty}\left(c_n \frac{u_n}{u_{n+1}} - c_{n+1}\right) = +\infty > 1.$$

于是由命题 3 知 $\sum u_n$ 收敛. 类似可证发散情形,从而命题 7 成立.

(ii)由 Kummer 判别法证明推广 Raabe 判别法.

设命题3中(3)式右边的极限值$R > 1$,并令$\lim\limits_{n\to\infty} \frac{\sum_{k=1}^{n} a_k}{a_n} = c_n$,则由假定$\sum a_n$发散及命题 6 知 $\sum \frac{1}{c_n}$ 发散. 而

$$c_n \frac{u_n}{u_{n+1}} - c_{n+1} = \frac{\sum_{k=1}^{n} a_k}{a_n} \frac{u_n}{u_{n+1}} - \frac{\sum_{k=1}^{n+1} a_k}{a_{n+1}} = \left(\frac{\sum_{k=1}^{n} a_k}{a_n} \frac{u_n}{u_{n+1}} - \frac{\sum_{k=1}^{n} a_k}{a_{n+1}}\right) - 1$$

$$= \frac{\sum_{k=1}^{n} a_k}{a_n}\left(\frac{u_n}{u_{n+1}} - \frac{a_n}{a_{n+1}}\right) - 1$$

因此

$$\lim_{n\to\infty}\left(c_n \frac{u_n}{u_{n+1}} - c_{n+1}\right) = \lim_{n\to\infty} \frac{\sum_{k=1}^{n} a_k}{a_n}\left(\frac{u_n}{u_{n+1}} - \frac{a_n}{a_{n+1}}\right) - 1 = R - 1 > 0,$$

于是由命题 7 知 $\sum u_n$ 收敛. 同理可证发散情形. 从而命题 3 成立.

由(i),(ii)知命题 8 成立.

参考文献

[1]唐翠娥. 级数敛散性的拉阿贝判别法的推广[J]. 大学数学,2005,21(2):132—134.

[2]菲赫金哥尔茨. 微积分学教程(二卷二分册)[M]. 北京:人民教育出版社,1954:261—286.

[3]华东师范大学数学系. 数学分析(第二版)[M]. 北京:高等教育出版社,1991.

(本文发表于 2007 年《大学数学》23 卷 2 期)

一个正项级数命题的另一种证明

杨钟玄*

贵刊1990年第1期与1994年第12期分别刊登了李铁烽同志的文[1]与高军同志的文[2].其中文[2]证明了文[1]中提出的正项级数敛散性的新比值判别法要强于拉贝判别法.这就是文[2]中的

命题1:若 $a_n>0$,且$\lim\limits_{n\to\infty}n(1-\frac{a_{n+1}}{a_n})=r$,则

$$\lim_{n\to\infty}\frac{a_{2n}}{a_n}=\lim_{n\to\infty}\frac{a_{2n+1}}{a_{n+1}}=\frac{1}{2^r}$$

证明中用到了泰勒中值公式与欧拉公式:

$$1+\frac{1}{2}+\ldots+\frac{1}{n}=C+\ln n+\varepsilon_n$$

其中 C 为欧拉常数,且$\lim\limits_{n\to\infty}\varepsilon_n=0$.

本文给出另一种证法,证明与命题1相当而更完全一些的.

命题2:设 $a_n>0$,如果

$$\lim_{n\to\infty}n(\frac{a_n}{a_{n+1}}-1)=\begin{cases}-\infty\\ r(-\infty<r<+\infty),\\ +\infty\end{cases}$$

则有$\lim\limits_{n\to\infty}\frac{a_{2n}}{a_n}=\lim\limits_{n\to\infty}\frac{a_{2n+1}}{a_{n+1}}=\begin{cases}+\infty\\ \frac{1}{2^r}\\ 0\end{cases}$

证明:当 $-\infty<r<+\infty$ 时,任意取定 $\varepsilon>0$,由命题条件,对一切充分大的 n 都有

---

* 作者简介:杨钟玄(1946—),男,甘肃天水人,天水师范学院数学与统计学院教授、学士,主要从事数学分析一研究。

$$r-\varepsilon < n(\frac{a_n}{a_{n+1}}-1) = r+\frac{\varepsilon}{2} \tag{1}$$

记 $r'=r-\varepsilon$,则易知,$\lim\limits_{n\to\infty}\dfrac{\left(1+\dfrac{1}{n}\right)^{r'}-1}{\dfrac{1}{n}}$ 等于函数 $(1+x)^{r'}$ 在点 $x=0$ 的导数,即等于 r'.

$\because r' < r-\frac{\varepsilon}{2}$,因此由极限保号性知,对充分大的 n,有

$$\frac{\left(1+\dfrac{1}{n}\right)^{r'}-1}{\dfrac{1}{n}} < r-\frac{\varepsilon}{2}$$

从而 $(1+\frac{1}{n})^{r-\varepsilon} = (1+\frac{1}{n})^{r'} < 1+\dfrac{r-\frac{\varepsilon}{2}}{n}$.

因此,由(1)式得

$$\frac{a_n}{a_{n+1}} > 1+\frac{r-\frac{\varepsilon}{2}}{n} > (1+\frac{1}{n})^{r-\varepsilon}$$

同理,当 n 充分大时,有

$$\frac{a_n}{a_{n+1}} < (1+\frac{1}{n})^{r+\varepsilon}$$

不妨设 $\varepsilon>0$ 充分小,由上述知有自然数 N,使对一切 $n>N$,有

$$(1+\frac{1}{n})^{r-\varepsilon} < \frac{a_n}{a_{n+1}} < (1+\frac{1}{n})^{r+\varepsilon}$$

$$(1+\frac{1}{n+1})^{r-\varepsilon} < \frac{a_{n+1}}{a_{n+2}} < (1+\frac{1}{n+1})^{r+\varepsilon}$$

……………………………………

$$(1+\frac{1}{2n-1})^{r-\varepsilon} < \frac{a_{2n-1}}{a_{2n}} < (1+\frac{1}{2n-1})^{r+\varepsilon}$$

以上 n 个不等式相乘后再取倒数得

$$\frac{1}{2^{r+\varepsilon}} < \frac{a_{2n}}{a_n} < \frac{1}{2^{r-\varepsilon}}$$

所以

$$\frac{1}{2^{r+\varepsilon}} \le \varliminf_{n\to\infty}\frac{a_{2n}}{a_n} \le \varlimsup_{n\to\infty}\frac{a_{2n}}{a_n} \le \frac{1}{2^{r-\varepsilon}}$$

再由 ε 的任意性,得 $\lim\limits_{n\to\infty}\dfrac{a_{2n}}{a_n}=\dfrac{1}{2^r}$.同理可推得,当 n 充分大时,又有

$$(1+\frac{1}{n+1})^{r-\varepsilon}<\frac{a_n}{a_{n+1}}<(1+\frac{1}{n+1})^{r+\varepsilon}$$

从而得到

$$(\frac{n+2}{2n+2})^{r+\varepsilon}<\frac{a_{2n+1}}{a_{n+1}}<(\frac{n+2}{2n+2})^{r-\varepsilon}$$

$$\therefore\quad (\frac{1}{2}+\frac{1}{2n+2})^{r+\varepsilon}<\frac{a_{2n+1}}{a_{n+1}}<(\frac{1}{2}+\frac{1}{2n+2})^{r-\varepsilon}.$$

$$\frac{1}{2^{r+\varepsilon}}\leq\underline{\lim_{n\to\infty}}\frac{a_{2n+1}}{a_{n+1}}\leq\overline{\lim_{n\to\infty}}\frac{a_{2n+1}}{a_{n+1}}\leq\frac{1}{2^{r-\varepsilon}}.$$

所以得$\lim\limits_{n\to\infty}\frac{a_{2n+1}}{a_{n+1}}=\frac{1}{2^r}$.

再设$\lim\limits_{n\to\infty}n(\frac{a_n}{a_{n+1}}-1)=-\infty$,则$M>0$,自然数$N_1>\frac{M}{2}$,当$n>N_1$时,有

$$n(\frac{a_n}{a_{n+1}}-1)<-M$$

$$\therefore\ \frac{a_n}{a_{n+1}}<1-\frac{M}{n}<1-\frac{M}{2n},\frac{a_{n+1}}{a_{n+2}}<1-\frac{M}{n+1}<1-\frac{M}{2n}$$

……………………………

$$\frac{a_{2n-1}}{a_{2n}}<1-\frac{M}{2n-1}<1-\frac{M}{2n}$$

又 $\because\ \lim\limits_{n\to\infty}\left(1-\frac{M}{2n}\right)^n=\lim\limits_{n\to\infty}\left[\left(1-\frac{M}{2n}\right)^{-\frac{2n}{M}}\right]^{-\frac{M}{2}}<2e^{-\frac{M}{2}}$

$\therefore\ \exists$ 自然数N_2,当$n>N_2$时有$\left(1-\frac{M}{2n}\right)^n<2e^{-\frac{M}{2}}$.这时,

$$0<\frac{a_n}{a_{2n}}<\left(1-\frac{M}{2n}\right)^n<2e^{-\frac{M}{2}}.$$

$\therefore\exists M>0$, 自然数 $N=\max\{N_1,N_2\}$,当$n>N$时,有$0<\frac{a_n}{a_{2n}}<\left(1-\frac{M}{2n}\right)^n<2e^{-\frac{M}{2}}$.再由$M$的任意性知$\lim\limits_{n\to\infty}\frac{a_n}{a_{2n}}=0$或$\lim\limits_{n\to\infty}\frac{a_{2n}}{a_n}=+\infty$.同理可证$\lim\limits_{n\to\infty}\frac{a_{2n+1}}{a_{n+1}}=+\infty$.

对于$\lim\limits_{n\to\infty}n(\frac{a_n}{a_{n+1}}-1)=+\infty$情形可以类似证明其结论也成立.于是可知命题2成立.

参考文献

[1]李铁烽.正项级数判敛的一种新的比值判别法[J].数学通报,1990,1.

[2]高军.谈谈几种正项级数敛散性判别法的比较[J].数学通报,1994,12.
[3]Г. М. 菲赫金哥尔茨,微积分学教程[M].二卷二分册.

(本文发表于 2001 年《数学通报》第 1 期)

2 类优美图

唐保祥　任韩*

1966 年，Rosa 猜想：所有的树都是优美图。这就是著名的优美树猜想。自 1972 年 Golomb. S. W 给出了优美图的定义[1]以来，人们发现了许多优美图。但是到底满足什么条件的图是优美图，即表征优美图至今仍是一个世界难题。

目前，虽然国内外已获得许多关于优美图的研究成果，但是由于缺乏一个系统和有力的工具，迄今，只能对一些特殊的图类探索其优美性[2-14]。然而优美图的研究成果已被广泛应用于射电天文学，雷达，通讯网络，X—射线衍射晶体学，密码设计，导弹控制码设计，同步机码设计等许多领域[15]。

本文通过构造的方法得到了 2 类新的优美图，对优美图的研究具有一定参考价值。

定义 1　对于一个图 $G=(V,E)$，如果对每个 $v \in V$，存在一个非负整数 $\theta(v)$（称为顶点 V 的标号）满足：

(1) $\forall u,v \in V$，如果 $u \neq v$，则 $\theta(u) \neq \theta(v)$；

(2) $\max\{\theta(v) \mid v \in V\} = |E|$；

(3) $\forall e_1,e_2 \in E$，如果 $e_1 \neq e_2$，则 $\theta'(e_1) \neq \theta'(e_2)$，其中 $\theta'(e) = |\theta(u)-\theta(v)|$，$e=uv$；

那么称 G 为优美图，θ 称为 G 的一个优美值，或称优美标号 θ' 称为由 θ 导出的边标号[2]。

定义 2　将宽和长分别为 m 和 n 个单位的矩形的长和宽分别 m 和 n 等分，所得图形中把交点看成图的顶点，把每个小正方形看成长为 4 的圈，这样所得的图称为一个 $m \times n$ 的棋盘，记作 $Q_{m\times n}$。

本文未给出的定义和记号见文献[15]和[16]。

---

* 作者简介：唐保祥（1961—），男，甘肃天水人，天水师范学院数学与统计学院教授、学士，主要从事图论方面的研究。

1 结果及其证明

定理1 对任意正整数 m 和 n，两个相同 $m\times n$ 的棋盘 $Q_{m\times n}$ 上，位置相同的对应顶点连一条边所得图记为 $Q^2_{m\times n}$，则 $Q^2_{m\times n}$ 是优美图。

证明 因为 $|V(Q_{m\times n})|=(m+1)(n+1)=mn+m+n+1$，

$$|E(Q_{m\times n})|=m(n+1)+n(m+1)=2mn+m+n,\text{所以}$$

$$|V(Q^2_{m\times n})|=2\cdot|V(Q_{m\times n})|=2mn+2m+2n+2,$$

$$|E(Q^2_{m\times n})|=2\cdot|E(Q_{m\times n})|+|V(Q_{m\times n})|=5mn+3m+3n+1。$$

把图 $Q^2_{m\times n}$ 看成 $n+1$ 个 $m\times 1$ 的棋盘 $Q_{m\times 1}$，从左至右顺次连结对应位置的顶点得到的。从左至右，设 $Q^2_{m\times n}$ 的第 i 个棋盘 $Q_{m\times 1}$ 从左上角起，从上到下“S”形地分别标识其顶点为：$v_{i1},v_{i2},v_{i3},v_{i4},\cdots,v_{i,2(m+1)}$，$i=1,2,\cdots,n+1$（如图 1 所示）。

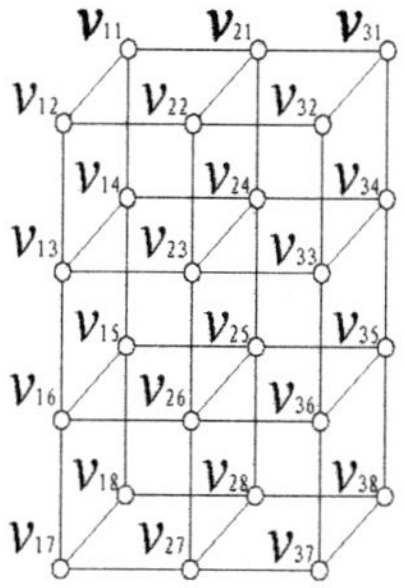

图 1 $Q^2_{3\times 2}$

首先证明：当 $m=1$ 时，对任意正整数 n，$Q^2_{1\times n}$ 是优美图。

定义 图 $Q^2_{1\times n}$ 的顶点标号 θ_1 如下（如图 2 所示。顶点 v_{ij} 是 $Q^2_{1\times n}$ 的第 i 个 4 圈 C_i 上的第 j 个顶点）：

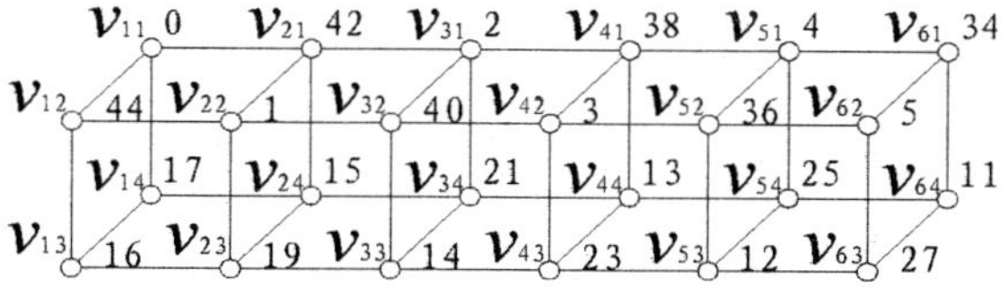

图 2 $Q^2_{1\times 5}$

(1) $\theta_1(v_{ij})=i-1,i=1,2,\cdots,n+1;j=\dfrac{3+(-1)^i}{2}$，即 $j=\begin{cases}1,i\equiv 1(\mathrm{mod}2),\\2,i\equiv 0(\mathrm{mod}2);\end{cases}$

(2)$\theta_1(v_{ij})=8n+6-2i, i=1,2,\cdots,n+1; j=\frac{3-(-1)^i}{2}$，即 $j=\begin{cases}2, i\equiv 1(\bmod 2),\\1, i\equiv 0(\bmod 2);\end{cases}$

(3)$\theta_1(v_{ij})=3n+2-i, i=1,2,\cdots,n+1; j=\frac{7+(-1)^i}{2}$，即 $j=\begin{cases}3, i\equiv 1(\bmod 2),\\4, i\equiv 0(\bmod 2);\end{cases}$

(4)$\theta_1(v_{ij})=3n+2i, i=1,2,\cdots,n+1; j=\frac{7-(-1)^i}{2}$，即 $j\begin{cases}4, i\equiv 1(\bmod 2),\\3, i\equiv 0(\bmod 2)。\end{cases}$

定义集合 S_1,S_2,S_3,S_4 如下：

$$S_1=\left\{\theta_1(v_{ij})\mid i=1,2,\cdots,n+1; j=\frac{3+(-1)^i}{2}\right\};$$

$$S_2=\left\{\theta_1(v_{ij})\mid i=1,2,\cdots,n+1; j=\frac{3-(-1)^i}{2}\right\};$$

$$S_3=\left\{\theta_1(v_{ij})\mid i=1,2,\cdots,n+1; j=\frac{7+(-1)^i}{2}\right\};$$

$$S_4=\left\{\theta_1(v_{ij})\mid i=1,2,\cdots,n+1; j=\frac{7-(-1)^i}{2}\right\};$$

下面证明当 $p\neq q$ 时，$S_p\cap S_q=\phi$，其中 $p,q\in\{1,2,3,4\}$。

(1) 若 $S_1\cap S_2\neq\phi$，则存在 $i_1,i_2\in\{1,2,\cdots,n+1\}$，使得 $i_1-1=8n+6-2i_2$，于是 $i_1=8n-2i_2+7$。当 $i_1,i_2\in\{1,2,\cdots,n+1\}$ 时，等式 $i_1=8n-2i_2+7$ 显然不可能成立。所以 $S_1\cap S_2=\phi$。

(2) 若 $S_1\cap S_2\neq\phi$，则存在 $i_1,i_2\in\{1,2,\cdots,n+1\}$，使得 $i_1-1=3n+2-i_2$，从而 $i_1=3n-i_2+3$。而当 $i_1,i_2\in\{1,2,\cdots,n+1\}$ 时，等式 $i_1=3n-i_2+3$ 显然不可能成立。所以 $S_1\cap S_3=\phi$。

(3) 若 $S_1\cap S_4\neq\phi$，则存在 $i_1,i_2\in\{1,2,\cdots,n+1\}$，使得 $i_1-1=3n+2i_2$，也就是 $i_1=3n+2i_2+1$。当 $i_1,i_2\in\{1,2,\cdots,n+1\}$ 时，等式 $i_1=3n+2i_2+1$ 显然不能成立。因此，$S_1\cap S_4=\phi$。

(4) 若 $S_2\cap S_3\neq\phi$，则存在 $i_1,i_2\in\{1,2,\cdots,n+1\}$，使得 $8n+6-2i_1=3n+2-i_2$，即 $i_2=2i_1-5n-4$。这不可能。故 $S_2\cap S_3=\phi$。

(5) 若 $S_2\cap S_4=\phi$，则存在 $i_1,i_2\in\{1,2,\cdots,n+1\}$，使 $8n+6-2i_1=3n+2i_2$，于是 $2i_2=5n-2i_1+6$。这不可能。故 $S_2\cap S_4=\phi$。

(6) 若 $S_3\cap S_4=\phi$，则存在 $i_1,i_2\in\{1,2,\cdots,n+1\}$，使得 $3n+2-i_1=3n+2i_2$，因此 $i_1+2i_2=2$。显然这个等式在给定条件下不成立。所以 $S_3\cap S_4=\phi$。

综上所述，当 $p \neq q$ 时，$S_p \cap S_q = \phi$，其中 $p,q \in \{1,2,3,4\}$。

对图 $Q^2_{1\times 1}$，容易验证由 θ_1 导出的边标号的集合恰为 $\{1,2,\cdots,12\}$。因为图 $Q^2_{1\times(n-1)}$ 有 $8(n-1)+4$ 条边，所以当 θ_1 是图 $Q^2_{1\times(n-1)}(n \geqslant 2)$ 的优美标号时，θ_1 导出的图 $Q^2_{1\times(n-1)}$ 的边标号集合恰为 $\{1,2,\cdots,8(n-1)+4\}$。把图 $Q^2_{1\times(n-1)}$ 看成 $Q^2_{1\times n}$ 的子图，则 $Q^2_{1\times n}$ 比 $Q^2_{1\times(n-1)}$ 多 4 个顶点，8 条边。上述定义的 θ_1 给图 $Q^2_{1\times(n-1)}$ 和 $Q^2_{1\times n}$ 的顶点标号有如下关系：

(1) 图 $Q^2_{1\times n}$ 和图 $Q^2_{1\times(n-1)}$ 的顶点集 $D_1 = \left\{v_{ij} \mid i = 1,2,\cdots,n; j = \dfrac{3+(-1)^i}{2}\right\}$ 上的标号相同；

(2) 图 $Q^2_{1\times n}$ 的顶点集 $D_2 = \left\{v_{ij} \mid i = 1,2,\cdots,n; j = \dfrac{3-(-1)^i}{2}\right\}$ 中每个顶点上的标号，恰好是其子图 $Q^2_{1\times(n-1)}$ 的对应顶点上的标号分别加 8 得到；

(3) 图 $Q^2_{1\times n}$ 的顶点集 $D_3 = \left\{v_{ij} \mid i = 1,2,\cdots,n; j = \dfrac{7-(-1)^i}{2}\right\}$ 和顶点集 $D_4 = \left\{v_{ij} \mid i = 1,2,\cdots,n; j = \dfrac{7+(-1)^i}{2}\right\}$ 中每个顶点上的标号，恰好是其子图 $Q^2_{1\times(n-1)}$ 的对应顶点上的标号分别加 3 得到；

(4) θ_1 给图 $Q^2_{1\times n}$ 的顶点集 $\{v_{n+1,1}, v_{n+1,2}, v_{n+1,3}, v_{n+1,4}\}$ 中顶点的标号是数集 $\{n, 2n+1, 5n+2, 6n+4\}$。

由(1)，(2)，(3) 和(4) 知道，θ_1 给图 $Q^2_{1\times n}$ 的子图 $Q^2_{1\times(n-1)}$ 的边集导出的边标号的集合为 $\{1,2,\cdots,8n+4\}\backslash\{3n-1,3n,3n+1,4n+2,4n+3,5n+4,5n+5,5n+6\}$。而 θ_1 导出的图 $Q^2_{1\times n}$ 的边集 $\{v_{n1}v_{n+1,1}, v_{n2}v_{n+1,2}, v_{n3}v_{n+1,3}, v_{n4}v_{n+1,4}, v_{n+1,1}v_{n+1,2}, v_{n+1,2}v_{n+1,3}, v_{n+1,3}v_{n+1,4}, v_{n+1,4}v_{n+1,1}\}$ 的标号恰好是 $\{3n-1, 3n, 3n+1, 4n+2, 4n+3, 5n+4, 5n+5, 5n+6\}$，所以 θ_1 导出的图 $Q^2_{1\times n}$ 的边边标号恰为 $\{1,2,\cdots,8n+4\}$。故 θ_1 是图 $Q^2_{1\times n}$ 的一个优美标号。

其次，给出图 $Q^2_{m\times n}(m \geqslant 2)$ 的如下一个迭代方式的优美标号(如图 3 所示)。

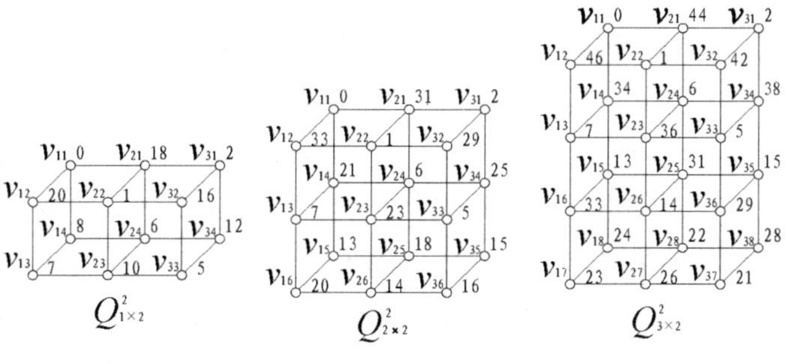

图 3

当 $m \geqslant 2$ 时，对任意正整数 n，图 $Q^2_{m\times n}$ 可以看成是，图 $Q^2_{(m-1)\times n}$ 中顶点 v_{ij}（其中 $i=1,2,\cdots,n+1;j=2m-1,2m$）与棋盘 $Q_{1\times n}$ 中位置对应的顶点间分别连结一条边所得。

故 $|E(Q^2_{m\times n})|=|E(Q^2_{(m-1)\times n})|+|E(Q_{1\times n})|+|V(Q_{1\times n})|$，$|E(Q^2_{m\times n})|=|E(Q^2_{(m-1)\times n})|+5n+3$。

当 $Q^2_{(m-1)\times n}$ 是优美图，$\theta_{(m-1)}$ 是它的优美标号时，定义图 $Q^2_{m\times n}$ 的顶点标号 θ_m 如下（如图 3 所示）：

(1) $\theta_m(v_{ij})=\theta_{(m-1)}(v_{ij}),i+j=0(\mathrm{mod}2),i=1,2,\cdots,n+1;j=1,2,\cdots,2m$。

(2) $\theta_m(v_{ij})=\theta_{(m-1)}(v_{ij})+5n+3,i+j\equiv 1(\mathrm{mod}2),i=1,2,\cdots,n+1;j=1,2,\cdots,2m$。

$$(3)\theta_m(v_{ij})=\begin{cases}\theta_{(m-1)}(v_{i,(j-2)})+4n+4-2i,m\equiv 1(\mathrm{mod}2),\\ \theta_{(m-1)}(v_{i,(j-2)})+2n+2i,m\equiv 0(\mathrm{mod}2),\\ \text{其中 } i+j\equiv 0(\mathrm{mod}2),\\ i=1,2,\cdots,n+1;j=2m+1,2m+2。\end{cases}$$

$$(4)\theta_m(v_{ij})=\begin{cases}\theta_{(m-1)}(v_{i,(j-2)})-4n-5+4i,m\equiv 1(\mathrm{mod}2),\\ \theta_{(m-1)})(v_{i,(j-2)})+3-4i,m\equiv 0(\mathrm{mod}2),\\ \text{其中 } i+j\equiv 1(\mathrm{mod}2),\\ i=1,2,\cdots,n+1;j=2m+1,2m+2。\end{cases}$$

把图 $Q^2_{(m-1)\times n}$ 看成 $Q^2_{m\times n}$ 的子图，则图 $Q^2_{m\times n}$ 比 $Q^2_{(m-1)\times n}$ 多 $2(n+1)$ 个顶点，边多 $5n+3$ 条。当 $Q^2_{(m-1)\times n}$ 是优美图，$\theta_{(m-1)}$ 是它的优美标号时，$\theta_{(m-1)}$ 是集合 $V(Q^2_{(m-1)\times n}$ 到数集 $\{0,1,2,\cdots,5(m-1)n+3(m-1)+3n+1\}$ 的一个单射，而且还是集合 $E(Q^2_{(m-1)\times n})$ 到数集 $\{1,2,\cdots,5(m-1)n+3(m-1)+3n+1\}$ 的双射。因此，由 θ_m 和图 $Q^2_{m\times n}$ 的定义知，θ_m 给 $Q^2_{m\times n}$ 的子图 $Q^2_{(m-1)\times n}$ 的顶点上的标号满足：

(1) 子图 $Q^2_{(m-1)\times n}$ 的顶点标号的最大值为 $\theta_m(v_{12})=m(5n+3)+3n+1$，且各个顶点的标号为互异的非负整数。设 θ_m 给 $Q^2_{(m-1)\times n}$ 的顶点标号的数集为 D_1。

(2) θ_m 在 $Q^2_{m\times n}$ 的子图 $Q^2_{(m-1)\times n}$ 上导出的边标号为数集 $S_1=\{5n+4,5n+5,\cdots,m(5n+3)+3n+1\}$。

设 $V=\{v_{1,2m+1},v_{1,2m+2},v_{2,2m+1},v_{2,2m+2},\cdots,v_{n+1,2m+1},v_{n+1,2m+2}\}$，$\theta_m$ 给图 $Q^2_{m\times n}$ 的顶点的子集的标号为数集 D_2，则 D_2 中元素均为小于 $m(5n+3)+3n+1$ 的互异正整数，且 $D_1\cap D_2=\phi$。θ_m 给 V 出的边边标号为数集 $S_2=\{1,2,\cdots,5n+3\}$，所以 $S_1\cup S_2=\{1,2,\cdots,m(5n+3)+3n+1)\}$。

综上所述，θ_m 是集合 $V(Q^2_{m\times n})$ 到数集 $\{0,1,2,\cdots,m(5n+3)+3n+1\}$ 的一个

单射,还是集合 $E(Q^2_{m\times n})$ 到数集$\{1,2,\cdots,m(5n+3)+3n+1\}$的双射。

故当$Q^2_{(m-1)\times n}$是优美图,$\theta_{(m-1)}$是它的优美标号时,则θ_m是图$Q^2_{m\times n}$的优美标号,所以对任意正整数 m 和 n,$Q^2_{m\times n}$ 是优美图。证毕。

定理 2　将给定顶点 u 与图 $Q^2_{m\times n}$ 中的顶点 v_{ij} 各连接一条边所得图记为 $u\cdot Q^2_{m\times n}$,则 $u\cdot Q^2_{m\times n}$ 是优美图,其中 $j=\dfrac{3+(-1)^i}{2}$,$i=1,2,\cdots,n+1$。

证明　显然 $|V(u\cdot Q^2_{m\times n})|=2mn+2m+2n+3$,$|E(u\cdot Q^2_{m\times n})|=5mn+3m+4n+2$. 设图 $Q^2_{m\times n}$ 由定理 1 给出的优美标号为 θ,定义图 $u\cdot Q^2_{m\times n}$ 的顶点标号 φ 如下(如图 4 所示):

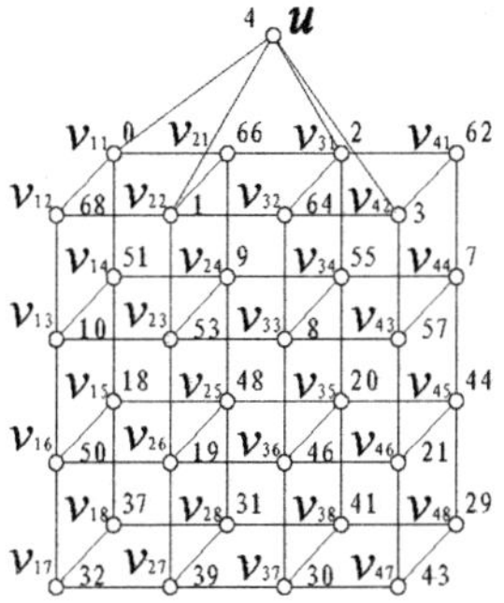

图 4　$u\cdot Q^2_{3\times 3}$

(1)$\varphi(v_{ij})=\theta(v_{ij})+n+1,i+j\equiv 0(\mathrm{mod}2),i=1,2,\cdots,n+1;j=1,2,\cdots,2m+2$;

(2)$\varphi(v_{ij})=\theta(v_{ij}),i+j\equiv 1(\mathrm{mod}2),i=1,2,\cdots,n+1;j=1,2,\cdots,2m+2$;

(3)$\varphi(u)=n+1$。

因为 θ 是图 $Q^2_{m\times n}$ 的优美标号,所以 θ 集合 $V(Q^2_{m\times n})$ 到数集数集$\{0,1,2,\cdots,m(5n+3)+3n+1\}$的一个单射,而且还是集合 $E(Q^2_{(m-1)\times n})$ 到数集$\{1,2,\cdots,m(5n+3)+3n+1\}$的双射。由 φ 图 $u\cdot Q^2_{m\times n}$ 的定义知,φ 是集合 $V(u\cdot Q^2_{m\times n})$ 到数集数集$\{0,1,2,\cdots,m(5n+3)+4n+1\}$的一个单射,也是集合 $E(u\cdot Q^2_{m\times n})$ 到数集$\{1,2,\cdots,m(5n+3)+4n+1\}$的双射。因此,$\varphi$ 是图 $u\cdot Q^2_{m\times n}$ 的一个有没标号,从而 $u\cdot Q^2_{m\times n}$ 是优美图。证毕。

参考文献

[1]Golomb S W, How to number a graph, Graph thery and computing[M], New York: Academic Press, 1972: 23-37.

[2]KATHIESAN K M. Two classes of graceful graphs[J]. Ars Combinatoria, 2000, 55:

129－132.

[3]WEI Lixia, YAN Shoufeng, ZHANG Kunlong. The researches on gracefilness of two kinds of unconnected graphs[J]. Journal of ShandongUniversity(Natural Science), 2008, 43(8):90－96.

[4]杨元生，容青，徐喜荣．一类优美图[J]．徐喜荣．数学研究与评论，2002，24(3)：520－524.

[5]Cai Hua, Wei Lixia, LU Xianrui. Gracefulness of Unconnected Graphs ($P_1 \vee P_2$) $\cup$ G_r, ($P_1 \vee P_n$) $\cup$ ($P_3 \vee \overline{K}_r$) and $W_n \cup St(m)$[J]. Journal of JilinUniversity(Science Edition), 2007, 45(4):539－543.

[6]Wei Lixia, JIA Zhizhong. The gracefulness of unconnected graphs $G_1 \cup G_2$ and $G_1 \cup G_2 \cup K_2$[J]. Acta Mathematicae Applicatae Sinica, 2005, 28(4):689－694.

[7]严谦泰．关于 $P_{2r,2m}$ 的优美标号[J]．系统科学与数学，2006，26(5)：513－517.

[8]魏丽侠，张昆龙．图 $K_1 \vee C_n$ 的非连通并图的优美性[J]．中山大学学报(自然科学版)，2007，46(4)：13－16.

[9]魏丽侠，张昆龙．几类并图的优美标号[J]．中山大学学报(自然科学版)，2008，47(3)：10－13.

[10]卞瑞玲．毛虫树的性质[J]．山东大学学报(理学版)，2002，37(6)：504－507.

[11]潘伟，路线．$K_2 \wedge K_{m,n}$ 的优美性[J]．吉林大学学报(理学版)，2004，42(3)：365－366.

[12]唐保祥．2 类包含 K_4 的优美图及其注记[J]．河北师范大学学报(自然科学版)，2001，25(3)：18－27.

[13]唐保祥．优美图的嵌入[J]．上海师范大学学报(自然科学版)，2007，36(2)：24－27.

[14]唐保祥．简单图都是优美图的子图[J]．海南师范大学学报(自然科学版)，2008，21(2)：126－128.

[15]马克杰．优美图[M]．北京：北京大学出版社，1991.

[16]Bondy J. A, Murty U. S. R. 图论及其应用[M]．吴望名，李念祖，吴兰芳，等译．北京：科学出版社，1984.

（本文发表于 2010 年《山东大学学报》第 10 期）

3 类特殊图完美对集数的计算

唐保祥　任韩*

1　引言

化学图论中,共轭分子图的 kekulé 结构称为完美对集.共轭分子图是否具有完美对集对芳香族系统的稳定性是极其重要的;图的完美对集数也是估计共振能量和 π—电子能量,计算鲍林键级的重要指标;图的对集计数理论与计晶体物理中的 dimmer 问题,组合论中棋盘的多米诺覆盖问题都有密切的关系[1—4].积和式在计算机科学,特别是计算复杂性理论中有重要的地位,二分图的完美对集的数目可以方便地表示为计算积和式的值.由于完美对集的计数理论在化学、物理学和计算机科学中有重要的理论价值和现实意义,所以引起人们对此问题的广泛研究[5—16].本文研究特殊图类完美对集的计数方法,给出了 3 类图完美对集数的显式表达式,其研究方法为得到一般有完美对集图的所有完美对集数目提供了借鉴.

定义　若图 G 的两个完美对集 M_1 和 M_2 中有一条边不同,则称 M_1 和 M_2 是 G 的两个不同的完美对集.

2　结果及其证明

定理 1　$4n$ 个长为 4 的圈为 C_4^{i1} : $u_{i1}u_{i4}u_{i6}u_{i3}u_{i1}$,$C_4^{i2}$: $u_{i2}u_{i5}u_{i7}u_{i4}u_{i2}$,$C_4^{i3}$: $v_{i1}v_{i4}v_{i6}v_{i3}v_{i1}$,$C_4^{i4}$: $v_{i2}v_{i5}v_{i7}v_{i4}v_{i2}$ $(i=1,2,\cdots,n)$.分别连接圈 C_4^{i1} 与 C_4^{i3},C_4^{i2} 与 C_4^{i4} 的顶点 u_{i6} 与 u_{i1},u_{i7} 与 v_{i2};再分别连接 C_4^{i2} 与 $C_4^{i+1,1}$,C^{i4} 与 $C_4^{i+1,3}$ 的顶点 u_{i5} 与 $u_{i+1,3}$,v_{i5} 与 $v_{i+1,3}$ $(i=1,2,\cdots,n-1)$.这样所得的图记为 $2-n(2-2C_4)$,如图 1 所示.$\lambda(n)$ 表示图 $2-n(2-2C_4)$ 的完美对集的数目.则

---

* 作者简介:唐保祥(1961—),男,甘肃天水人,天水师范学院数学与统计学院教授、学士,主要从事图论方面的研究。

$$\lambda(n)=\frac{17+4\sqrt{17}}{34}\cdot(4+\sqrt{17})^n+\frac{17-4\sqrt{17}}{34}\cdot(4-\sqrt{17})^n$$

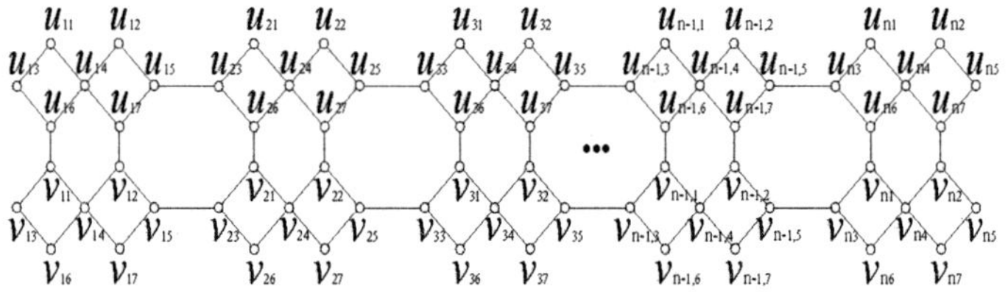

图1　$2-n(2-2C_4)$ 图

证明　显然图 $2-n(2-2C_4)$ 有完美对集. 设图 $2-n(2-2C_4)$ 的完美对集的集合为 $\mathscr{M}$,图 $2-n(2-2C_4)$ 含边 $u_{13}u_{11}$,$u_{13}u_{16}$ 的完美对集的集合分别为 $\mathscr{M}_1$,$\mathscr{M}_2$,则 $\mathscr{M}_1\cap\mathscr{M}_2=\phi$. 所以 $\mathscr{M}=\mathscr{M}_1\cup\mathscr{M}_2$,$\lambda(n)=|\mathscr{M}|=|\mathscr{M}_1|+|\mathscr{M}_2|$.

求 $|\mathscr{M}_1|$ 情形1　$\mathscr{M}_{11}\subseteq\mathscr{M}_1$,$\forall M_{11}\in\mathscr{M}_{11}$,$u_{13}u_{11}$,$u_{16}u_{14}$,$u_{12}u_{15}$,$u_{17}v_{12}$,$v_{13}v_{11}$,$v_{16}v_{14}$,$v_{17}v_{15}\in M_{11}$,由 $\lambda(n)$ 的定义知,$|\mathscr{M}_{11}|=\lambda(n-1)$.

情形2　$\mathscr{M}_{12}\subseteq\mathscr{M}_1$,$\forall M_{12}\in\mathscr{M}_{12}$,$u_{13}u_{11}$,$u_{16}u_{14}$,$u_{12}u_{15}$,$u_{17}v_{12}$,$v_{13}v_{16}$,$v_{11}v_{14}$,$v_{17}v_{15}\in M_{12}$,由 $\lambda(n)$ 的定义知,$|\mathscr{M}_{12}|=\lambda(n-1)$.

情形3　$\mathscr{M}_{13}\subseteq\mathscr{M}_1$,$\forall M_{13}\in\mathscr{M}_{13}$,$u_{13}u_{11}$,$u_{14}u_{12}$,$u_{17}u_{15}$,$u_{16}v_{11}$,$v_{13}v_{16}$,$v_{14}v_{12}$,$v_{17}v_{15}\in M_{13}$,由 $\lambda(n)$ 的定义知,$|\mathscr{M}_{13}|=\lambda(n-1)$.

情形4　$\mathscr{M}_{14}\subseteq\mathscr{M}_1$,$\forall M_{14}\in\mathscr{M}_{14}$,$u_{13}u_{11}$,$u_{14}u_{12}$,$u_{17}u_{15}$,$u_{16}v_{11}$,$v_{13}v_{16}$,$v_{14}v_{17}$,$v_{12}v_{15}\in M_{14}$,由 $\lambda(n)$ 的定义知,$|\mathscr{M}_{14}|=\lambda(n-1)$.

情形5　$\mathscr{M}_{15}\subseteq\mathscr{M}_1$,$\forall M_{15}\in\mathscr{M}_{15}$,$u_{13}u_{11}$,$u_{14}u_{12}$,$u_{16}u_{11}$,$u_{17}v_{12}$,$v_{13}v_{16}$,$v_{14}v_{17}$,$u_{15}u_{23}$,$v_{15}v_{23}$,$u_{21}u_{24}$,$u_{22}u_{25}$,$u_{26}v_{21}$,$u_{27}v_{22}$,$v_{26}v_{24}$,$v_{27}v_{25}\in M_{15}$,由 $\lambda(n)$ 的定义知,$|\mathscr{M}_{15}|=\lambda(n-2)$.

情形6　$\mathscr{M}_{16}\subseteq\mathscr{M}_1$,$\forall M_{16}\in\mathscr{M}_{16}$,$u_{13}u_{11}$,$u_{14}u_{17}$,$u_{12}u_{15}$,$u_{16}v_{11}$,$v_{13}v_{16}$,$v_{14}v_{12}$,$v_{17}v_{15}\in M_{16}$,由 $\lambda(n)$ 的定义知,$|\mathscr{M}_{16}|=\lambda(n-1)$.

情形7　$\mathscr{M}_{17}\subseteq\mathscr{M}_1$,$\forall M_{17}\in\mathscr{M}_1$,$u_{13}u_{11}$,$u_{14}u_{17}$,$u_{12}u_{15}$,$u_{16}v_{11}$,$v_{13}v_{16}$,$v_{14}v_{17}$,$v_{12}v_{15}\in M_{17}$,由 $\lambda(n)$ 的定义知,$|\mathscr{M}_{17}|=\lambda(n-1)$.

易知 $\mathscr{M}_1=\bigcup_{i=1}^{7}\mathscr{M}_{1i}$,$\mathscr{M}_{1i}\cap\mathscr{M}_{1j}=\phi(1\leqslant i<j\leqslant 7)$. 故 $|\mathscr{M}_1|=6\lambda(n-1)+2\lambda(n-2)$.

求 $|\mathscr{M}_2|$ 情形1　$\mathscr{M}_{21}\subseteq\mathscr{M}_2$,$\forall M_{21}\in\mathscr{M}_{21}$,$u_{13}u_{16}$,$u_{11}u_{14}$,$u_{12}u_{15}$,$u_{17}v_{12}$,$v_{13}v_{11}$,$v_{16}v_{14}$,$v_{17}v_{15}\in M_{21}$,由 $\lambda(n)$ 的定义知,$|\mathscr{M}_{21}|=\lambda(n-1)$.

情形2　$\mathscr{M}_{21}\subseteq\mathscr{M}_2$,$\forall M_{21}\in\mathscr{M}_{21}$,$u_{13}u_{16}$,$u_{11}u_{14}$,$u_{12}u_{15}$,$u_{17}v_{12}$,$v_{13}v_{16}$,$v_{11}v_{14}$,

$v_{17}v_{15} \in M_{22}$,由 $\lambda(n)$ 的定义知,$|\mathscr{M}_{22}| = \lambda(n-1)$.

易知 $\mathscr{M}_2 = \mathscr{M}_{21} \cup \mathscr{M}_{22}$,$\mathscr{M}_{21} \cap \mathscr{M}_{22} = \phi$,故 $|\mathscr{M}_2| = 2\lambda(n-1)$.

综上所述,$\lambda(n) = 8\lambda(n-1) + \lambda(n-2)$(1)

线性递推式(1)的特征方程的根为 $x = 4 \pm \sqrt{17}$. 易知 $\lambda(1) = 8$,$\lambda(2) = 65$. 故线性递推式(1)的通解为 $\lambda(n) = \dfrac{17+4\sqrt{17}}{34}\cdot(4+\sqrt{17})^n + \dfrac{17-4\sqrt{17}}{34}\cdot(4-\sqrt{17})^n$。

定理2　分别连接3圈 $u_{1i}u_{2i}u_{3i}$ 与 $u_{1,i+1}u_{2,i+1}u_{3,i+1}$ 的顶点 u_{ji} 与 $u_{j,i+1}(j=1,2,3)$ 得到的图称为3棱柱,记为 $Z_3^i = (i=0,1,2,\cdots,n-1)$. 将3棱柱 Z_3^i 的路 $u_{1i}u_{3i}$ 与 Z_3^{i+1} 的路 $u_{1,i+1}u_{3,i+1}$ 重合(顶点与边分别重合,$i=0,1,2,\cdots,n-1$),所得到的图记为 $2-nZ_3$,再对 $2-nZ_3$ 的顶点重新标识,如图2所示. $f(n)$ 表示图 $2-nZ_3$ 的完美对集的数目. 则

$$f(n) = \frac{13+7\sqrt{13}}{26}\cdot\left(\frac{1+\sqrt{13}}{2}\right)^n + \frac{13-7\sqrt{13}}{26}\cdot\left(\frac{1-\sqrt{13}}{2}\right)^n.$$

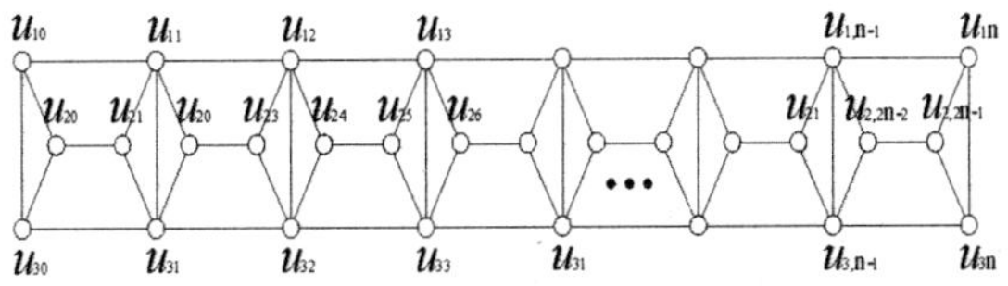

图2　$2-nZ_3$ 图

证明　显然图 $2-nZ_3$ 有完美对集. 设图 $2-nZ_3$ 的完美对集的集合为 $\mathscr{M}$,图 $2-nZ_3$ 含边 $u_{10}u_{11}$,$u_{10}u_{20}$,$u_{10}u_{30}$ 的完美对集的集合分别为 $\mathscr{M}_1$,$\mathscr{M}_2$,$\mathscr{M}_3$,则 $\mathscr{M}_i \cap \mathscr{M}_j = \phi(i \leqslant i < j \leqslant n)$. 所以 $\mathscr{M} = \mathscr{M}_1 \cup \mathscr{M}_2 \cup \mathscr{M}_3$,$f(n) = |\mathscr{M}| = |\mathscr{M}_1| + |\mathscr{M}_2| + |\mathscr{M}_3|$.

求 $|\mathscr{M}_1|$ 情形1　$\mathscr{M}_{11} \subseteq \mathscr{M}_1$,$\forall M_{11} \in \mathscr{M}_{11}$,$u_{10}u_{11}$,$u_{20}u_{21}$,$u_{30}u_{31}$,$u_{22}u_{23} \in M_{11}$,由 $f(n)$ 的定义知,$|\mathscr{M}_{11}| = f(n-2)$.

情形2　$\mathscr{M}_{12} \subseteq \mathscr{M}_1$,$\forall M_{12} \in \mathscr{M}_{12}$,$u_{10}u_{11}$,$u_{20}u_{30}$,$u_{21}u_{31}$,$u_{22}u_{23} \in M_{12}$,由 $f(n)$ 的定义知,$|\mathscr{M}_{12}| = f(n-2)$. 所以 $|\mathscr{M}_1| = 2f(n-2)$.

求 $|\mathscr{M}_2|$　$\forall M_2 \in \mathscr{M}_2$,因为 $u_{10}u_{20} \in M_2$,所以 $u_{30}u_{31}$,$u_{11}u_{21}$,$u_{22}u_{23} \in M_2$. 由 $f(n)$ 的定义知,$|\mathscr{M}_2| = f(n-2)$.

求 $|\mathscr{M}_3|$　$\forall M_3 \in \mathscr{M}_3$,因为 $u_{10}u_{30} \in M_3$,所以 $u_{20}u_{21} \in M_3$. 由 $f(n)$ 的定义知,$|\mathscr{M}_3| = f(n-1)$. 综上所述,$f(n) = f(n-1) + 3f(n-2)$(2)

递推式(2)的特征方程为 $x=\frac{1\pm\sqrt{13}}{2}$. 易知 $f(1)=4, f(2)=7$. 所以递推式(2)的通解为 $f(n)=\frac{13+7\sqrt{13}}{26}\cdot\left(\frac{1+\sqrt{13}}{2}\right)^n+\frac{13-7\sqrt{13}}{26}\cdot\left(\frac{1-\sqrt{13}}{2}\right)^n$.

定理 3　设完全图 K_6^i 的顶点集为 $V(K_6^i)=\{u_{i1},u_{i2},u_{i3},u_{i4},u_{i5},u_{i6}\}$，分别连接 K_6^i 于 K_6^{i+1} 的顶点 u_{ij} 与 $u_{i+1,7-j}(j=1,2,3;i=1,2,\cdots,n-1)$，这样得到的图记为 $3-nK_6$，如图 3 所示. $g(n)$ 表示图 $3-nK_6$ 的完美对集的数目. 则

$$g(n)=\frac{7+2\sqrt{7}}{14}\cdot(9+3\sqrt{7})^n+\frac{7-2\sqrt{7}}{14}\cdot(9-3\sqrt{7})^n.$$

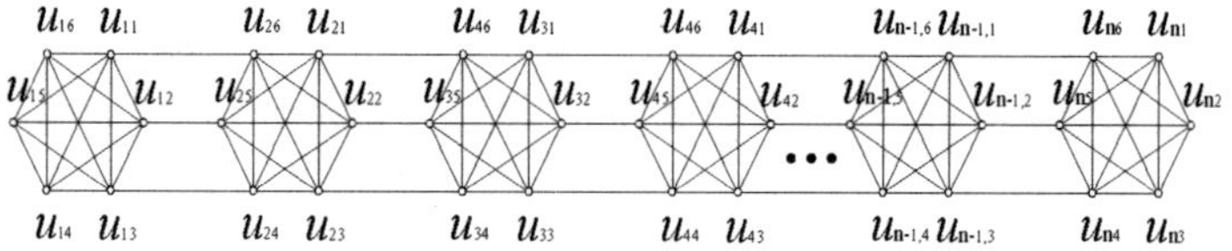

图 3　$3-nK_6$ 图

证明　欲求 $g(n)$，先定义 3 个图 G_1, G_2 和 G_3，并求其完美对集的数目. 将长为 1 的路 uv 的端点 u 和 v 分别与图 $3-nK_6$ 的顶点 u_{15} 和 u_{14}，u_{16} 和 u_{14}，u_{16} 和 u_{15}，各连接一条边，得到的图分别记为 G_1, G_2, G_3，如图 4,5,6 所示. 显然图 G_1, G_2, G_3 均有完美对集，且 $G_1\cong G_2\cong G_3$. $\alpha(n), \beta(n), \gamma(n)$ 分别表示图 G_1, G_2, G_3 的完美对集的数目，则 $\alpha(n)=\beta(n)=\gamma(n)$.

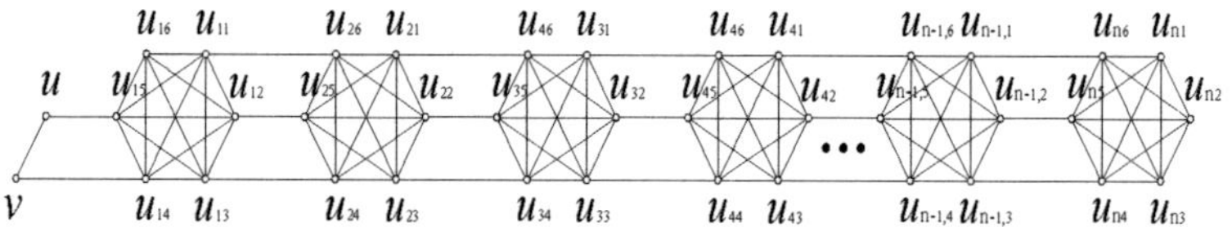

图 4　G_1 图

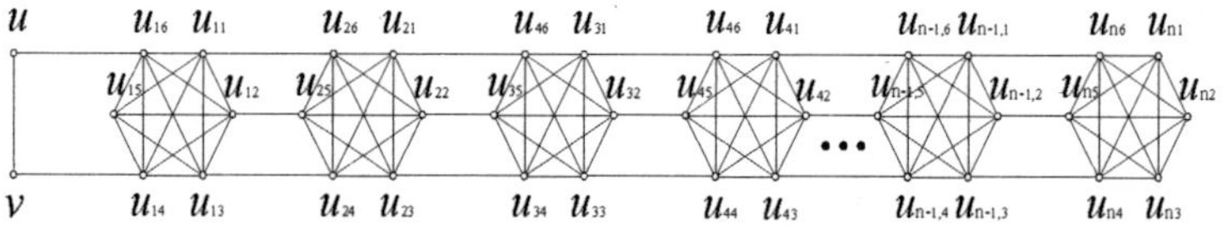

图 5　G_2 图

求 $\alpha(n)$. 设图 G_1 的完美对集的集合为 $\mathscr{M}$，G_1 含边 uv, uu_{15} 的完美对集的集合

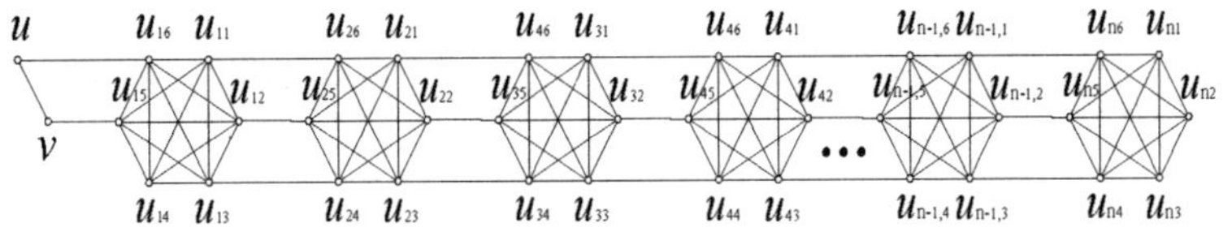

图 6　G_3 图

分别为 $\mathscr{M}_1$，$\mathscr{M}_2$，则 $\mathscr{M}_1 \cap \mathscr{M}_2 = \phi$. 所以 $\mathscr{M} = \mathscr{M}_1 \cup \mathscr{M}_2$，$\alpha(n) = |\mathscr{M}| = |\mathscr{M}_1| + |\mathscr{M}_2|$.

求 $|\mathscr{M}_1|$ $\forall M_1 \in \mathscr{M}_1$，因为 $uv \in M_1$，所以由 $g(n)$ 的定义知，$|\mathscr{M}_1| = g(n)$.

求 $|\mathscr{M}_2|$ 情形 1　$\mathscr{M}_{21} \subseteq \mathscr{M}_2$，$\forall M_{21} \in \mathscr{M}_{21}$，$uu_{15}$，$vu_{14}$，$u_{16}u_{11} \in M_{21}$，由 $\alpha(n)$ 的定义知，$|\mathscr{M}_{21}| = \alpha(n-1)$.

情形 2　$\mathscr{M}_{22} \subseteq \mathscr{M}_2$，$\forall M_{22} \in \mathscr{M}_{22}$，$uu_{15}$，$vu_{14}$，$u_{16}u_{12} \in M_{22}$，由 $\beta(n)$ 的定义知，$|\mathscr{M}_{22}| = \beta(n-1)$.

情形 3　$\mathscr{M}_{23} \subseteq \mathscr{M}_2$，$\forall M_{23} \in \mathscr{M}_{23}$，$uu_{15}$，$vu_{14}$，$u_{16}u_{13} \in M_{23}$，由 $\gamma(n)$ 的定义知，$|\mathscr{M}_{23}| = \gamma(n-1)$.

易知 $\mathscr{M}_2 = \mathscr{M}_{21} \cup \mathscr{M}_{22} \cup \mathscr{M}_{23}$，$\mathscr{M}_{2i} \cap \mathscr{M}_{2j} = \phi (1 \leqslant i < j \leqslant 3)$.

故 $|\mathscr{M}_2| = \alpha(n-1) + \beta(n-1) + \gamma(n-1)$.

综上所述，$\alpha(n) = g(n) + \alpha(n-1) + \beta(n-1) + \gamma(n-1)$(3)

求 $g(n)$. 显然图 $3-nK_6$ 有完美对集. 设 $2-nZ_3$ 的完美对集的集合为 $\mathscr{M}$，$2-nZ_3$ 含边 $u_{15}u_{16}$，$u_{15}u_{11}$，$u_{15}u_{12}$，$u_{15}u_{13}$，$u_{15}u_{14}$ 的完美对集的集合分别为 $\mathscr{M}_1$，$\mathscr{M}_2$，$\mathscr{M}_3$，$\mathscr{M}_4$，$\mathscr{M}_5$，则 $\mathscr{M}_i \cap \mathscr{M}_j = \phi (1 \leqslant i < j \leqslant 5)$.

所以 $\mathscr{M} = \bigcup_{i=1}^{5} \mathscr{M}_i$，$|\mathscr{M}| = |\mathscr{M}_1| + |\mathscr{M}_2| + |\mathscr{M}_3| + |\mathscr{M}_4| + |\mathscr{M}_5|$.

求 $|\mathscr{M}_1|$ 情形 1　$\mathscr{M}_{11} \subseteq \mathscr{M}_1$，$\forall M_{11} \in \mathscr{M}_{11}$，$u_{15}u_{16}$，$u_{14}u_{11} \in M_{11}$，由 $\alpha(n)$ 的定义知，$|\mathscr{M}_{11}| = \alpha(n-1)$.

情形 2　$\mathscr{M}_{12} \subseteq \mathscr{M}_1$，$\forall M_{12} \in \mathscr{M}_{22}$，$u_{15}u_{16}$，$u_{14}u_{12} \in M_{12}$，由 $\beta(n)$ 的定义知，$|\mathscr{M}_{12}| = \beta(n-1)$.

情形 3　$\mathscr{M}_{13} \subseteq \mathscr{M}_1$，$\forall M_{13} \in \mathscr{M}_{23}$，$u_{15}u_{16}$，$u_{14}u_{13} \in M_{13}$，由 $\gamma(n)$ 的定义知，$|\mathscr{M}_{23}| = \gamma(n-1)$.

易知 $\mathscr{M}_1 = \mathscr{M}_{11} \cup \mathscr{M}_{12} \cup \mathscr{M}_{13}$，$\mathscr{M}_{1i} \cap \mathscr{M}_{1j} = \phi (1 \leqslant i < j \leqslant 3)$.

故 $|\mathscr{M}_1| = \alpha(n-1) + \beta(n-1) + \gamma(n-1)$.

求 $|\mathscr{M}_2|$ 情形 1　$\mathscr{M}_{21} \subseteq \mathscr{M}_2$，$\forall M_{21} \in \mathscr{M}_{21}$，$u_{15}u_{11}$，$u_{16}u_{12}$，$u_{14}u_{13} \in M_{21}$，由 $g(n)$ 的定义知，$|\mathscr{M}_{21}| = g(n-1)$.

情形 2　$\mathscr{M}_{22} \subseteq \mathscr{M}_2$，$\forall M_{22} \in \mathscr{M}_{22}$，$u_{15}$，$u_{11}$，$u_{16}u_{12} \in M_{22}$，由 $g(n)$ 的定义知，|

$\mathscr{M}_{22}| = g(n-1)$.

情形 3 $\mathscr{M}_{23} \subseteq \mathscr{M}_2$, $\forall M_{23} \in \mathscr{M}_{23}$, u_{15}, u_{11}, $u_{16}u_{14} \in M_{23}$, 由 $\alpha(n)$ 的定义知, $|\mathscr{M}_{23}| = \alpha(n-1)$.

易知 $\mathscr{M}_2 = \mathscr{M}_{21} \cup \mathscr{M}_{22} \cup \mathscr{M}_{23}$, $\mathscr{M}_{2i} \cap \mathscr{M}_{2j} = \phi(1 \le i < j \le 3)$.

故 $|\mathscr{M}_2| = 2g(n-1) + \alpha(n-1)$.

求 $|\mathscr{M}_3|$ 情形 1 $\mathscr{M}_{31} \subseteq \mathscr{M}_3$, $\forall M_{31} \in \mathscr{M}_{31}$, $u_{15}u_{12}$, $u_{16}u_{11}$, $u_{14}u_{13} \in M_{31}$, 由 $g(n)$ 的定义知, $|\mathscr{M}_{31}| = g(n-1)$.

情形 2 $\mathscr{M}_{32} \subseteq \mathscr{M}_3$, $\forall M_{32} \in \mathscr{M}_{32}$, $u_{15}u_{12}$, $u_{16}u_{13}$, $u_{14}u_{11} \in M_{32}$, 由 $g(n)$ 的定义知, $|\mathscr{M}_{32}| = g(n-1)$.

情形 3 $\mathscr{M}_{33} \subseteq \mathscr{M}_2$, $\forall M_{33} \in \mathscr{M}_{33}$, $u_{15}u_{12}$, $u_{16}u_{14} \in M_{33}$, 由 $\beta(n)$ 的定义知, $|\mathscr{M}_{33}| = \beta(n-1)$.

易知 $\mathscr{M}_2 = \mathscr{M}_{21} \cup \mathscr{M}_{22} \cup \mathscr{M}_{23}$, $\mathscr{M}_{2i} \cap \mathscr{M}_{2j} = \phi(1 \le i < j \le 3)$.

故 $|\mathscr{M}_3| = 2g(n-1) + \beta(n-1)$.

求 $|\mathscr{M}_4|$ 情形 1 $\mathscr{M}_{41} \subseteq \mathscr{M}_4$, $\forall M_{41} \in \mathscr{M}_{41}$, $u_{15}u_{13}$, $u_{16}u_{11}$, $u_{14}u_{12} \in M_{41}$, 由 $g(n)$ 的定义知, $|\mathscr{M}_{31}| = g(n-1)$.

情形 2 $\mathscr{M}_{42} \subseteq \mathscr{M}_4$, $\forall M_{42} \in \mathscr{M}_{42}$, $u_{15}u_{13}$, $u_{16}u_{12}$, $u_{14}u_{11} \in M_{42}$, 由 $g(n)$ 的定义知, $|\mathscr{M}_{42}| = g(n-1)$.

情形 3 $\mathscr{M}_{43} \subseteq \mathscr{M}_4$, $\forall M_{43} \in \mathscr{M}_{33}$, $u_{15}u_{13}$, $u_{16}u_{14} \in M_{43}$, 由 $\gamma(n)$ 的定义知, $|\mathscr{M}_{43}| = \gamma(n-1)$.

易知 $\mathscr{M}_4 = \mathscr{M}_{41} \cup \mathscr{M}_{42} \cup \mathscr{M}_{43}$, $\mathscr{M}_{4i} \cap \mathscr{M}_{4j} = \phi(1 \le i < j \le 3)$.

故 $|\mathscr{M}_4| = 2g(n-1) + \gamma(n-1)$.

求 $|\mathscr{M}_5|$ 情形 1 $\mathscr{M}_{51} \subseteq \mathscr{M}_5$, $\forall M_{51} \in \mathscr{M}_{51}$, $u_{15}u_{14}$, $u_{16}u_{11} \in M_{51}$, 由 $\alpha(n)$ 的定义知, $|\mathscr{M}_{51}| = \alpha(n-1)$.

情形 2 $\mathscr{M}_{52} \subseteq \mathscr{M}_5$, $\forall M_{52} \in \mathscr{M}_{32}$, $u_{15}u_{14}$, $u_{16}u_{12} \in M_{52}$, 由 $\beta(n)$ 的定义知, $|\mathscr{M}_{52}| = \beta(n-1)$.

情形 3 $\mathscr{M}_{53} \subseteq \mathscr{M}_5$, $\forall M_{43} \in \mathscr{M}_{33}$, $u_{15}u_{14}$, $u_{16}u_{13} \in M_{53}$, 由 $\gamma(n)$ 的定义知, $|\mathscr{M}_{53}| = \gamma(n-1)$.

易知 $\mathscr{M}_5 = \mathscr{M}_{51} \cup \mathscr{M}_{52} \cup \mathscr{M}_{53}$, $\mathscr{M}_{5i} \cap \mathscr{M}_{5j} = \phi(1 \le i < j \le 3)$.

故 $|\mathscr{M}_5| = \alpha(n-1) + \beta(n-1) + \gamma(n-1)$.

综上所述,

$$g(n) = 6g(n-1) + 3\alpha(n-1) + 3\beta(n-1) + 3\gamma(n-1) \tag{4}$$

由(3) 和(4),得

$$g(n) = 15g(n-1) + 9\alpha(n-2) + 9\beta(n-2) + 9\gamma(n-2) \tag{5}$$

再由(4),得

$$g(n-1)=6g(n-1)+3\alpha(n-2)+3\beta(n-2)+3\gamma(n-2) \tag{6}$$

(5) − 3 × (6),得

$$g(n)=18g(n-1)-18g(n-2) \tag{7}$$

(7) 式的特征方程的根为 $x=9\pm 3\sqrt{7}$.

易得 $g(1)=15$,$\alpha(1)=\beta(1)=\gamma(1)=18$. 所以由(4) 式得,$g(2)=252$.

线性递推式(7) 的通解为

$$g(n)=\frac{7+2\sqrt{7}}{14}\cdot(9+3\sqrt{7})^n+\frac{7-2\sqrt{7}}{14}\cdot(9-3\sqrt{7})^n.$$

参考文献

[1]KasteleynPW. Dimmer statistics and phase transition[J]. Math Phys,1963,4:287−293.

[2]Valiant L G. The complexity of computing the permanent[J]. Theoretical Compute Science,1979,8(2):189−201.

[3] CyvinS J, GutmanI. Kekulé structures in Benzennoid hydrocarbons [M], Berlin: Springer Press,1988.

[4]Fournrei J C. Combinatorics of perfect matchings in planar bipartite graphs and application to tilings[J]. Theoretical Computer Science. 2003,303:333−351.

[5]ZhangH P. The connectivity of Z − transformation graphs of perfect matchings of polyominoes[J]. Discrete Mathematics,1996,158:257−272.

[6]ZhangHP,Zhang FJ. Perfectmatchings of polyomino graphs[J]. Graphs and Combinatorics,1997,13:259−304.

[7]张莲珠. 涉位四角系统完美匹配数的计算[J]. 厦门大学学报:自然科学版,1998,37(5):629−633.

[8]林泓,林晓霞. 若干四角系统完美匹配数的计算[J]. 福州大学学报:自然科学版,2005,33(6):704−710.

[9]Yan WG, Zhang Fu − ji. Enumeration of perfect matchings of a type of Cartesian products of graphs[J],Discrete Applied Mathematics,2006,154:145−157.

[10]唐保祥,任韩. 几类图完美匹配的数目[J]. 南京师大学报:自然科学版,2010,33(3):1−6.

[11]唐保祥,李刚,任韩. 3 类图完美匹配的数目[J]. 浙江大学学报:理学版,2011,38(4):16−19.

[12]唐保祥,任韩. 6 类图完美匹配的数目[J]. 中山大学学报:自然科学版,2012,51(2):40−44.

[13]唐保祥,任韩.四类图完美匹配的计数公式[J].福州大学学报:自然科学版,2012,40(4):437—440.

[14]唐保祥,任韩.4类图完美匹配的计数[J].武汉大学学报:理学版,2012,58(5):441—446.

[15]唐保祥,任韩.2类偶图完美匹配的数目[J].西南大学学报:自然科学版,2012,34(10):91—95.

[16]唐保祥,任韩.5类图完美匹配的数目[J].中山大学学报:自然科学版,2012,51(4):31—37.

(本文发表于2014年《南开大学学报》第47卷5期)

克尔媒质中耦合三能级T－C模型中原子信息熵的性质

董忠　尤良芳*

利用全量子理论，讨论了克尔媒质中耦合V型三能级原子与相干态光场相互作用过程中原子信息熵的演化特性。数值计算结果表明：原子信息熵的演化特性对系统初始时刻场平均光子数的大小很敏感。初始时刻场的平均光子数比较大时，原子信息熵的演化会出现明显的周期性崩塌和回复现象。初态中原子激发态概率幅，原子间耦合强度，失谐量与克尔系数等系统参量都会对原子信息熵的演化特性产生明显的影响。

1　引言

Jaynes－Cummings模型[JCM]是上世纪六十年代建立起来的描述单模量子化光场与单原子相互作用的理论模型。1968年，Tavis和Cummings在JCM的基础上又建立了描述两个两能级原子与量子化光场相互作用的Tavis－Cummings模型[TCM]。在此后的岁月中，科学家们经过推广，建立了考虑克尔效应在内的描述量子化光场与两个三能级原子相互作用的各种理论模型[1-5]。这些模型都是精确可解的。在这些模型中，光场和原子的行为呈现出许多非经典效应，如光场的反聚束和压缩，原子反转时间演化的周期崩溃和回复，原子偶极矩的压缩等[6-8]。但后来的研究表明：仅仅研究这些量子效应，还不能完全理解光场与原子相互作用的动力学行为。Phoenix和Knight等人用量子熵研究光场与二能级原子相互作用时的动力学特性[9-10]，显示出很大的优越性。熵是一个十分灵敏、有用的物理量，因为熵函数自动包含了量子系统密度矩阵的全部统计矩。特别值得一提的是，量子信息学是建立在量子信息熵概念基础之上的，而量子信息熵则

* 作者简介：董忠(1967—)，男，湖北武汉人，天水师范学院电子信息与电气工程学院教授，主要从事原子与分子物理学研究。

是用来确定场一原子间量子纠缠与量子关联程度、解释场一原子相互作用过程中各种量子效应的微观机制以及进行量子信息测量的理论根据。场一原子相互作用系统中原子的量子场熵和场一原子间的量子纠缠及其时间演化特性研究，由于在量子通信和量子光通信有着直接重要的应用，因而引起极大关注[11]。同时，研究表明，三能级原子在量子系统的密钥分发方案中比二能级原子更能有效地防止窃听，也具有更显著的量子非局域性，所以，研究三能级原子与光场相互作用系统的量子性质有着更大的意义和实用价值[12]。原子(光场)信息熵的时间演化在一定程度上反映了光场与原子关联程度的时间演化。原子(光场)信息熵越大，光场与原子关联的程度越高。于是，研究光场和原子相互作用过程中子系统信息熵的性质，对于更加深刻的了解原子与光场的相互作用过程有着十分重要的意义。

将近二十年来，科学家们对单个两能级原子、耦合的两能级原子和单个的三能级原子与光场相互作用过程中子系统信息熵随时间的演化特性做了比较系统的研究[13-19]。但是，对耦合的三能级原子，特别是对克尔媒质中耦合的V型三能级原子与相干态光场相互作用过程中，原子(光场)信息熵随时间演化特性的研究目前还没有见到相关的报道。本篇论文，首先对克尔媒质中耦合的全同V型三能级原子与相干态光场相互作用的动力学过程进行求解，然后对原子子系统信息熵的时间演化特性进行了数值计算并作了分析。

2 模型与求解

2.1 系统的哈密顿量

两个全同的V型三能级原子，其单个原子的能级构型如下图：

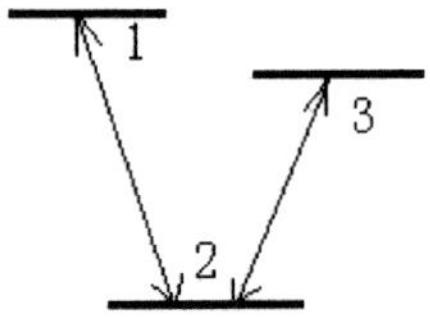

图1 V型三能级原子能级结构

能级$|1\rangle \leftrightarrow |2\rangle$和$|3\rangle \leftrightarrow |2\rangle$之间的跃迁是允许的，能级$|1\rangle \leftrightarrow |3\rangle$之间的跃迁是禁戒的。在相互作用绘景中，耦合的两个全同V型三能级原子与单模光场单光子相互作用的哈密顿量为(取 $\hbar=1$)：

$$H_1 = a[g_1(S_{12}^{(A)} + S_{12}^{(B)})\exp(-i\Delta_1 t) + g_3(S_{32}^{(A)} + S_{32}^{(B)})\exp(-i\delta_3 t)]$$

$$+\Omega[S_{12}^{(A)}+S_{12}^{+(B)}]+S_{32}^{(A)}S_{32}^{+(B)}+S_{12}^{(A)}S_{32}^{+(B)}\exp(i\Delta_{31}t)$$
$$+S_{32}^{(A)}S_{12}^{+(B)}\exp(-i\Delta_{31}t)]+H.C.+\chi a^{+2}a^{2} \tag{1}$$

其中 $\Delta_1=\omega-(\omega_1-\omega_2)$,$\Delta_3=\omega-(\omega_3-\omega_2)$,$\Delta_{31}=\omega_1-\omega_3=\Delta_3-\Delta_1$。$\omega$ 为光场的频率,ω_1 和 ω_3 分别是原子 $|1\rangle$,$|2\rangle$ 和 $|3\rangle$ 能级的能量频率;"$H.C.$"为共轭项;g_1 与 g_3 分别是光场和能级 $|1\rangle\leftrightarrow|2\rangle$,$|3\rangle\leftrightarrow|2\rangle$ 之间跃迁的耦合系数;Ω 为原子偶极矩之间的耦合强度;$S_{ij}^{(k)}$ 为表示第k个原子从能级j到能级i的跃迁算符;Δ_1 和 Δ_3 分别是 $|1\rangle\leftrightarrow|2\rangle$ 和 $|3\rangle\leftrightarrow|2\rangle$ 之间跃迁的失谐量;χ 为克尔媒质的三阶非线性系数;a 和 a^+ 分别是光子的湮灭和产生算符。

2.2 态函数的微分方程及其解

设原子子系统的初态为两个全同 V 型三能级原子都处于基态与激发态的相干叠加,则原子子系统的态函数为:

$$|\psi_a(0)\rangle=\cos\theta|2,2\rangle+\frac{1}{\sqrt{2}}\sin\theta\exp(i\varphi)(|1,1\rangle+|3,3\rangle) \tag{2}$$

其中 φ 为叠加相位。$\cos\theta$,$\frac{1}{\sqrt{2}}\sin\theta$ 为原子系统初始时刻处于两较低能态与激发态的概率幅。设系统初始时刻的光场为相干态光场:

$$|\psi_f(0)\rangle=\sum_{n=0}^{\infty}F_n|n\rangle=\sum_{n=0}^{\infty}\exp(-|\alpha|^2/2)\frac{a^n}{\sqrt{n!}}|n\rangle;(\alpha=|\alpha|\exp(i\phi)) \tag{3}$$

式中 $|\alpha|^2=\bar{n}$ 为平均光子数,$|\alpha|$ 越大,平均光子数越大,ϕ 为α 的相位角。则系统初始时刻的态函数可表示为:

$$|\Psi(0)\rangle=\sum_{n=0}^{\infty}F_n\left[\cos\theta|n,2,2\rangle+\frac{1}{\sqrt{2}}\sin\theta\exp(i\varphi)(|n,1,1\rangle+|n,3,3\rangle)\right] \tag{4}$$

设在时刻 t,波函数演化为:

$$|\psi(t)\rangle=\sum_{n=0}^{\infty}[C_1^n|n,1,1\rangle+C_2^n|n,1,2\rangle+C_3^n|n,1,3\rangle+C_4^n|n,2,1\rangle$$
$$+C_5^n|n,2,2\rangle+C_6^n|n,2,3\rangle+C_7^n|n,3,1\rangle+C_8^n|n,3,2\rangle$$
$$+C_9^n|n,3,3\rangle] \tag{5}$$

上式中的 C_1^n 都是时间 t 的函数,其初始值分别为:$C_1^n(0)=C_9^n(0)=F_n\sin\theta\exp(i\varphi)/\sqrt{2}$;$C_3^n(0)=C_7^n(0)=0$;$C_2^n(0)=C_4^n(0)=0$;$C_6^n(0)=C_8^n(0)=0$;$C_5^n(0)=F_n\cos\theta$。

考虑一种比较简单的情况:$\Delta_1=\Delta_3=\Delta$,即 $|1\rangle$,$|3\rangle$ 能级兼并,则 $\Delta_{31}=0$,且 $g_1=g_3=g$. 在这种情况下,依据相互作用绘景中的薛定谔方程,解方程得微分

方程组的解为：

$$C_1^n = C_9^n = (d^n + h^n)/2 \tag{6}$$

$$C_3^n = C_7^n = (d^n - h^n)/2 \tag{7}$$

$$\begin{aligned} d^n = \frac{\exp(-2i\Delta t)}{4D_3 D_4}\{[(D_2 - \Delta + \gamma_1)(D_1 + \gamma_1) - 4D_4^2]m_1 \exp(i\gamma_1 t) \\ + [(D_2 - \Delta + \gamma_2)(D_1 + \gamma_2) - 4D_4^2]m_2 \exp(i\gamma_2 t) \\ + [(D_2 - \Delta + \gamma_3)(D_1 + \gamma_3) - 4D_4^2]m_3 \exp(i\gamma_3 t)\} \end{aligned} \tag{8}$$

$$h^n = (F_n \cos\theta / \sqrt{2})\exp(-iD_5 \chi t) \tag{9}$$

$$C_2^0 = C_4^0 = C_6^0 = C_8^0 = \alpha_1 \exp(i\beta_1 t) + \alpha_2 \exp(i\beta_2 t) \tag{10a}$$

$$\begin{aligned} C_2^{n+1} = -\frac{\exp(-i\Delta t)}{4D_4}[(\gamma_1 + D_1)m_1 \exp(i\gamma_1 t)] + (\gamma_2 + D_1)m_2 \exp(i\gamma_2 t) \\ + (\gamma_3 + D_1)m_3 \exp(i\gamma_3 t)] \end{aligned} \tag{10b}$$

$$C_5^0 = F_0 \cos\theta \tag{11a}$$

$$C_5^1 = -\frac{1}{g}\exp(i\Delta t)[(\beta_1 + 2\Omega)\alpha_1 \exp(i\beta_1 t) + (\beta_2 + 2\Omega)\alpha_2 \exp(i\beta_2 t)] \tag{11b}$$

$$C_5^{n+2} = m_1 \exp(i\gamma_1 t) + m_2 \exp(i\gamma_2 t) + m_3 \exp(i\gamma_3 t) \tag{11c}$$

其中

$$\begin{gathered} D_1 = (n+1)(n+2)\chi; D_2 = 2\Omega + n(n+1)\chi; D_3 = g\sqrt{n+1}; \\ D_4 = g\sqrt{n+2}; D_5 = n(n-1)\chi \end{gathered} \tag{12}$$

β_1, β_2 是下面一元二次方程的两个根：

$$\beta^2 + (\Delta + 2\Omega)\beta + 2(\Delta\Omega - 2g^2) = 0$$

α_1, α_2 是下面二元一次方程组的两个根：

$$\alpha_1 + \alpha_2 = 0 \tag{14a}$$

$$\alpha_1\beta_1 + \alpha_2\beta_2 = -gF_1\cos\theta \tag{14b}$$

γ_1, γ_2 和 γ_3 是下面一元三次方程的三个根：

$$\begin{aligned} \gamma^3 + (D_1 + D_2 + D_5 - 3\Delta)\gamma^2 - [4(D_3^2 + D_4^2) - D_1 D_5 - D_1 D_2 - D_2 D_5 \\ + \Delta(D_1 + D_5) + 2\Delta(D_1 + D_2 - \Delta)]\gamma - [4D_5 D_4^2 + 4D_1 D_3^2 - D_1 D_2 D_5 \\ + \Delta(D_1 D_5 - 2\Delta D_1 + 2D_1 D_2 - 8D_4^2)] = 0 \end{aligned} \tag{15}$$

m_1, m_2, m_3 是下面三元一次方程组的三个根：

$$m_1 + m_2 + m_3 = F_{n+2}\cos\theta \tag{16a}$$

$$m_1\gamma_1 + m_2\gamma_2^2 + m_3\gamma_3^2 = -D_1 F_{n+2}\cos\theta \tag{16b}$$

$$m_1\gamma_1^2 + m_2\gamma_2^2 + m_3\gamma_3^2 = 2\sqrt{2}D_3 D_4 F_n \sin\theta \exp(i\varphi) + (4D_4^2 + D_1^2)F_{n+2}\cos\theta \tag{16c}$$

3 原子信息熵的性质

3.1 原子的约化密度算符和原子信息熵

依据(5)式,求得原子子系统的约化密度算符为:

$$\rho_a = Tr_f(|\psi(t)\rangle\langle\psi(t)|) \tag{17}$$

下标 f 表示对光场求迹。则原子子系统的信息熵为[16](取 $k=1$,k 为玻尔兹曼常数):

$$S_a = -Tr(\rho_a \text{In}\beta_a) = -\sum_{i=1}^{9} P_i \text{In} P_i \tag{18}$$

$$P_i = \sum_{n=0}^{\infty} C_i^n C_i^{n^*} \tag{19}$$

要得到原子信息熵的解析表达式不易,本文采用数值计算来分析原子信息熵的性质(以下取 $\phi=0$,$\varphi=0$,$g=1$)。

3.2 数值计算和讨论

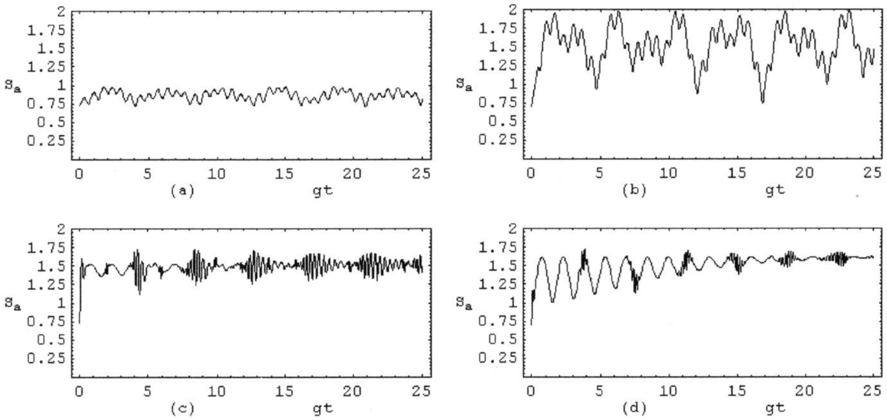

图 2 S_a 在 $\Delta=0.7$,$\Omega=3$,$\chi=0.8$ 和 (a)$\bar{n}=0.01$,$\theta=\pi/6$;(b)$\bar{n}=0.01$,$\theta=\pi/2$;(c)$\bar{n}=16$,$\theta=\pi/6$;(d)$\bar{n}=16$,$\theta=\pi/2$ 的时间演化曲线

从图 2 可以看出,原子信息熵随时间的变化总是呈现振荡的形态,这说明原子与场之间的关联总是处于加强和减弱的周期性循环过程中。原子信息熵随时间的演化规律与系统初始时刻平均光子数的大小有很大关系。

在初始时刻平均光子数比较小,原子处于激发态的概率幅从小变大时,原子信息熵随时间的演化从只是做简单的窄幅振荡(图 2(a))到其振荡幅度变大,且其均值上移(图 2(b))。这说明在系统初始时刻的平均光子数比较小时,原子处于激发态的概率幅越大,原子与场之间的关联越强。

在初始时刻平均光子数比较大，且原子处于激发态的概率幅比较小时，在原子和场相互作用的初期，原子信息熵随时间的演化体现出了明显的周期性崩塌和回复现象(图 2(c))；在初始时刻平均光子数比较大，且原子处于激发态的概率幅比较大时，原子信息熵振荡的最大值基本上没有变化，在原子和场相互作用的初期，原子信息熵振荡的最小值逐渐增大。随着时间的延长，原子信息熵的演化又体现出幅度很小的崩塌和回复现象，最终趋于振幅很小的稳定值(图 2(d))。

造成上述现象的原因不难分析。当光场比较弱时(即初始时刻光场的平均光子数比较小时)，光场与原子之间作用的主要动力学来源在于原子系统初始时刻的内部能量，因为初始时刻原子内部的能量比较小时，系统总的能量比较小，当然在系统演化的过程中，系统中光场的总能量也相应的比较小，自然场与原子的相互作用就弱，原子与光场在系统动力学演化的过程中相互的关联就弱。由此看来，当初始时刻场比较弱时，原子与光场的关联主要决定与初始时刻原子系统的能量。而当初始时刻光场比较强时，系统中总的能量主要来源于光场的能量，原子在与光场作用过程中辐射的光场能量在场的总能量中所占比例不大。实际上，也就是说，系统中光场总能量的变化幅度与系统总能量的比值变小。所以，在这种情况下，初始时刻原子处于激发态概率幅的变化对原子信息熵在演化过程中均值变化的影响不大。

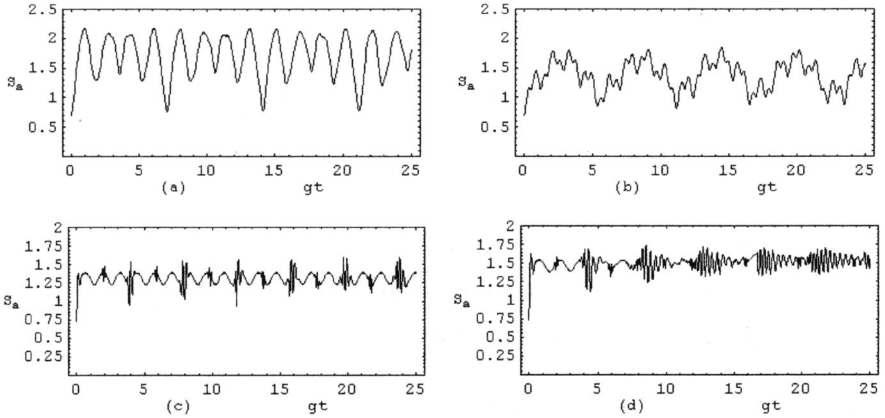

图 3　S_a 在 $\alpha=0.1,\theta=\pi/2,\Delta=2,\chi=0.8$ 和(a)$\Omega=0.4$;(b)$\Omega=3$ 与 S_a 在 $\alpha=4,\theta=\pi/6,\Delta=0.1,\chi=0.8$ 和(c)$\Omega=0.4$;(d)$\Omega=3$ 的时间演化曲线

从图 3 可以看出，系统初始时刻场的平均光子数比较小时，原子间耦合强度的增大会降低原子信息熵随时间演化的平均值，使原子信息熵演化曲线从周期比较短的振荡向周期比较长、噪声比较大的振荡形态过渡(图 3(a)(b))。在系统初始时刻场的平均光子数比较大时,原子间耦合强度的增大对原子信息熵随时间演化的平均值略有增大。同时,使原子信息熵演化呈现崩塌回复现象的时间变短(图 3(c)(d))。原子间耦合强度变化对原子信息熵随时间的演化当系统初始时刻平均光子数变化时会体现出不同的影响特性。

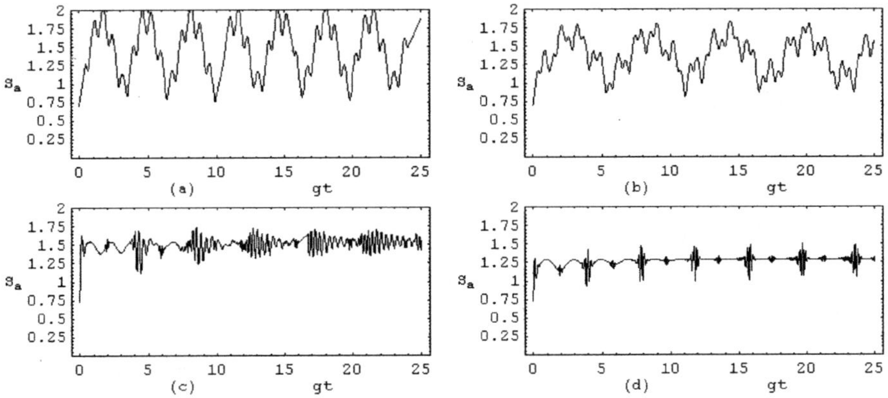

图 4　S_a 在 $\alpha=0.1,\theta=\pi/2,\Omega=3,\chi=0.8$ 和(a)$\Delta=0.1$;(b)$\Delta=2$ 与 S_a 在 $\alpha=4,\theta=\pi/6,\Omega=3,\chi=0.8$ 和(c)$\Delta=0.1$;(d)$\Delta=10$ 的时间演化曲线

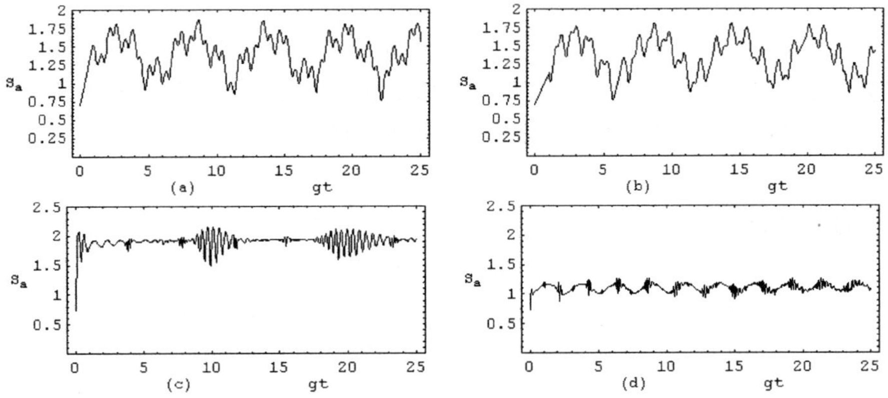

图 5　S_a 在 $\alpha=0.1,\theta=\pi/2,\Delta=2,\Omega=3$ 和(a)$\chi=0.4$;(b)$\chi=1.5$ 与 S_a 在 $\alpha=4,\theta=\pi/6,\Delta=0.1,\Omega=3$ 和(c)$\chi=0.4$;(d)$\chi=1.5$ 的时间演化曲线

从图 4 和图 5 可以看出,失谐量和克尔系数的增大,都会使原子信息熵随时间演化的平均值降低,特别是在系统初始时刻平均光子数比较大时尤其显著。在初始时刻平均光子数比较小时,失谐量和克尔系数的增大,会使原子信息熵振荡的周期变长(图 4(a)(b),图 5(a)(b))。而在初始时刻平均光子数比较大时,失谐量和克尔系数的增大，对原子信息熵随时间的演化特性体现出了不同的影响特性。失谐量增大，会使原子信息熵演化的周期性崩塌和回复现象更加明显（图4(c)(d)）。而克尔系数的增大，会使原子信息熵演化的崩塌和回复周期变短（图5(c)(d)）。但是，从总体上来讲，失谐量和克尔系数的增大，会降低原子与光场之间的关联。这是容易理解的，失谐量增大，会降低原子在场的作用下的跃起几率，从而降低原子与场之间的关联。克尔系数的增大，会增强场与克尔介质之间的作用，当然也就使场与原子之间的作用减小，同时使原子与场之间的关联降低。

4 结论

利用全量子理论，讨论了耦合V型三能级原子与相干态光场相互作用过程中原子信息熵的演化特性。数值计算结果表明：原子信息熵的演化特性对场的平均光子数的大小很敏感。

(1)初始时刻场的平均光子数比较小时，原子初态中激发态概率幅的增大，使原子信息熵的振荡幅度增大，平均值升高。原子间耦合的强度，失谐量与克尔系数的增大都会使原子信息熵的演化周期变长。

(2)初始时刻场的平均光子数比较大时，原子信息熵的演化会出现明显的周期性崩塌和回复现象。在这种情况下，原子初态中激发态概率幅的增大，不会使原子信息熵的平均值和最大值有明显的变化，而是使其振荡过程中的最小值逐渐升高。原子间耦合强度的增大，会使原子信息熵演化的平均值略有升高，失谐量与克尔系数的增大却会使原子信息熵的平均值降低。

参考文献

[1] JaynesET, CummingsFW. Comparisionofquantumandsemi-classicalradiationtheorieswithapplication to beam maser[J]. Proc IEEE,1963,51(1):89－109.

[2]Tavis M,CummingsFW. Exact solution for an N－Molecule－Radiation－Field Hamiltonian[J]. PhysRev,1968,170:379－384.

[3]B. W. Shore,P. L. Knght. The Jaynes－Cummings model,J. Modern Opt. 1993,40(7):1195－1238.

[4]Agarwal GS,PuriRR. Collapse and revival phenomenon in the evolution of a resonant field in a Kerr－like medium[J]. PhysRevA,1989,39(6):2969－2997.

[5]Dong C H. Dynamic behavior of a system of coupling three－level atoms interaction with light field[J]. Acta Optica Sinica,2003,23(2):142－149. (in Chinese)

[董传华.耦合三能级原子与光场相互作用中系统的动力学行为[J].光学学报,2003,23(2):142－149.]

[6]Fu－li Li, Shao－yan Gao. Controlling nonclassical properties of the Jaynes－Cummings model by an external coherent field[J]. Phys. Rev. A,2000,62:043809－043815

[7]LuisA, Korolkova N. Polarization squeezing and nonclassical properties of light[J]. Phys. Rev. A,2006,74:043817－043823

[8]CHEN Z H, Liao C G, Luo C L. Nonclassical properties in the resonant interaction of a three－levelΛ－type atom with pair coherent state[J]. Acta Physica Sinica,2010,59(9):6152－6158. (in Chinese)

[陈子翊,廖长庚,罗成立. Λ 型三能级原子与对相干态光场共振作用中的非经典性质[J]. 物理学报,2010,59(9):6152－6158.]

[9]Phoenix S J D, Barnett SM. Entropy as measure of quantum optical correlation[J]. PhysRev A,1989,40(5):2404－2409.

[10]Phoenix S J D, Knight P L. Establishment of an entangled atom－field state in the Jaynes－Cummings model[J]. PhysRevA,1991,44:6023－6029.

[11]Park K, JeongH. Entangled coherent states versus entangled photon pairs for practical quantum－information processing[J]. Phys. Rev. A,2010,82:062325－062332

[12]Kaszlikowski D, Gosal D, Ling E J, et al. Three－qutrit correlations violate local realism more strongly than those of three qubits[J]. Phy. Rev. A,2002,66(3):032103－032107

[13]SHAN Chuan－jia, Xia Yun－jie. The entanglement character of two entangled atoms in Tavis－Cummings model[J]. Acta Physica Sinica,2006,55(4):1585－1590. (in Chinese)

[单传家,夏云杰. Tavis－Cummings 模型中两纠缠原子纠缠的演化特性[J]. 物理学报,2006,55(4):1585－1590.]

[14]WEI Qiao, Yan Yan, Li Gao－Xiang. Sudden death and revival of entanglement in two V－typethree－level atoms[J]. Acta Physica Sinica,2010,59(7):4453－4459. (in Chinese)

[魏巧,鄢嫣,李高翔. 两个 V 型三能级原子系统的纠缠突然死亡与复苏[J]. 物理学报,2010,59(7):4453－4459.]

[15]WAN Lin, LIU San－qin, LIU Su－mei. The Field Entropy Evolution of Two Atoms Interacting with Light Field inthe KerrMedium[J]. Journal of Optoelectronics · Laser,2002,13(6):626－631. (in Chinese)

[万琳,刘三秋,刘素梅. Kerr 介质中双原子与光场相互作用的场熵演化[J]. 光电子·激光,2002,13(6):626－631.]

[16]LIU Wang－yun, YANG Zhi－yong, AN Yu－ying. Properties of Evolution of Quantum Field Entropy in a Single－mode Vacuum Field－Coupled Two－level Atoms' System[J]. Journal of Optoelectronics · Laser,2007,18(9):1124－1127. (in Chinese)

[刘王云,杨志勇,安毓英. 单模真空场－耦合双原子系统的量子场熵演化特性[J]. 光电

子·激光,2007,18(9):1124—1127.]

[17]CUI Ying—hua,Sachuerfu,GONG Yan—li. Quantumentanglement of the binomial field interacting with the moving atoms in tavis—cummings model[J]. Journal of Optoelectronics · Laser,2008,19(9):1265—1268.(in Chinese)

[崔英华,萨楚尔夫,宫艳丽. T—C 模型中运动原子与二项式光场互作用的量子纠缠[J]. 光电子·激光,2008,19(9):1265—1268.]

[18]CHEN xing. The scattering of two—mode squeezed vacuum state on one dimensional potential barrier[J]. Journal of Optoelectronics · Laser,2009,20(10):1410—1413.(in Chinese)

[陈星. 双模压缩真空态的一维势垒散射[J]. 光电子·激光,2009,20(10):1410—1413.]

[19]HAN Feng,MAN Zhong—xiao,XIA Yun—jie. The entanglement dynamics between tw o atoms in bimode cavity field[J]. Journal of Optoelectronics · Laser,2010,21(3):470—475.(in Chinese)

[韩峰,满忠晓,夏云杰. 双模腔场中两原子纠缠的动力学性质[J]. 光电子·激光,2010,21(3):470—475.]

(本文发表于 2011 年《光电子·激光》第 11 期)

Global attractivity of the difference equation $x_{n+1}=\alpha+(x_{n-k}/x_n)$

Wan-Sheng He　Wan-Tong Li　Xin-Xue Yan *

摘要:在本文中,我们研究了差分方程 $x_{n+1}=\alpha+\frac{x_{n-k}}{x_n}$ 的全部非负解的全局稳定性,其中,$\alpha<1$ 是个实数,$k\geq 1$ 是个整数,且初始条件 $x_{-k},\cdots,x_0$ 是任意的实数。我们证明了方程的唯一非负平衡是一个全局吸引子,其取决于系数的某些条件。

In this paper, we investigate the global stability of all negative solutions of thedifference equation $x_{n+1}=\alpha+\frac{x_{n-k}}{x_n}$, $n=0,1,\cdots$, where $\alpha<1$ is a real number, $k\geq 1$ is an integer, and the initial conditions $x_{-k},\cdots,x_0$ are arbitrary real numbers. We show that the unique negative equilibrium of the equation is a global attractor with a basin that depends on certain conditions of the coefficient.

1. Introduction

In this paper we consider the difference equation

$$x_{n+1}=\alpha+\frac{x_{n-k}}{x_n},n=0,1,\cdots, \tag{1}$$

where $\alpha\in(-\infty,1)$ is a real number, $k\geq 1$ is an integer, and the initial conditions $x_{-k},\cdots,x_0$ are arbitrary real numbers.

In 1999, Amleh et al. [1]studied the special case of $k=1$ of Eq. (1), that is, the recursive sequence

---

* 作者简介:何万生(1956—),男,甘肃通渭人,天水师范学院数学与统计学院教授、学士,主要从事数学建模、差分方程的理论研究。

$$x_{n+1} = \alpha + \frac{x_{n-1}}{x_n}, n = 0,1,\cdots,$$

where is a real number, and the initial conditions are arbitrary positive real numbers.

Recently, EI－Owaidy et al. [3]studied the difference equation

$$x_{n+1} = \alpha + \frac{x_{n-k}}{x_n}, n = 0,1,\cdots,$$

where $\alpha \in [1,\infty)$ is a real number, $k \geqslant 1$ is an integer, and the initial conditions $x_{-k},\cdots,x_0$ are arbitrary positive real numbers.

They investigated the global stability and the periodic character of all positive solutions of the equation.

Other related results can be found in[4－18].

Our aim in this paper is to study the periodic character, invariant intervalsand the global attractivity of all negative solutions of Eq. (1).

We show that theunique negative equilibrium of Eq. (1)is a global attractor with a basin that depends on certain conditions of the coefficient.

Here, we recall some notations and results which will be useful in our proofs.

Let I be some interval of real numbers and let F be continuous function defined on I^{K+1}. Then, for initial conditions $x_{-k},\cdots,x_0 \in I$, it is easy to see that the difference equation

$$x_{n+1} = F(x_n, x_{n-1},\cdots,x_{n-k}), n = 0,1,\cdots \qquad (2)$$

has a unique solution f $\{x_n\}_{n=-k}^{\infty}$.

A point $\overline{x}$ is called an equilibrium of Eq. (2)if

$$\overline{x} = F(\overline{x},\overline{x},\cdots,\overline{x}),$$

That is, $x_n = \overline{x}$ for $n \geqslant 0$, is a solution of Eq. (2), or equivalently, $\overline{x}$ is a fixed point of F.

An interval $J \subseteq I$ is called an invariant interval for Eq. (2)if $x_{-k},\cdots,x_0 \in J \Rightarrow x_n \in J$ for all $n > 0$.

That is, every solution of Eq. (2)with initial conditions in J remains in J.

The linearized equation associated with Eq. (2)about the equilibrium $\overline{x}$ is

$$y_{n+1} = \sum_{0}^{k} \frac{\partial F}{\partial u_i}(\overline{x},\overline{x},\cdots\overline{x}) y_{n-1}, n = 0,1,\cdots \qquad (3)$$

Its characteristic equation is

$$\lambda^{n+1} = \sum_{0}^{k} \frac{\partial F}{\partial u_i}(\overline{x},\overline{x},\cdots\overline{x})\lambda^{n-1}. \tag{4}$$

Definition 1. 1. The equilibrium $\overline{x}$ of Eq. (1)is said to be

(i)Locally stable if for every $\varepsilon>0$, there exists $\delta>0$ such that for all x_{-k}, $\cdots, x_0 \in I$ with $\sum_{i=-k}^{0} | x_i - \overline{x} | < \delta$ implies $|x_i - \overline{x}| < \varepsilon$ for all $n \geqslant -k$.

(ii)Locally asymptotically stable if it is locally stable, and if there exist $\gamma>0$ such that for all $x_{-k}, \cdots, x_0 \in I$ with $\sum_{i=-k}^{0} | x_i - \overline{x} | < \gamma$ implies $\lim_{n\to\infty} x_n = \overline{x}$.

(iii)Global attractor if for all $x_{-k}, \cdots, x_0 \in I$ implies $\lim_{n\to\infty} x_n = \overline{x}$.

(iv)Global asymptotically stable if $\overline{x}$ is locally stable and $\overline{x}$ is also a global attractor.

(v)Unstable if x is not locally stable.

(vi)Saddle point if some of the roots of Eq. (4)are larger and some are lass than one in absolute value.

Theorem A(Kocic, Ladas, 1993[4]). Assume that $p, q \in R$ and $k \in \{0, 1, \cdots\}$.

Then

$$| p | + | q | < 1 \tag{5}$$

is a sufficient condition for asymptotic stability of the difference

$$x_{n+1} + px_n + qx_{n-k} = 0, n = 0, 1, \cdots, \tag{6}$$

Theorem B(Cunningham et al. , 2001[2]). Consider the difference equation.

$$x_{n+1} = f(x_n, x_{n-k}), n = 0, 1, \cdots, \tag{7}$$

where $k \geqslant 1$ is an integer. Let $I = [a, b]$ be some interval of real numbers, and assumethat

$$f: [a, b] \times [a, b] \rightarrow [a, b]$$

is a continuous function satisfying the following properties:

(a)$f(u, v)$ is a nondecreasing function in u, and a nonincreasing function in v.

(b)If $(m, M) \in [a, b] \times [a, b]$ is a solution of the system

$$m = f(m, M) \text{ and}, M = f(M, m) \tag{8}$$

then $m = M$.

Then Eq. (7)has a unique equilibrium point $\overline{x}$ and every solution of Eq. (7) converges to $\overline{x}$.

2. Main results

The unique negative equilibrium of Eq. (1) is

$$\bar{x} = \alpha + 1$$

The linearized equation associated with Eq. (1) about the equilibrium $\bar{x}$ is

$$y_{n+1} + \frac{1}{\alpha+1} y_n - \frac{1}{\alpha+1} y_{n-k} = 0, n = 0, 1, \cdots$$

Its characteristic equation is

$$\lambda^{k+1} + \frac{1}{\alpha+1}\lambda^k - \frac{1}{\alpha+1} = 0$$

By Theorem A and Definition 1. 1, we have the following results.

Theorem 2. 1

(i) If $\alpha < -3$, then the equilibrium $\bar{x}$ of Eq. (1) is locally asymptotically stable.

(ii) If $\alpha \in [-2, -1] \cup (-1, 0)$, then the equilibrium $\bar{x}$ of Eq. (1) is unstable.

(iii) If $\alpha \in [-3, -2] \cup (0, 1)$, then the equilibrium $\bar{x}$ of Eq. (1) is a saddle point.

Theorem 2. 2. Eq. (1) has no negative solution with prime period two for $\alpha \neq 1$.

Proof. Assume for the sake of contradiction that there exist distinctive negative real numbers ϕ and ψ, such that $\cdots, \phi, \psi, \cdots$ is a prime period two solution of Eq. (1). Then, there are two cases to be considered.

Case I. k is odd.

In this case $x_{n+1} = x_{n-k}$, ϕ and ψ satisfy the system

$$\phi\psi = \alpha\psi + \phi,$$

$$\phi\psi = \alpha\phi + \psi,$$

Subtracting these two equations, we get

$$(\phi - \psi)(\alpha - 1) = 0.$$

In view of $\alpha \neq 1$ then $\phi = \psi$, which contradicts the hypothesis of $\phi \neq \psi$.

Case II. If k is even, then $\phi = \psi = \alpha + 1$ which is also a contradiction. The proof is complete.

The following lemma shows that the function $f(u, v)$ is strictly monotonic.

Lemma2. 1. Assume $\alpha \in (-\infty, -3)$, and let $f(u, v) = \alpha + \frac{u}{v}$. Then the following statements are true.

(i) $-\infty<\bar{x}=\alpha+1<\alpha+2$.

(ii) If $u,v\in(-\infty,\alpha+1)$, then $f(u,v)$ is a strictly increasing function in u, and a strictly decreasing function in v. The proof is simple and omitted.

Theorem 2.3. Let $\tau>0$ is an arbitrary positive real number. Assume that $\alpha\in(-\infty,-(\tau+5)]$, and the initial values $x_{-k},\cdots,x_0\in I(I=[\alpha-\tau,\alpha+2]$. Then the Eq. (1) is permanent, that is, there exists constants P and Q, such that $P\leqslant x_n\leqslant Q$ for $n\geqslant 1$.

Proof. Set $Q=\alpha+2$, $P=\alpha-\tau$, then

$$f(\alpha+2,\alpha-\tau)=\alpha+\frac{\alpha-\tau}{\alpha+2}\leqslant\alpha+2=Q,$$

$$f(\alpha-\tau,\alpha+2)=\alpha+\frac{\alpha+2}{\alpha-\tau}>\alpha>\alpha-\tau=P$$

Hence $P=\alpha-\tau\leqslant f(u,v)\leqslant\alpha+2=Q$ for all $u,v\in I$. The proof is complete.

By Theorem 2.3 and definition of invariant interval we see that the interval $|\alpha-\tau,\alpha+2|$ is an invariant interval of Eq. (1).

Theorem 2.4. Assume $\alpha\in(-\infty,-(\tau+5)]$, Then the unique negative equilibrium $\bar{x}$ of Eq. (1) is a global attractor with a basin $S=[\alpha-\tau,\alpha+2]^{k+1}$.

Proof. Let $\{x_n\}$ be a solution of Eq. (1) with initial conditions $(x_{-k},\cdots,x_0)\in S$ for $u,v\in I=[\alpha-\tau,\alpha+2]$. Let

$$f(u,v)=\alpha+\frac{v}{u}.$$

Then $f: I\times I\to I$ is a continuous function and nondecreasing in u and nonincreasingin v. Let $m,M\in I$. is a solution of the system

$$m=f(m,M),M=f(M,m)$$

then

$$(M-m)(M+m-\alpha+1)=0$$

Since $M+m-\alpha+1<0$, $m=M$

By Theorem B, we have

$$\lim_{n\to\infty}x_n=\bar{x}$$

The proof is complete.

References

[1]A. M. Amleh, E. A. Grove, D. A. Georgiou, G. Ladas, On the recursive sequenc $x_{n+1}=$

$\alpha+(x_{n-1}/x_n)$, J. Math. Anal. Appl. 233(1999)790—798.

[2]K. C. Cunningham, M. R. S. Kulenovic, G. Ladas, S. V. Valicenti, On the recursive Sequence $x_n=(\alpha+\beta x_n)/(Bx_n+Cx_{n-1})$, Nonlinear Anal. TMA 47(2001)4603—4614.

[3]H. M. El—Owaidy, A. M. Ahmed, M. S. Mousa, On asymptotic behaviour of the Differenceequation $x_{n+1}=\alpha+(x_{n-k}/x_n)$, Appl. Math. Comput. (in press).

[4]V. L. Kocic, G. Ladas, Global Behavior of Nonlinear Difference Equations of Higher Orderwith Application, Kluwer Academic Publishers, Dordrecht, 1993.

[5]S. A. Kuruklis, The asymptotic stability of $x_{n+1}-\alpha x_n+bx_{n-k}=0$, J. Math. Anal. Appl. 188(1994)719—731.

[6]W. T. Li, H. R. Sun, Global Attractivity of a Rational Recursive Sequence, Dynam. Syst. Applicat. 11(3)(2002)339—345.

[7]C. H. Gibbons, M. R. S. Kulenovic, G. Ladas, H. D. Voulov, On the trichotomy character of $x_{n+1}=(\alpha+\beta x_n+\gamma x_{n-1})/(A+x_n)$, J. Differ. Equat. Appl. 8(1)(2002)75—92.

[8]M. T. Aboutaleb, M. A. El—Sayed, A. E. Hamza, Stability of the recursive sequence $x_n=(\alpha-\beta x_n)/(\gamma+x_{n-1})$, J. Math. Anal. Appl. 261(2001)126—133.

[9]C. Darwen, W. T. Patula, Properties of a certain Lyness equation, J. Math. Anal. Appl. 218(1998)458—478.

[10]R. DeVault, W. Kosmala, G. Ladas, S. W. Schultz, Global behavior of $y_{n+1}=(p+y_{n-k})/(qy_n+y_{n-k})$, Nonlinear Anal. TMA 47(2001)4743—4751.

[11]H. M. El—Owaidy, M. M. El—Afifi, A note on the periodic cycle of $x_{n+2}=(1+x_{n+1})/x_n$, Appl. Math. Comput. 109(2000)301—306.

[12]J. Feuer, E. J. Janowski, G. Ladas, Lyness—type equations in the third quadrant, NonlinearAnal. TMA 30(1997)1183—1189.

[13]V. L. Kocic, G. Ladas, Global attractivity in a second—order nonlinear difference equation, J. Math. Anal. Appl. 180(1993)144—150.

[14]V. L. Kocic, G. Ladas, I. W. Rodrigues, On rational recursive sequences, J. Math. Anal. Appl. 173(1993)127—157.

[15]M. R. S. Kulenovic, G. Ladas, N. R. Prokup, A rational difference equation, Comput. Math. Applic. 41(2001)671—678.

[16]X. X. Yan, W. T. Li, Global attractivity in the recursive sequence $x_{n+1}=(\alpha-\beta x_n)/(\gamma-x_{n-1})$, Appl. Math. Comput. 128(2—3)(2003)415—423.

[17]X. X. Yan, W. T. Li, Global attractivity in a rational recursive sequence, Appl. Math. Comput. (in press).

[18]X. X. Yan, W. T. Li, H. R. Sun, Global attractivity in a higher order nonlinear Differenceequation, Appl. Math. E—Notes 2(2002)51—58.

(本文发表于 2005 年《大学数学》第 2 期)

Multiple Solutions for a Class of Fractional Equations with Combined Nonlinearities

Hongming Xia　Ruichang Pei *

摘要:本文研究了一类具有组合非线性项的分数阶 Laplacian 方程,在共振与非共振情形下,运用山路理论、Morse 理论、Ekeland 变分原理,建立了 5 个非平凡解的存在性结果.

In this paper, we study a class of fractional Laplace equations with combined nonlinearities. The existence results of five nontrivial solutions under the resonance and non-resonance conditions are established by using the minimax method, Ekeland variational principle and Morse theory.

1. INTRODUCTION

In this article, we are interested in the following non-local fractional equations:

$$\begin{cases}(-\Delta)^s u = a(x)\,|u|^{q-2}u + f(x,u), & \text{in } \Omega,\\ u = 0, & \text{in } \mathbb{R}^N\backslash\Omega,\end{cases} \tag{1.1}$$

where $s \in (0,1)$ is fixed parameter, Ω is a bounded domain in $\mathbb{R}^N$ with smooth boundary $\partial\Omega$, $N > 2s$, $1 < q < 2$, and $(-\Delta)^s$ is the fractional Laplace operator.

In recent years, a great attention has been focused on the study of fractional and nonlocal operators of elliptic type, both for the pure mathematical research and for concrete real-world applications. Fractional and nonlocal operators appear in many fields such as, among the others, optimization, finance, phase transi-

---

* 作者简介:夏鸣鸣(1968—),男,甘肃天水人,天水师范学院数学与统计学院教授、硕士,主要从事非线性数学物理方程的研究。

tions, stratified materials, anomalous diffusion, crystal dislocation, soft thin films, semipermeable membranes, flame propagation, conservation laws, ultra−relativistic limits of quantum mechanics, quasi−geostrophic flows, multiple scattering, minimal surfaces, materials science and water waves. Just to mention a few, we recall, for instance, the following papers and the references therein: [4, 16] for regularity results, [2,5,17,22] for the existence of solutions and [20,23] for multiplicity of solutions. Particularly, when a(x) = 0, Fiscella et al. [8] studied problem (1.1) with asymptotically linear right−hand side and obtained some existence results by using saddle point theorem; Iannizzotto et al. [9] also studied fractional p−Laplacian equations with asymptotically p−linear and obtained two nontrivial solutions by the use of mountain pass theorem.

There are many interesting problems in the standard framework of theLaplacian (or higher order Laplacian), widely studied in the literature. A natural question is whether or not the existence results of multiple solutions obtained in the classical context can be extended to the non−local framework of the fractional Laplacian operator. LI et al. [12] showed the existence of three nontrivial solutions for semilinear elliptic problems with combined nonlinearities via mountain pass theorem and Ekeland variational principle. PU et al. [15] did similar work for fourth−order Navier boundary value problem with combined nonlinearities.

Motivated by their work, we study the following non−local problem with homogeneousDirichlet boundary conditions investigated by Servadei et al. [19] when $a(x) = 0$ and the related works[18,20]:

$$\begin{cases} -\mathscr{L}_k u = a(x)\,|u|^{q-2}u + f(x,u), & \text{in}\,\Omega, \\ u = 0, & \text{in}\,\mathbb{R}^N\setminus\Omega, \end{cases} \tag{1.2}$$

where $\mathscr{L}_k$ is the integro−differential operator defined as follows:

$$\mathscr{L}_k u(x) = \int_{\mathbb{R}^N} (u(x+y) + u(x-y) - 2u(x))K(y)dy, x \in \mathbb{R}^N, \tag{1.3}$$

with the kernel $K: \mathbb{R}^N\setminus 0 \to (0, +\infty)$ such that

($B1$) $mK \in L^1(\mathbb{R}^N)$, where $m(x) = \min\{|x|^2, 1\}$,

($B2$) There exists $\theta > 0$ such that $K(x) \geq \theta |x|^{-(N+2s)}$ for any $x \in \mathbb{R}^N\setminus\{0\}$,

($B3$) $K(x) = K(-x)$ for any $x \in \mathbb{R}^N\setminus\{0\}$.

For narrative convenience, in this paper, we only consider the particular case of problem (1.2), i. e. , we let K being given by the singular kernel $K(x) = |x$

$|^{-(N+2s)}$ which leads to the fractional Laplace operator $-(-\Delta)^s$, which, up to normalization factors, may be defined as

$$-(-\Delta)^s u(x) = \int_{\mathbb{R}^N} \frac{u(x+y)+u(x-y)-2u(x)}{|y|^{N+2s}} dy, x \in \mathbb{R}^N. \quad (1.4)$$

In fact, our methods and results in this paper also adapt for the general problem(1.2).

The conditions imposed on $a(x)$ and $f(x,t)$ are as follows:

(a) $a(x) \in L^\infty(\Omega)$, $a(x) > 0$ and $\inf_{x\in\Omega} a(x) > 0$;

(h_1) $f \in C^1(\overline{\Omega} \times \mathbb{R}, \mathbb{R})$, $f(x,0) = 0$, $f(x,t)t \geq 0$ for all $x \in \Omega$, $t \in \mathbb{R}$;

(h_2) $\lim_{|t|\to 0} \frac{f(x,t)}{t} = f_0$, $\lim_{|t|\to\infty} \frac{f(x,t)}{t} = l$ uniformly for $x \in \Omega$, where f_0 and l are constants;

(h_3) f' is subcritical in t, i. e. there is a constant $p \in (2, 2^*)$, $2^* = \frac{2N}{N-2s}$ such that

$$\lim_{t\to\infty} \frac{f_t(x,t)}{|t|^{p-1}} = 0 \text{ uniformly for } x \in \overline{\Omega},$$

(h_4) $\lim_{|t|\to\infty} [f(x,t)t - 2F(x,t)] = -\infty$ uniformly in $x \in \Omega$, where $F(x,t) = \int_0^t f(x,s)ds$;

(h_5) $2F(x,t) - f(x,t)t > 0$ for all $x \in \Omega$ and $t \neq 0$.

Now, we give our main results.

Theorem 1.1 Under conditions (a), (h_1), (h_2), (h_3) and (h_5). If $f_0 < \lambda_1$ and $l \in (\lambda_k, \lambda_{k+1})$ for some $k \geq 2$, there exists $m = m(f_0, q, f, N, \Omega)$ such that for all $a(x) \in L^\infty(\Omega)$ and $a(x) > 0$ with $|a|_\infty < m$, problem (1.1) has at least five nontrivial solutions.

Theorem 1.2 Under conditions (a), (h_1), (h_2), (h_3), (h_4) and (h_5). If $f_0 < \lambda_1$ and $l = \lambda_k$ for some $k \geq 2$, there exists $m = m(f_0, q, f, N, \Omega)$ such that for all $a(x) \in L^\infty(\Omega)$ and $a(x) > 0$ with $|a|_\infty < m$, problem (1.1) has at least five nontrivial solutions.

Here, $0 < \lambda_1 < \lambda_2 < \cdots < \lambda_k < \cdots$ be the eigenvalues of $(-\Delta)^s$ with homogeneous Dirichlet boundary data and $\varphi_1(x) > 0$ be the eigenfunction corresponding to λ_1.

Inview of the condition (h_2), problem (1.1) is called asymptotically inear at infinity which means that sual Ambrosetti－Rabinowitz condition (see [1]) is not atisfied. Moreover, if l in the above condition (h_2) is an eigenvalue of $(-\Delta)^s$, hen the problem (1.1) is called resonance at infinity. Otherwise, we call it non－resonance. This will bring some difficulty if the mountain pass theorem is used to seek nontrivial solutions of problem (1.1). For standard Laplacian Dirichlet problem, Zhou [24] have overcome it by using some monotonicity condition. Novelties of our this paper are as following:

We consider multiple solutions of problem(1.1) in the resonance and non－resonance by using the mountain pass theorem, the Ekeland variational principle and the Morse theory. At first, we use the truncated technique, the Ekeland variational principle and the mountain pass theorem to obtain two positive solutions and two negative solutions of problem (1.1) under our more general conditions (h_1) and (h_2) with respect to the conditions (H_1) and (H_3) in [24]. In the course of proving existence of positive solutions and negative solutions, our conditions are general, but the proof of our compact condition is more simple than that in [24]. Furthermore, we can obtain a nontrivial solution when the nonlinear term g is resonance or non－resonance at infinity and concave near the origin by computing the critical groups and Morse theory.

The paper is organized as follows. In section 2, we present some necessary preliminary knowledge about working space. In section 3, we prove some lemmas in order to prove our main results. In section 4, we give the proofs for our main results.

2. PRELIMINARIES

In this section, we give some preliminary results which will be used in the sequel. We briefly recall the related definition and notes for functional space X_0 introduced in [19].

The functional space X denotes the linear space of Lebesgue measurable functions from $\mathbb{R}^N$ to $\mathbb{R}$ such that the restriction to Ω of any function g in X belongs to $L^2(\Omega)$ and the map $(x,y)\mapsto(g(x)-g(y))\sqrt{K(x-y)}$ is in $L^2(\mathbb{R}^N\times\mathbb{R}^N)\backslash(C\Omega\times C\Omega), dxdy)$ (here $C\Omega=\mathbb{R}^N\backslash\Omega$). Also, we denote by X_0 the following linear subspace of X

$$X_0:=\{g\in X: g=0 \text{ a.e. in } \mathbb{R}^N\backslash\Omega\}.$$

Note that X and X_0 are non-empty, since $C_0^2(\Omega) \subseteq X_0$ by[19]. Moreover, the space X is endowed with the norm defined as

$$\| g \|_X = | g |_{L^2(\Omega)} + \left(\int_Q \left| g(x) - g(y) \right|^2 K(x-y) dxdy \right)^{\frac{1}{2}}, \quad (2.1)$$

where $Q = (\mathbb{R}^N \times \mathbb{R}^N) \backslash O$ and $O = (C\Omega) \times (C\Omega) \subset \mathbb{R}^N \times \mathbb{R}^N$. We equip X_0 with the following norm

$$\| g \|_{X_0} = \left(\int_Q \left| g(x) - g(y) \right|^2 K(x-y) dxdy \right)^{\frac{1}{2}}, \quad (2.2)$$

which is equivalent to the usual one defined in (2.1) (see[18]). It is easy to know that $(X_0, \| \cdot \|_{X_0})$ is a Hilbert space with scalar product

$$\langle u, v \rangle_{X_0} = \int_Q (u(x) - u(y))(v(x) - v(y)) K(x-y) dxdy. \quad (2.3)$$

Denote by $H^s(\Omega)$ the usual fractional Sobolev space with respect to the Gagliardo norm

$$\| g \|_{H^s(\Omega)} = | g |_{L^2(\Omega)} + \left(\int_{\Omega \times \Omega} \frac{| g(x) - g(y) |^2}{| x - y |^{N+2s}} dxdy \right)^{\frac{1}{2}}. \quad (2.4)$$

Now, we give a basic fact which will be used later.

Lemma 2.1[18] The embedding $j : X_0$

$L^v(\Omega)$ is continuous for any $v \in [1, 2^*]$, while it is compact whenever $v \in [1, 2^*)$.

Next, we state some used Propositions for operator $(-\Delta)^s$. Let $\lambda_1 < \lambda_2 \leq \lambda_3 \leq \cdots \leq \lambda_k \leq \cdots$ be the sequence of the eigenvalues of $(-\Delta)^s$ (see [8] and φ_k be the k-th eigenfunction corresponding to the eigenvalues λ_k. Moreover, we will set

$$P_{k+1} = \{ u \in X_0 : \langle u, \varphi_j \rangle_{X_0} = 0, \forall j = 1, 2, \cdots, k \}$$

and

$$H_k = \mathrm{span}\{ \varphi_1, \ldots, \varphi_k \}.$$

Proposition 2.1[8] The following inequality holds true

$$\| u \|_{X_0}^2 \leq \lambda_k | u |_{L^2(\Omega)}^2$$

for all $u \in H_k$ and $k \in N$.

Proposition 2.2[8] The following inequality holds true

$$\| u \|_{X_0}^2 \geq \lambda_{k+1} | u |_{L^2(\Omega)}^2$$

for all $u \in P_{k+1}$ and any $k \in N$.

Next, we recall some definitions for compactness condition and a version of mountain pass theorem.

Definition 2.1 Let $(X_0, \|\cdot\|_{X_0})$ be a real Banach space with its dual space $(X_0^*, \|\cdot\|_{X_0^*})$ and $J \in C^1(X_0, \mathbb{R})$. For $c \in \mathbb{R}$, we say that J satisfies the $(PS)_c$ condition if for any sequence $\{x_n\} \subset X_0$ with

$$J(x_n) \to c, DJ(x_n) \to 0 \text{ in } X_0^*,$$

there is a subsequence $\{x_{n_k}\}$ such that $\{x_{n_k}\}$ converges strongly in X_0. Also, we say that J satisfy the $(C)_c$ condition stated in [7] if for any sequence $\{x_n\} \subset X_0$ with

$$J(x_n) \to c, \ \|DJ(x_n)\|_{X_0^*}(1 + \|x_n\|_{X_0}) \to 0,$$

there is subsequence $\{x_{n_k}\}$ such that $\{x_{n_k}\}$ converges strongly in X_0.

3. SOME LEMMAS

First, we observe that problem (1.1) has a variational structure, indeed it is the Euler—Lagrange equation of the functional $J: X_0 \to \mathbb{R}$ defined as follows:

$$J(u) = \frac{1}{2}\int_{\mathbb{R}^N \times \mathbb{R}^N} \left| u(x) - u(y) \right|^2 K(x-y)dxdy - \frac{1}{q}\int_{\Omega} a(x) \left| u \right|^q dx$$

$$-\int_{\Omega} F(x, u(x))dx.$$

It is well know that the functional J is Frechet differentiable in X_0 and for any $\varphi \in X_0$

$$\langle J'(u), \varphi \rangle = \int_{\mathbb{R}^N \times \mathbb{R}^N} (u(x) - u(y))(\varphi(x) - \varphi(y))K(x-y)dxdy$$

$$-\int_{\Omega} a(x) \mid u \mid^{q-2} u\varphi(x)dx - \int_{\Omega} f(x, u(x))\varphi(x)dx.$$

Thus, critical points of J are solutions of problem (1.1).

Consider the following problem

$$\begin{cases} (-\Delta)^s u = a(x)u^{q-2}u^+ + f_+(x,u), & \text{in } \Omega; \\ u = 0, & \text{in } \mathbb{R}^N \backslash \Omega \end{cases}$$

Where

$$f_+(x,t) = \begin{cases} f(x,t), & t > 0, \\ 0, & t \le 0. \end{cases}$$

Define a functional $J_+ : X_0 \to \mathbb{R}$ by

$$J_+(u) = \frac{1}{2}\int_{\mathbb{R}^N \times \mathbb{R}^N} |u(x) - u(y)|^2 K(x-y)dxdy$$

$$-\frac{1}{q}\int_{\Omega} a(x)(u^+)^q dx - \int_{\Omega} F_+(x, u(x))dx,$$

where $F_+(x,t)=\int_0^t f_+(x,s)ds$, then $J_+ \in C^{200}(X_0,\mathbb{R})$.

Lemma 3.1 J_+ satisfies the (PS) condition.

Proof Let $\{u_n\}\subset X_0$ be a sequence such that $|J_+'(u_n)|\le c$, $\langle J_+'(u_n),\varphi\rangle \to 0$ as $n\to\infty$. Note that

$$J_+'(u_n),\varphi\rangle=\int_{\mathbb{R}^N\times\mathbb{R}^N}(u_n(x)-u_n(y))(\varphi(x)-\varphi(y))K(x-y)dxdy$$

$$-\int_\Omega a(x)(' q-1\varphi dx-\int_\Omega f_+(x,u_n)\varphi dx=o(\|\varphi\|_{X_0}) \quad (3.1)$$

for all $\varphi\in X_0$. Assume that $|u_n|_{L^2(\Omega)}$ is bounded. Taking $\varphi=u_n$ in (3.1), by (h_2), there exists $c>0$ such that $|f_+(x,u_n(x))|\le c|u_n(x)|$, a.e. $x\in\Omega$. So u_n is bounded in X_0. If $|u_n|_{L^2(\Omega)}\to+\infty$, as $n\to\infty$, set $v_n=\frac{u_n}{|u_n|_{L^2(\Omega)}}$, then $|v_n|_{L^2(\Omega)}=1$. Taking $\varphi=v_n$ in (3.1), it follows that $\|v_n\|_{X_0}$ is bounded. Without loss of generality, we assume that $v_n\rightharpoonup v$ in X_0, then $v_n\to v$ in $L^2(\Omega)$. Hence, $v_n\to v$ a.e. in Ω. Dividing both sides of \eqref{3.1} by $|u_n|_{L^2(\Omega)}$, we get

$$\int_{\mathbb{R}^N\times\mathbb{R}^N}(v_n(x)-v_n(y))(\varphi(x)-\varphi(y))K(x-y)dxdy-$$

$$\int_\Omega a(x)(' q-1(|u_n|_{L^2(\Omega)})q-2\varphi dx$$

$$-\int_\Omega\frac{f_+(x,u_n)}{|u_n|_{L^2(\Omega)}}\varphi dx=o\left(\frac{\|\varphi\|_{X_0}}{|u_n|_{L^2(\Omega)}}\right),\quad\forall\varphi\in X_0. \quad (3.2)$$

Then for a.e. $x\in\Omega$, we deduce that $\frac{f_+(x,u_n)}{|u_n|_{L^2(\Omega)}}\to lv_+$ as $n\to\infty$, where $v_+=\max\{v,0\}$. In fact, when $v(x)>0$, by (h_2) we have

$$u_n(x)=v_n(x)|u_n|_{L^2(\Omega)}\to+\infty$$

and

$$\frac{f_+(x,u_n)}{|u_n|_{L^2(\Omega)}}=\frac{f_+(x,u_n)}{u_n}v_n\to lv.$$

When $v(x)=0$, we have

$$\frac{f_+(x,u_n)}{|u_n|_{L^2(\Omega)}}\le c|v_n|\to 0.$$

When $v(x)<0$, we have

$$u_n(x)=v_n(x)|u_n|_{L^2(\Omega)}\to-\infty$$

and

$$\frac{f_+(x,u_n)}{|u_n|_{L^2(\Omega)}}=0.$$

Since $\frac{f_+(x,u_n)}{|u_n|_{L^2(\Omega)}} \le c|v_n|$, by (3.2) and the Lebesgue dominated convergence theorem, we arrive at

$$\int_{\mathbb{R}^N\times\mathbb{R}^N}(v(x)-v(y))(\varphi(x)-\varphi(y))K(x-y)dxdy-\int_{\Omega} lv_+\varphi dx$$
$$=0, \text{ for any } \varphi\in X_0. \qquad (3.3)$$

From the strong maximum principle (see[9]) we deduce that $v>0$. Choosing $\varphi=\varphi_1$ in (3.3), we obtain

$$l\int_{\Omega} v\varphi_1 dx=\lambda_1\int_{\Omega} v\varphi_1 dx.$$

This is a contradiction.

Lemma 3.2 Let φ_1 be the eigenfunction corresponding to λ_1 with $||\varphi_1||=1$. If $f_0<\lambda_1<l$, then there exists $m=m(f_0,q,f,N,\Omega)$ such that for all $a(x)\in L^{\infty}(\Omega)$ and $a(x)\ge 0$ with $|a|_{\infty}<m$ we have

(a) There exist $\rho,\beta>0$ such that $J_+(u)\ge\beta$ for all $u\in X_0$ with $||u||=\rho$;

(b) $J_+(t\varphi_1)=-\infty$ as $t\to+\infty$.

Proof By (h_1) and (h_2), if $l\in(\lambda_1,+\infty)$, for any $\varepsilon>0$, there exist $A=A(\varepsilon)\ge 0$ and $B=B(\varepsilon)$ such that for all $(x,s)\in\Omega\times\mathbb{R}$,

$$F_+(x,s)\le\frac{1}{2}(f_0+\varepsilon)s^2+As^{p+1}, \qquad (3.4)$$

$$F_+(x,s)\ge\frac{1}{2}(l-\varepsilon)s^2-B, \qquad (3.5)$$

where $p\in(1,\frac{N+s}{N-s})$.

Choose $\varepsilon>0$ such that $f_0+\varepsilon<\lambda_1$. By (3.4) and Lemma 2.1, we get

$$\begin{aligned} J_+(u)&=\frac{1}{2}\|u\|_{X_0}^2-\frac{||a||_{\infty}}{q}\int_{\Omega}|u|^q dx-\int_{\Omega}F_+(x,u)dx\\ &\ge\frac{1}{2}\|u\|_{X_0}^2-\frac{||a||_{\infty}}{q}\int_{\Omega}|u|^q dx-\frac{1}{2}\int_{\Omega}[(f_0+\varepsilon)u^2+A|u|^{p+1}]dx\\ &\ge\frac{1}{2}\left(1-\frac{f_0+\varepsilon}{\lambda_1}\right)||u||_{X_0}^2-\frac{||a||_{\infty}K^q}{q}||u||_{X_0}^q-AK^{p+1}||u||_{X_0}^{p+1}.\end{aligned}$$

Set

$$\rho=\left(\frac{\lambda_1-f_0-\varepsilon}{4\lambda_1 AK^{p+1}}\right)^{\frac{1}{p-1}},\ m=\frac{\lambda_1-f_0-\varepsilon}{8\lambda_1 K^q\rho^{q-2}}.$$

So, part (a) holds if we choose $||u||=\rho>0$ and $||a||_{\infty}\le m$.

On the other hand, if $l\in(\lambda_1,+\infty)$, take $\varepsilon>0$ such that $l-\varepsilon>\lambda_1$. By

(3.5), we have

$$J_{+}(u) \leq \frac{1}{2}\|u\|_{X_0}^2 - \frac{t^q}{q}\int_{\Omega} a(x)(\varphi_1)^q dx - \frac{l-\varepsilon}{2}|u|_{L^2(\Omega)}^2 + B|\Omega|.$$

Since $l-\varepsilon > \lambda_1$ and $\|\varphi_1\| = 1$, it is easy to see that

$$J_{+}(t\varphi_1) \leq \frac{1}{2}\left(1 - \frac{l-\varepsilon}{\lambda_1}\right)t^2 - \frac{t^q}{q}\int_{\Omega} a(x)(\varphi_1)^q dx + B|\Omega| \to -\infty \ as \ t \to +\infty$$

and part (b) is proved.

Lemma 3.3 Let $X_0 = H_k \oplus W$. If f satisfies (h_1), (h_2) and (h_4), then

(i) the functional J is coercive on W, that is

$$J(u) \to +\infty \text{as } \|u\| \to +\infty, \ u \in W$$

and bounded from below on W,

(ii) the functional J is anti-coercive on H_k.

Proof (i) For $u \in W$, by (h_2), for any $\varepsilon > 0$, there exists $B_1 = B_1(\varepsilon)$ such that for all $(x,s) \in \Omega \times \mathbb{R}$,

$$F(x,s) \leq \frac{1}{2}(l+\varepsilon)s^2 + B_1. \tag{3.6}$$

So we have

$$J(u) = \frac{1}{2}\|u\|_{X_0}^2 - \frac{1}{q}\int_{\Omega} a(x)|u|^q dx - \int_{\Omega} F(x,u)dx$$

$$\geq \frac{1}{2}\|u\|_{X_0}^2 - \frac{1}{2}(l+\varepsilon)|u|_{L^2(\Omega)}^2 - \|a\|_{\infty} K \|u\|_{X_0}^q - B_1|\Omega|$$

$$\geq \frac{1}{2}\left(1 - \frac{l+\varepsilon}{\lambda_{k+1}}\right)\|u\|_{X_0}^2 - \|a\|_{\infty} K \|u\|_{X_0}^q - B_1|\Omega|.$$

Choose $\varepsilon > 0$ such that $l+\varepsilon < \lambda_{k+1}$. This proves (i).

(ii) We firstly consider thecase $l = \lambda_k$.

Write $G(x,t) = F(x,t) + \frac{a(x)|t|^q}{q} - \frac{1}{2}\lambda_k t^2$, $g(x,t) = [a(x)|t|^{q-2}t + f(x,t)] - \lambda_k t$. Then (h_2) and (h_4) imply that

$$\lim_{|t|\to\infty}[g(x,t)t - 2G(x,t)] = -\infty \tag{3.7}$$

and

$$\lim_{|t|\to\infty} \frac{2G(x,t)}{t^2} = 0. \tag{3.8}$$

It follows from(3.7) that for every $M > 0$, there exists a constant $T > 0$ such that

$$g(x,t)t - 2G(x,t) \leq -M, \forall t \in \mathbb{R}, |t| \geq T, \text{ a.e. } x \in \Omega. \tag{3.9}$$

For $\tau > 0$, we have

$$\frac{d}{d\tau}\frac{G(x,\tau)}{\tau^2} = \frac{g(x,\tau)\tau - 2G(x,\tau)}{\tau^3}. \tag{3.10}$$

Integrating(3.10) over $[t,s] \subset [T,+\infty)$, we deduce that

$$\frac{G(x,s)}{s^2} - \frac{G(x,t)}{t^2} \leq \frac{M}{2}\left(\frac{1}{s^2} - \frac{1}{t^2}\right). \tag{3.11}$$

Letting $s \to +\infty$ and using (3.8), we see that $G(x,t) \geq \frac{M}{2}$, for $t \in \mathbb{R}$, $t \geq T$, a. e. $x \in \Omega$. A similar argument shows that $G(x,t) \geq \frac{M}{2}$, for $t \in \mathbb{R}$, $t \leq -T$, a. e. $x \in \Omega$. Hence

$$\lim_{|t| \to \infty} G(x,t) \to +\infty, \text{ a. e. } x \in \Omega. \tag{3.12}$$

By (3.12), we get

$$\begin{aligned} J(v) &= \frac{1}{2}\|v\|_{X_0}^2 - \int_\Omega G(x,v)dx \\ &= \frac{1}{2}\|v\|_{X_0}^2 - \frac{1}{2}\lambda_k\int_\Omega v^2 dx - \int_\Omega G(x,v)dx \\ &\leq -\delta\|v^-\|_{X_0}^2 - \int_\Omega G(x,v)dx \to -\infty \end{aligned}$$

for $v \in V$ with $\|v\| \to +\infty$, where $v^- \in H_{k-1}$.

In the caseof $\lambda_k < l < \lambda_{k+1}$, we needn't the assumption (h_3) and it is easy to see that the conclusion also holds.

Lemma 3.4 If $\lambda_k < l < \lambda_{k+1}$, then J satisfies the (PS) condition.

Proof Let $\{u_n\} \subset X_0$ be a sequence such that $|J(u_n)| \leq c$, $\langle J'(u_n),\varphi\rangle \to 0$. Since

$$\begin{aligned} \langle J'(u_n),\varphi\rangle &= \int_{\mathbb{R}^N\times\mathbb{R}^N}(u_n(x)-u_n(y))(\varphi(x)-\varphi(y))K(x-y)dxdy \\ &\quad - \int_\Omega a(x)|u_n|^{q-2}u_n\varphi dx - \int_\Omega f(x,u_n)\varphi dx \\ &= o(\|\varphi\|_{X_0}). \end{aligned}$$

for all $\varphi \in X_0$. If $|u_n|_{L^2(\Omega)}$ is bounded, we can take $\varphi = u_n$. By (h_2), there exists a constant $c > 0$ such that $|f(x,u_n(x))| \leq c|u_n(x)|$, a. e. $x \in \Omega$. So u_n is bounded in X_0. If $|u_n|_{L^2(\Omega)} \to +\infty$, as $n \to \infty$, set $v_n = \frac{u_n}{|u_n|_{L^2(\Omega)}}$, then $|v_n|_{L^2(\Omega)} = 1$. Taking $\varphi = v_n$ in (3.13), it follows that $\|v_n\|_{X_0}$ is bounded. Without loss of generality, we assume $v_n \rightharpoonup v$ in X_0, then $v_n \to v$ in $L^2(\Omega)$. Hence, $v_n \to v$ a. e. in

Ω. Dividing both sides of (3.13) by $|u_n|_{L^2(\Omega)}$, we get

$$\int_{\mathbb{R}^N\times\mathbb{R}^N}(v_n(x)-v_n(y))(\varphi(x)-\varphi(y))K(x-y)dxdy-\int_\Omega a(x)|u_n|^{q-2}v_n\varphi dx$$

$$-\int_\Omega \frac{f(x,u_n)}{|u_n|_{L^2(\Omega)}}\varphi dx=o\left(\frac{\|\varphi\|_{X_0}}{|u_n|_{L^2(\Omega)}}\right),\quad \forall\varphi\in X_0. \tag{3.14}$$

Then for a. e. $x\in\Omega$, we have $\frac{f(x,u_n)}{|u_n|_{L^2(\Omega)}}\to lv$ as $n\to\infty$. In fact, if $v(x)\neq 0$, by (h_2), we have

$$|u_n(x)|=|v_n(x)||u_n|_{L^2(\Omega)}\to+\infty$$

and

$$\frac{f(x,u_n)}{|u_n|_{L^2(\Omega)}}=\frac{f(x,u_n)}{u_n}v_n\to lv.$$

If $v(x)=0$, we have

$$\frac{|f(x,u_n)|}{|u_n|_{L^2(\Omega)}}\leq c|v_n|\to 0.$$

Since $\frac{|f(x,u_n)|}{|u_n|_{L^2(\Omega)}}\leq c|v_n|$, by (3.14) and the Lebesgue dominated convergence theorem, we arrive at

$$\int_{\mathbb{R}^N\times\mathbb{R}^N}(v(x)-v(y))(\varphi(x)-\varphi(y))K(x-y)dxdy-\int_\Omega lv\varphi dx=0,$$

$$\text{for any }\varphi\in X_0.$$

Obviously $v\neq 0$, hence, l is an eigenvalue of $(-\Delta)^s$. This contradicts our assumption.

Lemma 3.5　Suppose $l=\lambda_k$ and f satisfies (h_4). Then the functional J satisfies the (C) condition.

Proof Suppose $u_n\in X_0$ satisfies

$$J(u_n)\to c\in\mathbb{R},(1+\|u_n\|)\|J'(u_n)\|\to 0\text{ as }n\to\infty. \tag{3.15}$$

In viewof (h_2), it suffices to prove that u_n is bounded in X_0. Similar to the proof of Lemma 3.4, we have

$$\int_{\mathbb{R}^N\times\mathbb{R}^N}(v(x)-v(y))(\varphi(x)-\varphi(y))K(x-y)dxdy-\int_\Omega lv\varphi dx=0 \tag{3.16}$$

for any $\varphi\in X_0$. Therefore　$v\neq 0$ is an eigenfunction of λ_k, then $|u_n(x)|\to\infty$ for a. e. $x\in\Omega$. It follows from (h_4) that

$$\lim_{n\to+\infty}[f(x,u_n(x))u_n(x)-2F(x,u_n(x))]=-\infty$$

holds uniformly in $x \in \Omega$, which implies that

$$\int_{\Omega}(f(x,u_n)u_n - 2F(x,u_n))dx \to -\infty \text{ as } n \to \infty. \qquad (3.17)$$

On the other hand, (3.15) implies that

$$2J(u_n) - \langle J'(u_n), u_n \rangle \to 2c \text{ as } n \to \infty.$$

Thus

$$\int_{\Omega}(f(x,u_n)u_n - 2F(x,u_n))dx \to +\infty \text{ as } n \to \infty,$$

Which contradicts to (3.17). Hence u_n is bounded.

It is well known that critical groups and Morse theory are the useful tools in studying elliptic partial differential equations. Let us recall some results which will be used later. We refer the readers to the book [6] for further understanding on Morse theory.

Let X be a Hilbert space and $J \in C^1(X, \mathbb{R})$ be a functional satisfying the (PS) condition or (C) condition, and $H_q(X,Y)$ be the q −th singular relative homology group with integer coefficients. Let u_0 be an isolated critical point of J with $J(u_0) = c, c \in \mathbb{R}$, and U be a neighborhood of u_0. The group

$$C_q(J, u_0) := H_q(J^c \cap U, J^c \cap U \backslash \{u_0\}), \quad q \in Z$$

is said to be the q −th critical group of J at u_0, where $J^c = \{u \in X : J(u) \leq c\}$.

Let $K := \{u \in X : J'(u) = 0\}$ be the set of critical points of J and $a < \inf J(K)$, the critical groups of J at infinity are formally defined by (see [3])

$$C_q(J, \infty) := H_q(X, J^a), \quad q \in Z.$$

The following result comes from[3,6] and will be used to prove the results in this paper.

Proposition 3.1[3] Assume that $X = V \oplus W$, J is bounded from below on W and $J(u) \to -\infty$ as $\| u \| \to \infty$ with $u \in V$. Then

$$C_k(J, \infty) \not\cong 0, if k = \dim V < \infty. \qquad (3.18)$$

Next, we prove auseful result for calculating the critical group of local minimum points of the functional J.

Lemma 3.6 Under conditions (a), (h_1), (h_2), then there exists $m = m(f_0, q, f, N, \Omega)$ such that for all $a(x) \in L^{\infty}(\Omega)$ and $a(x) > 0$ with $|a|_{\infty} < m$, problem (1.1) has two solutions $u^*, u^{**} \in X_0$, $u^* > 0, u^{**} < 0$ and

$$C_q(J, u^*) = \delta_{q,0} Z$$

and

$$C_q(J, u^{**}) = \delta_{q,0} Z.$$

Proof Under our conditions, we still canprove it by using the Ekeland variational principle. Sine the proof is completely similar to that of Li shujie, Wu shaoping and Zhouhuan-song's Theorem 1.1 [12] and Perera Kanishka's Lemma 2.1 [14], we omitit.

4. PROOF OF THE MAIN RESULTS

Proof of Theorem 1.1 By Lemmas 3.1-3.2 and the mountain pass theorem, the functional J_+ has a critical point u_1 satisfying $J_+(u_1) \geq \beta$. Since $J_+(0) = 0$, $u_1 \neq 0$ and by the maximum principle (see [9]), we get $u_1 > 0$. Hence u_1 is a positive solution of the problem (1.1) and satisfies

$$C_1(J_+, u_1) \neq 0, \quad u_1 > 0. \tag{4.1}$$

By (h_3), the functional J is $C^2(X_0 \backslash \{0\})$. Using the results in [6, 11, 13], we obtain

$$C_q(J, u_1) = C_q(J_{C_d^0(\overline{\Omega})}, u_1) = C_q(J_+|_{C_d^0(\overline{\Omega})}, u_1) = C_q(J_+, u_1) = \delta_{q1} Z. \tag{4.2}$$

Here

$$C_d^0(\overline{\Omega}) = \{u \in C^0(\overline{\Omega}) : ud^{-\gamma} \in C^0(\overline{\Omega})\},$$

where $d(x) = \text{dist}(x, \partial\Omega)$ for all $x \in \overline{\Omega}$ and $0 < \gamma < 1$. More detailed topology knowledge will be seen in [9] and we omit it.

Similarly, we can obtain another negative critical point u_2 of J satisfying

$$C_q(J, u_2) = \delta_{q,1} Z. \tag{4.3}$$

Now, we claim that

$$C_q(J, 0) = 0. \tag{4.4}$$

In fact, for $u \in X_0$, by (a), (h_2), we have

$$\lim_{|\sigma| \to 0} \frac{J(\sigma u)}{|\sigma|^2} = -\infty. \tag{4.5}$$

Using (4.5), we see that for any $u \in X_0 \backslash \{0\}$, there exists $\sigma_0 \in (0,1)$ such that

$$J(\sigma u) < 0 \text{ for } \sigma \in (0, \sigma_0). \tag{4.6}$$

Assume that $J(u) \geq 0$, that is

$$||u||_{X_0}^2 \geq \int_\Omega G(x, u) dx.$$

Then accordingto (h_5), for $u \in J^{-1}[0, +\infty)\setminus\{0\}$ we obtain

$$\frac{d}{d\sigma}\Big|_{\sigma=1} J(\sigma u) > 0.$$

Next, we adjust the argument in the proof of Proposition 2.1 in [10] slightly. From [4.6] and (4.7), we see that for any $u \in J^{-1}[0, +\infty)\setminus\{0\}$, there exists a unique $\sigma^* = \sigma^*(u) > 0$ such that $J(\sigma^* u) = 0$. Moreover, since $J(\sigma^* u) = 0$, from (4.7), we deduce

$$\frac{d}{d\sigma}\Big|_{\sigma=\sigma^*} J(\sigma u) > 0.$$

Thus, by the Implicit Function Theorem we see that σ^* is continuous on $J^{-1}[0, +\infty)\setminus\{0\}$. If $J(u) \leq 0$, we set $\sigma^*(u) = 1$. Then $\sigma^*: X_0 \to \mathbb{R}$ continuous.

We now define $\eta: [0,1] \times X_0 \to X_0$ by

$$\eta(\sigma, u) = (1-\sigma)u + \sigma\sigma^*(u)u, \ (\sigma, u) \in [0,1] \times X_0.$$

It is easy to see that η is a continuous deformation from $X_0, X_0\setminus\{0\})$ to $(J_0, J_0\setminus\{0\})$, hence

$$C_k(J,0) \cong H_k((J_0, J_0\setminus\{0\})) = 0, \ k \in N.$$

Thus, our claim holds.

On the other hand, by Lemma 3.3, Lemma 3.4 and the Proposition 3.6, we have

$$C_k(J, \infty) \not\cong 0. \tag{4.8}$$

Hence I has a critical point u_3 satisfying

$$C_k(J, u_3) \not\cong 0. \tag{4.9}$$

Since$k \geq 2$, it follows from Lemma 3.7, (4.1)-(4.4), (4.8) and (4.9) thatu^*, u^{**}, u_1, u_2 and u_3 are five different nontrivial solutions of the problem (1.1).

Proof of Theorem 1.2　By Lemma 3.3, Lemma 3.5 and the Proposition 3.6, we can prove the conclusion (4.8). The other proof is similar to that of Theorem 1.1.

Acknowledgments

This study was supported by the National NSF (Grant No. 11571176) and Natural Science Foundation of Gansu Province China (Grant No. 1506RJZE114).

References

[1]AMBROSETTI A, RABINOWITZ P H, Dualvariational methods in critical point the-

ory and applications,J. Funct. Anal. 14(1973)349—381.

[2]BARRIOS B,COLORADO E,PABLO A,et al. ,On some critical problems for the fractional Laplacian operator,J. Differential Equations 252(2012)6133—6162.

[3]BARTSCH T,LI S J,Critical point theory forasymptotically quadratic functionals and applications to problemswith resonance,Nonlinear Anal. TMA 28(1997)419—441.

[4]CAFFARELLI L,SILVESTRE L,Regularity theory for fully nonlinearintegro—differentialequations,Comm. Pure Appl. Math. 62(2009)597—638.

[5]CABRÉ,TAN,Positivesolutions of nonlinear problems involving the square root of theLaplacian,Adv. Math. 42(2009)2052—2093.

[6]CHANG K C,Infinite Dimesional Morse Theoryand Multiple Solutions Problems, Progress in Nonlinear Differential Equations and their Applications,Vol. 6. Birkh ä user,Boston,1993.

[7]CERAMI G,On the existence of eigenvalues for anonlinear boundary value problem, Ann. Mat. Pura Appl. 124(1980)161—179.

[8]FISCELLA A,SERVADEI R,VALDINOCI E,Asymptotically linear problems driven by fractional Laplacianoperators,Math. Methods Appl. Sci. 38(2015)3551—3563.

[9]IANNIZZOTTO A,LIU S B,PERERA K et al. ,Existence results for fractional p—Laplacian problems via Morse theory,Adv. Calc. Var. 9(2014) 101—125.

[10] JIU Q S,LIU S B,Existence and multiplicity results for Dirichlet problem with p—Laplacian,J. Math. Anal. Appl. 281 (2003) 587—601.

[11]LIU J Q,WU S P,Calculating critical groups of solutions for elliptic problem with jumping nonlinearity,Nonlinear Anal. TMA 49 (2002) 779—797.

[12] LI S J,WU S P,and ZHOU H S,Solutions to semilinear elliptic problems with combined nonlinearities,Journal of Differential Equations 485 (2002) 200—224.

[13] LI C,LI S J,LIU J Q,Splitting theorem,Poincar\'{e}—Hopf theorem and jumping nonlinear problems,J. Functional Analysis 221 (2005) 439—455.

[14] PERERA K,Multiplicity results for some elliptic problems with concave nonlinearities,Journal of Differential Equations 140 (1997) 133—141.

[15] PU Y,WU X P,TANG C L,Fourth—order Navier boundary value problem with combined nonlinearities,Nonlinear Anal. TMA 398 (2013) 798—813.

[16] ROS—OTON X,SERRA J,The Dirichlet problem for the fractional Laplacian:regularity up to the boundary,J. Math. Pures Appl. 101 (2014) 275—302.

[17]SERVADEI R,A critical fractional Laplace equation in the resonant case,Topol. Methods Nonlinear Anal. 43 (2014) 251—267.

[18]SERVADEI R,VALDINOCI E,Mountain Pass solutions for non—local elliptic operators,J. Math. Anal. Appl. 389 (2012) 887—898.

[19]SERVADEI R, VALDINOCI E, Lewy－Stampacchia type estimates for variational inequalities driven by (non)local operators, Rev. Mat. Iberoam. 29 (2013) 1091－1126.

[20]SERVADEI R, Infinitely many solutions for fractional Laplace equations with subcritical nonlinearity, Contemp. Math. 595 (2013) 317－340.

[21] SERVADEI R, VALDINOCI E, Weak and viscosity solutions of the fractional Laplace equation, Publ. Mat. 58(2014) 133－154.

[22]TAN J, The Brezis－Nirenberg type problem involving the square root of the fractional Laplacian, Calc. Var. Partial Differential Equations 36 (2011) 21－41.

[23]ZHANG B L, FERRARA M, Multiplicity of solutions for a class of superlinear non－local fractional equations, Complex Var. Elliptic Equ. 60 (2015) 583－595.

[24]ZHOU H S, Existence of asymptotically linear Dirichlet problem, Nonlinear Anal. TMA 44 (2001) 909－918.

（本文发表于 2017 年《应用数学》第 30 卷第 4 期）

Existence Results for Asymmetric Fractional $p-$Laplacian Problem

Ruichang Pei Caochuan Ma Jihui Zhang *

摘要:我们研究了一类拟线性非局部问题,非线性项在正无穷远处超线性增长,但在负无穷远处线性增长,并且不满足(AR)条件.运用变分法结合 Moser—Trudinger 不等式获得非平凡解的存在性结果.

We investigated a class of quasi—linear nonlocal problems with aright—hand side nonlinearity which exhibits an asymmetric grow that $+\infty$ and $-\infty$. Namely, it is linear at $-\infty$ and superlinear at $+\infty$. However, it needs not satisfy the Ambrosetti—Rabinowitz condition on the positive semiaxis. Someexistence results for nontrivial solution are established byusing variational methods combined with the Moser—Trudingerinequality.

1. INTRODUCTION

For $p\in(1,\infty)$, $s\in(0,1)$ and n smooth enough, the fractional $p-$Laplacian is the nonlinear nonlocal operator defined by

$$(-\Delta)_p^s u(x)=2\lim_{\epsilon\searrow 0}\int_{\mathbb{R}^N\setminus B_\epsilon(x)}\frac{|u(x)-u(y)|^{p-2}(u(x)-u(y))}{|x-y|^{N+sp}}dy,x\in R^n.$$

This definition is consistent, up to a normalization constant depending on$N$ and s, with the linearfractional Laplacian $(-\Delta)^s$ in the case $p=2$. Some motivations that have led to the study of these kind of operators can be found in[5]. This operator, known as the fractional $p-$Laplacian, leads naturally to the study

* 作者简介:裴瑞昌(1975—),男,甘肃秦安人,天水师范学院数统学院教授、博士,主要从事偏微分方程研究。

of the quasi－linear problem

$$\begin{cases}(-\Delta)_p^s u(x) = f(x,u), & \text{in } \Omega, \\ u = 0, & \text{in } \mathbb{R}^N \backslash \Omega\end{cases} \tag{1.1}$$

In recent years, a great attention has been focused on the studyof fractional and non－local operators of elliptic type, both forthe pure mathematical research and for concrete real－world applications, we refer the interested reader to[19, 18, 10, 17]and the references therein. Fractional and nonlocal operators appear in manyfields such as, among the others, optimization, finance, phasetransitions, stratified materials, anomalous diffusion, crystaldislocation, soft thin films, semi-permeable membranes, flamepropagation, conservation laws, ultra－relativistic limits ofquantum mechanics, quasi－geostrophicflows, multiple scattering, minimal surfaces, materials science and water waves. For anelementary introduction to this topic and for a－still notexhaustive－list of related references see, e. g., [7]. For some more concrete applications, one may alsorefer to[4]. Very recently, there is a rapidly growing literature on this problem(1.1) when Ω is bounded with Lipschitz boundary. Inparticular, fractional p－eigenvalue problems have been studiedin[9, 13], regularity theory in[3, 12], existence theory in the subcritical case in[11, 23, 26, 25, 24, 27], and the critical case in[22].

The main purpose of this paper is to establish existence results of nontrivial solutions for problem(1.1) with $1<p\leqslant\frac{N}{s}$ when the nonlinearity term $f(x,.)$ exhibits an asymmetric behavior as $t\in\mathbb{R}$ approaches $+\infty$ and $-\infty$. More precisely, we assume that for a. e. $x\in\Omega$, $f(x,.)$ grows superlinear at $+\infty$, while at $-\infty$ it has a linear growth. Incase of $1<p<N$, the classical p－Laplacian Dirichlet problems with nonlinearities which are superlinear in one direction and linear in the other were investigated by Arcoya and Villegas[2], de Figueiredo and Ruf[8], Perera[20]. All three works express the superlinear growth at $+\infty$ using the Ambrosetti－Rabinowitz condition(AR－condition, for short). Recall that a function $f:\Omega\times\mathbb{R}\to\mathbb{R}$ is said to satisfy the(AR)condition in the positive direction if there exist $\mu>p$ and $M>0$ such that

$$0<\mu F(x,t)\leqslant tf(x,t) \text{ for all } t\geqslant M \text{ and a. e. } x\in\Omega,$$

where $F(x,t)=\int_0^t f(x,s)ds$.

Under almost all of the works mentioned above and the motivation of Lam and Lu[14], our first main results will be to study problem(1.1)inthe improved subcritical polynomial growth

$$(SCPI): \lim_{t\to\infty}\frac{f(x,t)}{t^{p^*-1}}=0$$

uniformly for all $x\in\Omega$ which is weaker than usualsubcritical growth condition, i.e. there is a constant $q\in(p,p^*)\left(p^*=\frac{Np}{N-sp}\right)$ such that

$$\lim_{t\to\infty}\frac{f(x,t)}{|t|^{q-1}}=0$$

uniformlyfor all $x\in\Omega$ Note that in this case, we do not have theSobolev compact embedding anymore. Our work is to study asymmetric problem(1.1)without the (AR)—condition in thepositive semi—axis. Our results have an essential difference compared with[14]. Lam} since our nonlinearitieshave different growth behavior at $+\infty$ and $-\infty$. Infact, this condition was studied by Liu and Wang in [15]in the case of Laplacian(i. e., $p=2$)by the Nehari manifoldapproach. In this paper, by using the mountain pass theorem and its suitable version, we try to get the nontrivial solutions to problem(1.1) with $1<p<\frac{N}{s}$. Our results and the proof of compactnesscondition are different from those in[14].

Let us now state our results: Suppose that $f(x,t)\in C(\overline{\Omega}\times\mathbb{R})$ and satisfies:

$(H_1)\lim\limits_{t\to 0}\frac{f(x,t)}{|t|^{p-2}t}=f_0$ uniformly for a. e. $x\in\Omega$ where $f_0\in[0,+\infty)$;

$(H_2)\lim\limits_{t\to-\infty}\frac{f(x,t)}{|t|^{p-2}t}=l$ uniformly for a. e. $x\in\Omega$, where $l\in[0,+\infty]$;

$(H_3)\lim\limits_{t\to+\infty}\frac{f(x,t)}{t^{p-1}}=+\infty$ uniformly for a. e. $x\in\Omega$;

(H_4)There exist $\theta\geqslant 1$ and $C_*>0$ such that $\theta F(x,t)\geqslant F(x,st)-C_*$ for $(x,t)\in\Omega\times(-\infty,\mathbb{R})$ and $s\in[0,1]$, where $F(x,t)=f(x,t)t-pF(x,t)$.

Let λ_1 bethe first eigenvalue of $(-\Delta)_p^s$ and $\varphi_1(x)>0$ for every $x\in\Omega$ be the λ_1 eigenfunction. More detailed definitions may refer to[11].

Theorem 1.1. Let $1<p<\frac{N}{s}$ andassume that f has the improved subcritical polynomial growth on Ω(condition(SCPI)) and satisfies $(H_1)-(H_3)$. If $f_0<\lambda_1<l<\infty$, then problem(1.1)has at least onenontrivial solution when l is not any of the eigenvalues of $(-\Delta)_p^s$.

Remark 1. 1. In view of the conditions (SCPI), (H_2) and (H_3), problem (1. 1) with the improved subcritical polynomial growth is called asymmetric. Hence, Theorem 1. 1 is completely different from the results contained in[11,23, 26,24,27].

Theorem 1. 2. Let $1<p<\frac{N}{s}$ and assume that f has the improved subcritical-polynomial growth on Ω(condition(SCPI)) and satisfies (H_1)−(H_3). If $f_0<\lambda_1=l$ and $\lim\limits_{t\to-\infty}[f(x,t)t-pF(x,t)]=-\infty$ uniformly for a. e. $x\in\Omega$. then problem (1. 1)has at leastone nontrivial solution.

Remark 1. 2. When $l=\lambda_1$, problem(1. 1)is called resonant at negative infinity. This caseis new and completely different from results contained in[11]. Here, we also give an example for $f(x,t)$. It satisfies our conditions (H_1)−(H_3) and (SCPI):

Example A. Define

$$f(x,t)=\begin{cases} g(t)\mid t\mid^{p-2}t, & t\leqslant 0,\\ g(t)\mid t\mid^{p-2}t+h(t), & t>0,\end{cases}$$

where $g(t)\in C(R)$, $g(0)=0$; $g(t)\geqslant 0$, $t\in\mathbb{R}$; $h(t)\in C[0,+\infty)$; $\lim\limits_{t\to+0}\frac{h(t)}{t^{p-1}}=0$; $\lim\limits_{t\to+\infty}\frac{h(t)}{t^{p^*-1}}=0$; $\lim\limits_{t\to+\infty}\frac{h(t)}{t^{p-1}}=+\infty$. Moreover, there exists $t_0>0$ such that $g(t)\equiv\lambda_1$ for all $|t|\geqslant t_0$.

Theorem 1. 3. Let $1<p<\frac{N}{s}$ and assume that f has theimproved subcritical polynomial growth on Ω(condition(SCPI)) and satisfies (H_1)−(H_4). If $f_0<\lambda_1$ and $l=+\infty$, then problem(1. 1)has at least one nontrivial solution.

Remark 1. 3. When $l=+\infty$, problem(1. 1)is generalized superlinear at negative infinity. Thisextends results of[23,24].

Incase of $ps=N$, we have $p^*=+\infty$. In this case, we mustredefine the subcritical(exponential)growth as follows:

(SCE): f has subcritical(exponential)growth on Ω, i. e.,

$$\lim_{t\to\infty}\frac{\mid f(x,t)\mid}{\exp(\alpha\mid t\mid^{\frac{N}{N-s}})}=0$$

uniformly on $x=\Omega$ for all $\alpha>0$.

When $ps=N$ and f has the subcritical(exponential)growth(SCE), our work is still to study asymmetric problem(1. 1)without the(AR)—condition inthe positive semi—axis. To our knowledge, this problem is rarely studied by other people for fractional p—Laplacian. Hence, ourresults are new and our methods are skillful since we skillfully combined mountain pass theorem with the Moser—Trudinger inequality. Our results are as follows:

Theorem 1. 4. Let $sp=N$ andassume that f has thesubcritical exponential growth on Ω(condition(SCE))and satisfies $(H_1)-(H_3)$. If $f_0<\lambda_1<l<\infty$, then problem(1. 1)has at least one nontrivial solution when l is not any of the eigenvalues of $(-\Delta)^{s}_{p}$.

Remark 1. 4. In view of the conditions (H_2), (H_3) and(SCE), problem(1. 1) is called asymmetric subcritical exponential problem. Hence, this result is new.

Theorem 1. 5. Let $sp=N$ andassume that f has the subcritical exponential growth on Ω(condition(SCE))and satisfies $(H_1)-(H_3)$. If $f_0<\lambda_1=l$ and $\lim\limits_{t\to-\infty}\left[f(x,t)t-\frac{N}{s}F(x,t)\right]=-\infty$ uniformly for a. e. $x=\Omega$, then problem(1. 1)has at least one nontrivial solution.

Remark 1. 5. When $l=\lambda_1$, problem(1. 1)is called resonant at negative infinity. This caseis completely new.

Theorem 1. 6. Let $sp=N$ and assume that $f$has the subcritical exponential growth on Ω(condition(SCE)) and satisfies $(H_1)-(H_4)$. If $f_0<\lambda_1$ and $l=+\infty$, then problem(1. 1)has at least one nontrivial solution.

Remark 1. 6. Probably, some more credit should be given to References[23] and[24], since they built the functional framework to prove theresults in Theorems 1. 4, 1. 5 and 1. 6. To our knowledge, for the case $p=2$, all ofour results are also new.

The paper is organized as follows. In Section 2, we present somenecessary preliminary knowledge about working space. In Section 3, we prove some lemmas in order to prove our main results. In Section 4, we give the proofs for our main results.

2. PRELIMINARIES

In this section, we give some preliminary results which will be used in the sequel. We briefly recall the variation formulation of theproblem introduced in [11]. Let $\Omega \subset \mathbb{R}^N$ be a bounded domain with smooth boundary $\partial\Omega$, and for all $1 \leqslant r \leqslant \infty$ denote by $|\cdot|_r$ thenorm of $L^r(\Omega)$. Moreover, let $0<s<1<p<\infty$ be real-numbers, and the fractional critical exponent be defined as $p^* = \frac{Np}{N-sp}$ if $sp<N$ and $p^* = \infty$ if $sp \geqslant N$. First we introduce a variational setting for problem(1.1). The gagliardo seminorm is defined for all measurable function $u: \mathbb{R}^N \to \mathbb{R}$ by

$$[u]_{s,p} = \left(\int_{\mathbb{R}^{2N}} \frac{|u(x)-u(y)|^p}{|x-y|^{N+sp}} dxdy\right)^{\frac{1}{p}}.$$

We define the fractional Sobolev space

$$W^{s,p}(\mathbb{R}^N) = \{u \in L^p(\mathbb{R}^N): u \text{ measurable}, [u]_{s,p} < \infty\}$$

endowed with the norm

$$\|u\|_{s,p} = (\|u\|_p^p + [u]_{s,p}^p)^{\frac{1}{p}}.$$

For a detailed account on the properties of $W^{s,p}(\mathbb{R}^N)$ we refer the reader to [7]. Weshall work in the closed linear subspace

$$X = \{u \in W^{s,p}(\mathbb{R}^N): u(x) = 0 \text{ a. e. in } \mathbb{R}^N \backslash \Omega\},$$

which can be equivalently renormed by setting $\|\cdot\| = [\cdot]_{s,p}$ (see Theorem 7.1 in[7]). It is readily seen that $(X, \|\cdot\|)$ is a uniforml yconvex Banach space and that the embedding $X \quad L^r(\Omega)$ is continuous for all $1 \leqslant r \leqslant p^*$ and compact for all $1 \leqslant r < p^*$ (see Theorem 6.5,7.1 in[7]). Thedual space of $(X, \|\cdot\|)$ is denoted by $(X^*, \|\cdot\|)$. We rephrase variationally the fractional $p-$Laplacian as thenonlinear operator $A: X \to X^*$ defined for all $u, x \in X$ by

$$\langle A(u), v\rangle = \int_{\mathbb{R}^{2N}} \frac{|u(x)-u(7)|^{p-2}(u(x)-u(y))(v(x)-v(y))}{|x-y|^{N+sp}} dxdy.$$

It can be seen that, if u is smooth enough, this definitioncoincides with that of $(-\Delta)_p^s u(x)$ in our introduction. Aweak solution of problem (1.1) is a function $u \in X$ such that

$$< A(u), v > = \int_\Omega f(x,u) v dx \tag{2.1}$$

for all $v \in X$. Clearly, A is odd, $(p-1)-$homogeneous, and satisfies for all $u \in X$

$$< A(u), u > = \|u\|^p, \|A(u)\|_* \leqslant \|u\|^{p-1}.$$

Since X is uniformly convex in X, by Proposition 1.3 in[21], A satisfiesthe following compactness condition: If (u_n) is a sequence in X such that $u_n \to u$ in X and $< A(u_n), u_n - u > \to 0$, then $u_n \to u$ in X. Fromcondition(SCPI), it is easy to know that(2. 1)is theEuler－Lagrange equation of the functional

$$J(u)=\frac{1}{p}\|u\|^{p}-\int_{\Omega}F(x,u)dx.$$

Next,we recall some definitions for compactness condition and aversion of mountain pass theorem. Definition 2. 1. Let$(X,\|\cdot\|_{X})$ bea real Banach space with its dual space $(X^{*},\|\cdot\|_{X^{*}})$ and $J\in C^{1}(X,\mathbb{R})$. For $c\in\mathbb{R}$,we say that J satisfies the $(PS)_{c}$ condition if forany sequence $\{x_{n}\}\subset E$ with

$$J(x_{n})\to c,DJ(x_{n})\to 0 \text{ in } X^{*},$$

there is a subsequence $\{x_{n_k}\}$ such that $\{x_{n_k}\}$ converges strongly in X. Also,we say that J satisfythe $(C)c$ condition if for any sequence $\{x_{n}\}\subset X$ with

$$J(x_{n})\to c,\backslash\|DJ(x_{n})\|_{X^{*}}(1+\|x_{n}\|_{X})\to 0,$$

there issubsequence $\{x_{n_k}\}$ such that $\{x_{n_k}\}$ convergesstrongly in X.

We have the following version of the mountain pass theorem(see[1,6]):

Proposition 2. 1. Let X be areal Banach space and suppose that $J\in C^{1}(X,R)$ satisfies the condition

$$\max\{J(0),J(u_{1})\}\leqslant\alpha\leqslant\beta\leqslant\inf_{\|u\|=\rho}J(u),$$

for some $\alpha<\beta,\rho>0$ and $u_{1}\in X$ with $\|u_{1}\|>\rho$. Let $c\geqslant\beta$ be characterized by

$$c=\inf_{\gamma\in\Gamma}\max_{0\leqslant t\leqslant 1}J(\gamma(t)),$$

where $\Gamma=\{\gamma\in C([0,1],X),\gamma(0)=0,\gamma(1)=u_{1}\}$ is the set of continuous paths joining 0 and u_{1}. Then,there exists a sequence $\{u_{n}\}\subset X$ such that

$$J(u_{n})\to c\geqslant\beta \text{ and } (1+\|u_{n}\|)\|J'(u_{n})\|_{X^{*}}\to 0 \text{ as } n\to\infty.$$

3. SOME LEMMAS

Lemma 3. 1. Let $1<p<\frac{N}{s}$ and $\varphi_{1}>0$ be a λ_{1}－eigenfunction with $\|\varphi_{1}\|=1$ and assume that $(H_{1})-(H_{3})$ and(SCPI)hold. If $f_{0}<\lambda_{1}<l\leqslant+\infty$, then:

(i)There exist $\beta,\alpha>0$ such that $J(u)\geqslant\alpha$ for all $u\in X$ with $\|u\|=\rho$,

(ii)$J(t\varphi_{1})\to-\infty$ as $t\to+\infty$.

Proof. By (SCPI) and $(H_{1})-(H_{3})$, if $l\in(\lambda_{1},+\infty)$,for any $\varepsilon>0$, there exist $A_{1}=A_{1}(\varepsilon),B_{1}-B_{1}(\varepsilon)$ such that for all $(x,s)\in\Omega\times\mathbb{R}$,

$$F(x,s)\leqslant\frac{1}{p}(f_{0}+\varepsilon)|s|^{p}+A_{1}|s|^{p^{*}},\qquad(3.1)$$

$$F(x,s)\geqslant\frac{1}{p}(1-\varepsilon)|s|^{p}-B_{1}\quad\text{if}\quad l\in(\lambda_{1},+\infty).\qquad(3.2)$$

Choose $\varepsilon>0$ such that $(f_0+\varepsilon)<\lambda_1$. By(3. 1), the Poincaré inequality and the Sobolev inequality: $|u|_p^{p^*}\leqslant K\|u\|^{p^*}$, we get

$$J(u)\geqslant\frac{1}{p}\|u\|^p-\frac{f_0+\varepsilon}{p}|u|_p^p-A_1|u|_p^{p^*}$$

$$\geqslant\frac{1}{p}\left(1-\frac{f_0+\varepsilon}{\lambda_1}\right)\|u\|^p-A_1K\|u\|^{p^*}.$$

So, part (i) is proved if we choose $\|u\|=\rho>0$ small enough.

On the other hand, if $l\in(\lambda_1+\infty)$, taking $\varepsilon>0$ such that $l-\varepsilon>\lambda_1$ and using (3. 2), we have

$$J(t\varphi_1)\leqslant\frac{1}{p}\left(1-\frac{l-\varepsilon}{\lambda_1}\right)|t|^p+B_1|\Omega|\rightarrow-\infty \text{ as } t\rightarrow+\infty.$$

Thus part (ii) is proved. By a exactly sligh tmodification of the proof above, we can prove (ii) if $l=+\infty$.

Lemma 3. 2. (see[16])In case of $sp=N$, let $u\in X$, then we have

$$\sup_{u\in X,\ \|u\|\leqslant 1}\int_\Omega\exp(\alpha|u|^{\frac{N}{N-s}})dx\leqslant C(\Omega) \text{ for } \alpha\leqslant\alpha_{N,s}.$$

The inequality is optimal: for any growth $\exp(\alpha|u|^{\frac{N}{N-s}})$ with $\alpha>\alpha_{N,s}$, the corresponding supremum is $+\infty$.

Lemma 3. 3. Let $ps=N$ and $\varphi_1>0$ be a λ_1-eigenfunction with $\|\varphi_1\|=1$ and assume $(H_1)-(H_3)$ and(SCE)hold. If $f_0<\lambda_1<l\leqslant+\infty$, then:

(i)There exist $\rho,\alpha>0$ such that $J(u)\geqslant\alpha$ for all $u\in X$ with $\|u\|=\rho$,

(ii)$J(t\varphi_1)\rightarrow-\infty$ as $t\rightarrow+\infty$.

Proof. By (SCE) and $(H_1)-(H_3)$, if $l\in(\lambda_1,+\infty)$ for any $\varepsilon>0$, there exist $A_1=A_1(\varepsilon)$, $B_1=B_1(\varepsilon)$, $\kappa>0$ and $q>\dfrac{N}{s}$ such that for all $(x,s)\in\Omega\times\mathbb{R}$,

$$F(x,s)\leqslant\frac{s}{N}(f_0+\varepsilon)|s|^{\frac{N}{s}}+A_1\exp(\kappa|s|^{\frac{N}{N-s}})|s|^q,\tag{3. 3}$$

$$F(x,s)\geqslant\frac{s}{N}(l-\varepsilon)|s|^{\frac{N}{s}}-B_1 \quad\text{if}\quad l\in(\lambda_1,+\infty).\tag{3. 4}$$

Choose $\varepsilon>0$ such that $(f_0+\varepsilon)<\lambda_1$. By(3. 3), the H\\"older inequality and the Moser-Trudinger embedding inequality, we get

$$J(u)\geqslant\frac{s}{N}\|u\|^{\frac{N}{s}}-\frac{f_0+\varepsilon}{\frac{N}{s}}|u|_{\frac{N}{s}}^{\frac{N}{s}}-A_1\int_\Omega\exp(\kappa|u|^{\frac{N}{N-s}})|u|^q dx$$

$$\geqslant\frac{s}{N}\left(1-\frac{f_0+\varepsilon}{\lambda_1}\right)\|u\|^{\frac{N}{s}}-A_1\left(\int_\Omega\exp\left(\kappa r\|u\|^{\frac{N}{N-s}}\left(\frac{|u|}{\|u\|}\right)^{\frac{N}{N-s}}\right)dx\right)^{\frac{1}{r}}$$

$$\left(\int_{\Omega} |u|^{r'q}dx\right)^{\frac{1}{r}} \geqslant \frac{s}{N}\left(1-\frac{f_0+\varepsilon}{\lambda_1}\right)\|u\|^{\frac{N}{s}} - C\|u\|^{q},$$

where $r>1$ sufficiently close to 1, $\|u\| \leqslant \sigma$ and $\kappa r \sigma^{\frac{N}{N-s}} < \alpha_{N,s}$. So, part (i) is proved if we choose $\|u\| = \rho > 0$ small enough.

On the other hand, if $l \in (\lambda_1, +\infty)$, taking $\varepsilon > 0$ such that $l-\varepsilon > \lambda_1$ and using (3.4), we have

$$J(t\varphi_1) \leqslant \frac{s}{N}\left(1-\frac{l-\varepsilon}{\lambda_1}\right)|t|^{\frac{N}{s}} + B_1|\Omega| \to -\infty \text{ as } t \to +\infty.$$

Thus part (ii) is proved. By exactly slight modification of the proof above, we can prove (ii) if $l=+\infty$.

4. PROOFS OF THE MAIN RESULTS

Here, we only prove Theorem 1.1 — Theorem 1.4. Others followthese results.

Proof of Theorem 1.1. By Lemma 3.1, the geometry conditions of mountain pass theorem hold. So, we only need toverify condition (PS). Let $\{u_n \subset X\}$ be a (PS) sequence such that for every $n \in \mathbb{N}$,

$$\left|\frac{\|u_n\|^{p}}{p} - \int_{\Omega} F(x,u_n)dx\right| \leqslant c, \tag{4.1}$$

$$\left|< A(u_n), v > - \int_{\Omega} f(x,u_n)v dx\right| \leqslant \varepsilon_n \|v\|, v \in X, \tag{4.2}$$

where $c>0$ is a positive constant and $\{\varepsilon_n\} \subset \mathbb{R}^{+}$ is a sequence which converges to zero.

Step 1. In order to prove that $\{u_n\}$ has a convergence subsequence, we first show that it is a bounded sequence. To dothis, we argue by contradiction assuming that for a subsequence, which we denote by $\{u_n\}$, we have

$$\|u_n\| \to +\infty \text{ as } n \to \infty.$$

Without loss of generality, we can assume $\|u_n\| > 1$ for all $n \in \mathbb{N}$ and define $z_n = \frac{u_n}{\|u_n\|}$. Obviously, $\|z_n\| = 1\ \forall n \in \mathbb{N}$ and then, it is possible to extract a subsequence(denoted also by $\{z_n\}$) such that

$$z_n \rightharpoonup z_0 \quad \text{in } X, \tag{4.3}$$

$$z_n \to z_0 \quad \text{in } L^{p}(\Omega), \tag{4.4}$$

$$z_n(x) \to z_0(x) \text{ a.e. } x \in \Omega, \tag{4.5}$$

$$|z_n(x)| \leqslant q(x) \text{ a. e. } x \in \Omega, \tag{4.6}$$

where $z_0 \in X$ and $q \in L^p(\Omega)$. Dividing both sides of(4.2)by $\|u_n\|^{p-1}$, we obtain

$$\left| < A(z_n), v > - \int_\Omega \frac{f(x,u_n)}{\|u_n\|^{p-1}} v dx \right| \leqslant \frac{\varepsilon_n}{\|u_n\|^{p-1}} \|v\| \quad \textit{for all} \quad v \in X.$$

Passing to the limit we deduce from(4.3)that

$$\lim_{n\to\infty}\int_\Omega \frac{f(x,u_n)}{\|u_n\|^{p-1}} v dx = < A(z_0), v > \tag{4.7}$$

for all $v \in X$.

Now we claim that $z_0(x) \leqslant 0$ for a. e. $x \in \Omega$ To verify this, let us observe that by choosing $v = z_0^+ = \max\{z_0, 0\}$ in(4.7)we have

$$\lim_{n\to\infty}\int_{\Omega^+} \frac{f(x,u_n)}{\|u_n\|^{p-1}} z_0 dx = < A(z_0), z_0 > < +\infty, \tag{4.8}$$

where $\Omega^+ = \{x \in \Omega | z_0(x) > 0\}$. But, on the other hand, from (H_2) and (H_3), it implies

$$\frac{f(x,u_n(x))}{\|u_n\|^{p-1}} z_0(x) \geqslant (-l|q(x)|^{p-2} q(x) - K_1) z_0(x), \text{a. e. } x \in \Omega$$

for some positive constant $K_1 > 0$. Moreover, using $\lim_{n\to\infty} u_n(x) = +\infty$ for a. e. $x \in \Omega^+$, (4.5)and the superlinearity of f(see (H_3)), we alsodeduce

$$\lim_{n\to\infty} \frac{f(x,u_n(x))}{\|u_n\|^{p-1}} z_0(x) = \lim_{n\to\infty} \frac{f(x,u_n(x))}{u_n^{p-1}} z_n(x)^{p-1} z_0(x) = +\infty, \text{a. e. } x \in \Omega^+.$$

Therefore, if $|\Omega^+| > 0$, by the Fatou's Lemma, we will obtain

$$\lim_{n\to\infty}\int_{\Omega^+} \frac{f(x,u_n(x))}{\|u_n\|^{p-1}} z_0(x) dx = +\infty,$$

which contradicts(4.8). Thus $|\Omega^+| = 0$ and the claim is proved.

Clearly, $z_0(x) \not\equiv 0$. By (H_2), there exists $c > 0$ suchthat $\frac{|f(x,u_n)|}{|u_n|^{p-1}} \leqslant c$ for a. e. $x \in \Omega$. By using the Lebesgue dominated convergence theoremin(4.7), we have

$$< A(z_0), v > - \int_\Omega l |z_0|^{p-2} z_0 v dx = 0 \tag{4.9}$$

for all $v \in X$. This contradicts our assumption, i. e. , l is not any of theeigenvalues of $(-\Delta)_p^s$ on X.

Step 2. Now, we prove that $\{u_n\}$ has a convergence subsequence. In fact, we can suppose that

$$u_n \rightharpoonup u \quad \text{in } X,$$

$$u_n \to u \quad \text{in } L^q(\Omega), \forall 1 \leqslant q < p^*,$$

$$u_n(x) \to u(x) \text{ a. e. } x \in \Omega.$$

Now, since f has the subcritical growth on Ω, for every $\varepsilon > 0$, we can find a constant $C(\varepsilon) > 0$ such that

$$f(x,s) \leqslant C(\varepsilon) + \varepsilon \mid s \mid^{p^{*}-1}, \forall (x,s) \in \Omega \times \mathbb{R},$$

then

$$\left| \int_{\Omega} f(x,u_n)(u_n - u)dx \right|$$

$$\leqslant C(\varepsilon)\int_{\Omega} \mid u_n - u \mid dx + \varepsilon\int_{\Omega} \mid u_n - u \mid \mid u_n \mid^{p^{*}-1} dx$$

$$\leqslant C(\varepsilon)\int_{\Omega} \mid u_n - u \mid dx + \varepsilon\left(\int_{\Omega} (\mid u_n \mid^{p^{*}-1})^{\frac{p^{*}}{p^{*}-1}} dx\right)^{\frac{p^{*}-1}{p^{*}}} \left(\int_{\Omega} \mid u_n - u \mid^{p^{*}}\right)^{\frac{1}{p^{*}}}$$

$$\leqslant C(\varepsilon)\int_{\Omega} \mid u_n - u \mid dx + \varepsilon C(\Omega).$$

Similarly, since $u_n \rightharpoonup u$ in $X \int_{\Omega} \mid u_n - u \mid dx \to 0$. Since $\varepsilon > 0$ is arbitrary, we canconclude that

$$\int_{\Omega} (f(x,u_n) - f(x,u))(u_n - u)dx \to 0 \text{ as } n \to \infty.$$

By(4. 2)and the formula above, wehave

$$< A(u_n), u_n - u > \to 0, \text{as } n \to \infty.$$

So, from property of operator A, we have $u_n \to u$ in X which means that J satisfies(PS).

Proof of Theorem 1. 2. Since $l = \lambda_1$, obviously, Lemma 3. 1 (i) holds. We only need to show that Lemma 3. 1 (ii) holds. Let $u = -t\varphi_1$, then

$$J(-t\varphi_1) = \frac{1}{p}t^p \parallel \varphi_1 \parallel^p - \int_{\Omega} F(x, -t\varphi_1)dx$$

$$= \frac{1}{p}t^p \parallel \varphi_1 \parallel^p - \frac{1}{p}\int_{\Omega} f(x, -t\varphi_1)(-t\varphi_1)dx$$

$$- \int_{\Omega} \left\{ F(x, -t\varphi_1) + \frac{f(x, -t\varphi_1)t\varphi_1}{p} \right\} dx$$

By $f(x,s) = \lambda_1 |s|^{p-2}s + \circ(s)$ as $s \to -\infty$, we have

$$J(-t\varphi_1) \to -\infty \quad \text{as} \quad t \to +\infty$$

and the claim is proved. By Proposition 2. 1, there exists a sequence $\{u_n\} \subset X$ such that

$$J(u_n) = \frac{1}{p} \parallel u_n \parallel^p - \int_{\Omega} F(x,u_n)dx = c + \circ(1), \tag{4. 11}$$

$$(1+\|u_n\|)\|J'(u_n)\|_{X^*}\to 0 \quad \text{as} \quad n\to\infty. \tag{4.12}$$

Clearly,(4.12)implies that

$$<J'(u_n),u_n>=\|u_n\|^p-\int_\Omega f(x,u_n(x))u_n dx=\circ(1). \tag{4.13}$$

To complete our proof, we firstly need to verify that $\{u_n\}$ is bounded in X. Similar to the proof of Theorem 1.1, we have $z_0(x)\leqslant 0, x\in\Omega, z_0(x)\not\equiv 0$ and

$$<A(z_0),v>-\int_\Omega l\,|z_0|^{p-2}z_0 v dx=0$$

for all $v\in X$. By the maximum principle(see[11]), $z_0<0$ is an eigenfunction of λ_1 then $|u_n(x)|\to\infty$ fora. e. $x\in\Omega$. By our assumptions, we have

$$\lim_{n\to\infty}(f(x,u_n(x))u_n(x)-pF(x,u_n(x)))=-\infty$$

uniformly in $x\in\Omega$, which implies that

$$\int_\Omega (f(x,u_n(x))u_n(x)-pF(x,u_n(x)))dx\to\infty \quad \text{as} \quad n\to\infty. \tag{4.14}$$

On the otherhand,(4.13)implies that

$$pJ(u_n)-<J'(u_n),u_n>\to pc \quad \text{as} \quad n\to\infty.$$

Thus

$$\int_\Omega (f(x,u_n)u_n-pF(x,u_n))dx\to pc \quad \text{as} \quad n\to\infty,$$

which contradicts(4.14). Hence $\{u_n\}$ is bounded. According to Step 2 of the proof of Theorem 1.1, we have $u_n\to u$ in X which means that J satisfies $(C)_c$.

Proof of Theorem 1.3. By Lemma 3.1and Proposition 2.1, then(4.11)-(4.13)hold. We still can provethat $\{u_n\}$ is bounded in X. Assume $\|u_n\|\to+\infty$ as $n\to\infty$. Similar tothe proof of Theorem 1.1, we have $z_0(x)\leqslant 0$ and when $z_0(x)<0, u_n=z_n\|u_n\|\to-\infty$ as $n\to\infty$. For any $c>0$, let

$$s_n=\frac{\sqrt[p]{2pc}}{\|u_n\|}, w_n=s_n u_n=\frac{\sqrt[p]{2pc}\,u_n}{\|u_n\|}. \tag{4.15}$$

Since $\{w_n\}$ is bounded in$X$, it is possiblet oextract as ubsequence(denotedal-soby $\{w_n\}$)suchthat

$$w_n\rightharpoonup w_0 \quad \text{in } X,$$

$$w_n\to w_0 \quad \text{in } L^p(\Omega),$$

$$w_n(x)\to w_0(x) \text{ a. e. } x\in\Omega,$$

$$|w_n(x)|\leqslant h(x) \text{ a. e. } x\in\Omega,$$

where $w_0\in X$ and $h\in L^p(\Omega)$.

If $\| u_n \| \to +\infty$ as $n \to \infty$, then $w_0(x) \equiv 0$. In fact, letting $\Omega^- = \{x \in \Omega : w_0(x) < 0\}$ and noticing $l = +\infty$, it follows from (H_3) that

$$\frac{f(x, u_n)}{|u_n|^{p-2} u_n} \geqslant M \quad \text{uniformly for all } x \in \Omega^-,$$

where M is large enough constant. Therefore, by(4. 13)and(4. 15), we have

$$\begin{aligned} 2pc &= \lim_{n \to \infty} \| w_n \|^p \\ &= \lim_{n \to \infty} \int_\Omega \frac{f(x, u_n)}{|u_n|^{p-2} u_n} |w_n|^p dx \\ &\geqslant \lim_{n \to \infty} \int_{\Omega^-} \frac{f(x, u_n)}{|u_n|^{p-2} u_n} |w_n|^p dx \\ &\geqslant M \lim_{n \to \infty} \int_{\Omega^-} |w_0|^p dx \end{aligned}$$

So $w_0 = 0$ for a. e. $x \in \Omega$. But, if $w_0 \equiv 0$, then $\int_\Omega F(x, w_n) dx \to 0$. Hence

$$J(w_n) = \frac{1}{p} \| w_n \|^p + o(1) = 2c + o(1). \tag{4.16}$$

On the other hand, we let $\{t_n\} \subset \mathbb{R}$ such that

$$J(t_n u_n^-) = \max_{t \in [0,1]} J(t u_n^-).$$

Then we have

$$J(t_n u_n^-) \geqslant J(w_n) = 2c - \int_\Omega F(x, w_n) dx \geqslant c,$$

which deduces that

$$J(t_n u_n^-) \to \infty, \quad \text{as} \quad n \to \infty. \tag{4.17}$$

According to $J(0) = 0$ and $J(u_n^-) \to c$ we have $t_n \in (0,1)$, then

$$< J' t_n u_n^-), t_n u_n^- >= t_n \frac{d}{dt}\Big|_{t = t_n} J(t u_n^-) = 0.$$

Then, from (H_4) it follows that

$$\begin{aligned} \frac{1}{\theta} J(t_n u_n^-) &= \frac{1}{\theta} \left(J(t_n u_n^-) - \frac{1}{p} < J'(t_n u_n^-), t_n u_n^- > \right) \\ &= \frac{1}{p\theta} \int_\Omega F(x, t_n u_n^-) dx \\ &\leqslant \frac{1}{p} \int_\Omega F(x, u_n^-) dx + \frac{1}{p\theta} |\Omega| C_* \\ &= J(u_n^-) - \frac{1}{p} < J'(u_n^-), u_n^- > + c \to C \end{aligned}$$

This contradicts the fact that $J(t_n u_n^-) \to \infty$. So $\{u_n\}$ is bounded in X. According to Step 2 of the proof of Theorem 1. 1, we have $u_n \to u$ in X which means that J satisfies $(C)_c$.

Proof of Theorem 1. 4. By Lemma 3. 3, the geometry conditions of mountain pass theorem hold. So, we onlyneed to verify condition (PS). Similar to Step 1 ofthe proof of Theorem 1. 1, we easily know that (PS) sequence $\{u_n\}$ is bounded in X. Next, we prove that $\{u_n\}$ has a convergence subsequence. Without loss ofgenerality, suppose that

$$\begin{cases} \| u_n \| \leqslant \beta, \\ u_n \rightharpoonup u \quad \text{in } X, \\ u_n \to u \quad \text{in } L^q(\Omega), \forall q \geqslant 1, \\ u_n(x) \to u(x) \text{ a. e. } x \in \Omega. \end{cases}$$

Now, since f has the subcritical exponential growth (SCE) on Ω, we can find a constant $C_\beta > 0$ such that

$$| f(x,t) | \leqslant C_\beta \exp\left(\frac{\alpha_{N,s}}{2\beta^{\frac{N}{N-s}}} | t |^{\frac{N}{N-s}}\right), \forall (x,t) \in \Omega \times \mathbb{R}.$$

Thus, by the fractional case of the Moser—Trudinger inequality(see Lemma 3. 2),

$$\begin{aligned} & \left| \int_\Omega f(x,u_n)(u_n - u) dx \right| \\ & \leqslant C\left(\int_\Omega \exp\left(\frac{\alpha_{N,s}}{\beta^{\frac{N}{N-s}}} | u_n |^{\frac{N}{N-s}}\right) dx\right)^{\frac{1}{2}} | u_n - u |_2 \\ & \leqslant C\left(\int_\Omega \exp\left(\frac{\alpha_{N,s}}{\beta^{\frac{N}{N-s}}} \| u_n \|^{\frac{N}{N-s}} \left| \frac{u_n}{\| u_n \|} \right|^{\frac{N}{N-s}}\right) dx\right)^{\frac{1}{2}} | u_n - u |_2 \\ & \leqslant C | u_n - u |_2 \to 0 \end{aligned}$$

Similar to the last proof of Theorem 1. 1, we have $u_n \to u$ in X which means that J satisfies (PS).

Acknowledgements

This research was supported by the NSFC(Nos. 11661070 and 11571176), NSF of Gansu Province (Nos. 1506RJZE114 and 1606RJYE237), TSNC (No. TSA1406) and Scientific Research Foundation of the Higher Education Institutions of Gansu Province(No. 2015A—131).

References

[1]A. Ambrosetti and P. H. Rabinowitz, Dualvariational methods in critical point theory

and applications, J. Funct. Anal. 14, 349—381(1973).

[2]D. Arcoya and S. Villegas, Nontrivialsolutions for a Neumann problem with a nonlinear termasymptotically linear at $-\infty$ and superlinear at $+\infty$, Math. Z. 219, 499—513(1995).

[3]L. Brasco and E. Parini, The second eigenvalue of thefractional p—Laplacian, preprint.

[4]C. Bucur, E. Valdinoci, Nonlocal Diffusion and Applica—tions, Lecture Notes of the Unione Matematica Italiana Series, Volume 20, (Springer Verlag, 2016).

[5]L. Caffarelli, Non—local diffusions, drifts andgames. In Nonlinear Paritial Differential Equations, Abel Symposia 7, 37—52(2012).

[6]D. G. Costa and O. H. Miyagaki, Nontrivial solutions for perturbations of the p—Laplacianon unbounded domains, J. Math. Anal. Appl. 193, 737—755(1995).

[7]E. Di Nezza, G. Palatucci, and E. Valdinoci, Hitchhiker's guide to the fractional Sobolev spaces, Bull. Sci. Math. 136, 521—573(2012).

[8]D. G. de Figueiredo and B. Ruf, On asuperlinear Sturm—Liouville equation and a related bouncingproblem, J. Reine Angew. Math. 421, 1—22(1991).

[9]G. Franzina and G. Palatucci. Fractional p—eigenvalues, Riv. Mat. Univ. Parma, \\textbf{5}, 373—386(2014).

[10]A. Fiscella, G. Molica Bisci, and R. Servadei, Bifurcation and multiplicity results for critical nonlocal fractionalproblems, Bull. Sci. Math 140, 14—35(2016).

[11]A. Iannizzotto, S. B. Liu, K. Perera and M. Squassina, Existence results for fractional p—Laplacian problems via Morse

theory, Advances in Calculus of Variations(2014)doi: 10.1515/acv—2014—0024.

[12]A. Iannizzotto, S. Mosconi, and M. Squassina, GlobalHölder regularity for the fractional p—Laplacian, preprint, arXiv: 1411.2956.

[13]A. Iannizzotto and M. Squassina, Weyl—type laws forfractional p—eigenvalue problems, Asymptot. Anal. 88, 233—245(2014).

[14]N. Lam and Guozhen Lu, N—Laplacian equations in $\mathbb{R}^N$ withsubcritical and critical growth without the Ambrosetti—Rabinowitzcondition, Adv. Nonlinear Stud. \\textbf{13}, 289—308(2013).

[15]Z. L. Liu and Z. Q. Wang, On the Ambrosetti—Rabinowitz superlinear condition, Adv. Nonlinear Stud. 4, 563—574(2004).

[16]L. Martinazzi, Fractional Adams—Moser—Trudinger typeinequalities, preprint, arXiv: 1506.00489.

[17]G. Molica Bisci, Sequence of weak solutions for fractional equations, Math. Res. Lett. 21, 241—253(2014).

[18]G. Molica Bisci and D. V. Rădulescu, Ground state solutions of scalar field fractional

Schrödinger equations,Calc. Var. Partial Differential Equations 54,2985—3008(2015).

[19]G. Molica Bisci,D. V. Rădulescu,and R. Servadei,Variational methods for nonlocal fractional problems,Encyclopedia of Mathematics and its Applications,162. (Cambridge University Press,Cambridge,2016).

[20]K. Perera,Existence and multiplicity resultsfor a Sturm—Liouville equation asymptotically linear at $-\infty$ and superlinear at $+\infty$,Nonlinear Anal. 39,669—684(2000).

[21]K. Perera,R. P. Agarwal,and D. O'Regan,Morse theoreticaspects of p—Laplacian operators,volume 161 of Mathematical

Surveys and Monographs,American mathematical Society,Providence,RI,2010.

[22]K. Perera,M. Squassina,and Y. Yang,Bifurcation andmultiplicity results for critical fractional p—laplacianproblems,preprint,arXiv:1407. 8061.

[23]R. Servadei and E. Valdinoci,Mountain Pass solutionsfor non—local elliptic operators,J. Math. Anal. Appl. 389,887—898(2012).

[24]R. Servadei and E. Valdinoci,Variational methods fornon—local operators of elliptic type,Discrete and ContinuousDynamical Systems 33,2105—2137(2013).

[25]M. Q. Xiang,G. Molica Bisci,G. H. Tian,and B. L. Zhang,Infinitely many solutions for the stationaryKirchhoff problems involving the fractional p—Laplacian,Nonlinearity 29,357—374(2015).

[26]B. L. Zhang and M. Ferrara,Multiplicity of soutions for a class of superlinear non—local fractional equations,Complex

Variables and Elliptic Equations60,583—595(2015).

[27]B. L. Zhang,G. Molica Bisci and R. Servadei,Superlinear nonlocal fractional problems with infinitely manysolutions,Nonlinearity 28,2247—2264(2015).

(本文发表于 2017 年德国杂志《数学通讯》290 卷 16 期)

General Padé Approximation Method for Time Space Fractional Diffusion Equation

Hengfei Ding *

摘要:近年来,分数阶微分方程的研究引起了人们的兴趣。在这篇文章中,我们提出了一个新的数值方法去求解时间空间分数阶微分方程,其中应用一个四阶的分数阶紧致格式去离散 Riesz 导数,则此方程变成一个分数阶的常微分方程系统。再利用拉普拉斯和反拉普拉斯变换,得到分数阶微分方程的解析解,进一步,我们利用广义的 Padé 逼近去近似 Mittag—Leffler 函数,得到所求方程的数值解。最近,我们给出两个数值例子,进一步验证了数值结果和我们的理论分析是一致的。

In recent years, fractional differential equations have attracted much attention due to their wide application. In this paper, we present a novel numerical method for the space — time Riesz — Caputo fractional diffusion equation, which discrete the Riesz derivative by a fourth — order fractional — compact difference scheme, then the above space — time Riesz — Caputo fractional diffusion equation change into a fractional ordinary differential equation (FODE) system. Again using Laplace and inverse Laplace transforms, one can get the analytical solution of the FODE system, furthermore, we approximate the Mittag — Leffler function by the global Padé approximation and obtain the numerical method for the space — time Riesz — Caputo fractional diffusion equation. Finally, two numerical examples are presented to show that the numerical results are in good line with our theoretical analysis.

---

* 作者简介:丁恒飞(1979—),男,甘肃天水,理学博士,天水师范学院数学与统计学院副教授(校聘教授),主要从事分数阶微分方程的建模及其高阶快速算法研究。

1. Introduction

Recently, it is found that some phenomena (in engineering, control theory, chemical physics, stochastic processes, anomalous diffusion, rheology, biology, medical and other science) can be described more accurately by using fractional differential equations [1,2]. So, great attention has been given to finding solutions of theirs. However, it is usually very difficult or even impossible to obtain the analytical solution. As a consequence, it is necessary to search for new numerical methods for the fractional differential equations, at present, there have some numerical methods for them[3−6], etc.

In this paper, our aim is to numerically study the following space−time Riesz−Caputo fractional diffusion equation

$$ {}_{C}D_{0,t}^{\alpha}u(x,t) = k_{\beta}\frac{\partial^{\beta}u(x,t)}{\partial|x|^{\beta}}, 0 < x < L, 0 \leqslant t \leqslant T, \tag{1}$$

together with the corresponding initial and boundary value conditions

$$u(x,0) = g(x), 0 \leqslant t \leqslant L,$$
$$u(0,t) = u(L,t) = 0, 0 \leqslant t \leqslant T,$$

where ${}_{C}D_{0,t}^{\alpha}u(x,t)$ denotes Caputo fractional derivative of order α with respect to time t for $0<\alpha \leqslant 1$, $\frac{\partial^{\beta}}{\partial|x|^{\beta}}$ denotes Riesz fractional derivative of order β with respect to space variable x for $1<\beta<2$, respectively. The coefficient K_{β} is a positive constant.

The paper is organized as follows. In Section2, some preliminaries are listed. In Section 3, numerical method for solving the above space−time Riesz−Caputo fractional diffusion equation is established. In Section 4, two numerical examples are given to support our theoretical analysis. Some concluding remarks are given in the final section.

2. Preliminaries

In this section, we give two important definitions and one lemma as follows, which can be found from [2]:

Definition 1. The Riesz fractional operator for $n-1<\beta\leqslant n$ on a finite interval $0<x<L$ is defined as

$$\frac{\partial^{\beta}}{\partial \mid x \mid^{\beta}}(x \cdot t) = - C_{\beta}({}_{RL}D_{0,x}^{\beta} + {}_{RL}D_{x,L}^{\beta})u(x,t),$$

Where

$$C_{\beta} = \frac{1}{2\cos\frac{\pi\beta}{2}}, \beta \neq 1,$$

$${}_{RL}D_{0,x}^{\beta}u(x,t) = \frac{1}{\Gamma(n-\beta)} \frac{\partial n}{\partial x^{n}} \int_{0}^{x} \frac{u(\xi,t)}{(x-\xi)^{\beta+1-n}} d\xi,$$

and

$${}_{RL}D_{x,L}^{\beta}u(x,t) = \frac{(-1)^{n}}{\Gamma(n-\beta)} \frac{\partial^{n}}{\partial x^{n}} \int_{x}^{L} \frac{u(\xi,t)}{(\xi-x)^{\beta+1-n}} d\xi,$$

Definition 2. The Caputo fractional derivative operator ${}_{C}D_{0,t}^{\alpha}$ of order α is defined in the following form:

$${}_{C}D_{0,t}^{\alpha}u(x,t) = \frac{1}{\gamma(m-\alpha)} \int_{0}^{t} \frac{\partial^{m}u(x,\eta)}{\partial \eta^{m}} (t-\eta)^{m-\alpha-1} d\eta,$$

where $m-1<\alpha\leqslant m, m\in N$,

Lemma 1. The Laplace transform of the Caputo derivative ${}_{C}D_{0,t}^{\alpha}u(x,t)$ with respect to t is

$$L_{t}\{{}_{C}D_{0,t}^{\alpha}u(x,t)\} = s^{\alpha}U(x,s) - \sum_{j=0}^{m-1} s^{\alpha-j-1} \frac{\partial^{j}u(x,t)}{\partial t^{j}}\bigg|_{t=0},$$

where $U(x,s)=L_{t}\{u(x,t)\}, m-1<\alpha\leqslant m, m\in N$.

3. Construction of the numerical method

Let $t_{k}=k\tau, k=0,1,\cdots,M, x_{j}=jh, j=0,1,\cdots,N$, where $\tau=T/M$ and $h=L/N$ arethe time and space steps, respectively. For numerical treatment of Eq. (1), we usually assume the same values of the function $u(x,t)$ outside thedomain(0, L).

Firstly, we discrete the Riesz derivative by the following fourth-order fractional-compact difference scheme [7]

$$\frac{\partial^{\beta}u(x,t)}{\partial \mid x \mid^{\beta}} = -\frac{1}{h^{\beta}}\left(1+\frac{\beta}{24}\delta_{x}^{2}\right)^{-1}\Delta_{h}^{\beta}u(x,t) + \vartheta(h^{4}),$$

where $\delta_{x}{}^{2}$ is the second-order central difference operator with respect to x and defined by $\delta_{x}{}^{2}u(x,t)=u(x-h,t)-2u(x,t)+u(x+h,t)$. And Δ_{h}^{β} is the following fractional centered difference operator

$$\Delta_{h}^{\beta}u(x,t) = \sum_{l=-\infty}^{\infty} g_{l}^{(\beta)}u(x-lh,t),$$

in which

$$g_l^{(\beta)} = \frac{(-1)^l \Gamma(\beta+1)}{\Gamma\left(\frac{\beta}{2}-l+1\right)\left(\frac{\beta}{2}+l+1\right)}$$

Substituting (2) into (1) yields

$${}_{C}D_{0,t}^{\alpha}u(x_j,t) = -\frac{1}{h^{\beta}}\left(1+\frac{\beta}{24}\delta_x^2\right)^{-1}\sum_{l=-(N-j)}^{j} g_k^{(\beta)}u(x_{j-l},t) + \vartheta(h^4).$$

Neglecting the high-order term $\vartheta(\mathrm{h}^4)$ and denoting u_j^k as the numerical solution of $u(x_j,t_k)$, then one has

$${}_{C}D_{0,t_k}^{\alpha}u_j^k = -\frac{1}{h^{\beta}}\left(1+\frac{\beta}{24}\delta_x^2\right)^{-1}\sum_{l=-(N-j)}^{j} g_k^{\beta}u_{j-l}^k$$

and its matrix form is

$${}_{C}D_{0,t_k}^{\alpha}U(t_k) = \mu A^{-1}BU(t_k),$$

where

$$\mu = -\frac{k_{\beta}}{h^{\beta}}, U(t_k) = (u_1^k, u_2^k, \cdots, u_{N-1}^k)^T,$$

$$A = \begin{pmatrix} \frac{\beta}{24} & 1-\frac{\beta}{24} & 0 & \cdots & 0 & 0 \\ 1-\frac{\beta}{24} & \frac{\beta}{24} & 1-\frac{\beta}{24} & 0 & \cdots & 0 \\ 0 & 1-\frac{\beta}{24} & \frac{\beta}{24} & 1-\frac{\beta}{24} & 0 & \cdots \\ \vdots & \ddots & 1-\frac{\beta}{24} & \ddots & \ddots & \vdots \\ \cdots & 0 & \ddots & \frac{\beta}{24} & 1-\frac{\beta}{24} & 0 \\ 0 & \cdots & 0 & 1-\frac{\beta}{24} & \frac{\beta}{24} & 1-\frac{\beta}{24} \\ 0 & 0 & \cdots & 0 & 1-\frac{\beta}{24} & \frac{\beta}{24} \end{pmatrix}$$

and

$$B=\begin{pmatrix} g_0^{(\beta)} & g_{-1}^{(\beta)} & g_{-2}^{(\beta)} & \cdots & g_{4-M}^{(\beta)} & g_{3-M}^{(\beta)} & g_{2-M}^{(\beta)} \\ g_1^{(\beta)} & g_0^{(\beta)} & g_{-1}^{(\beta)} & g_{-2}^{(\beta)} & \cdots & g_{4-M}^{(\beta)} & g_{3-M}^{(\beta)} \\ g_2^{(\beta)} & g_1^{(\beta)} & g_0^{(\beta)} & g_{-1}^{(\beta)} & g_{-2}^{(\beta)} & \cdots & g_{4-M}^{(\beta)} \\ \vdots & \ddots & g_1^{(\beta)} & \ddots & \ddots & \ddots & \vdots \\ g_{4-M}^{(\beta)} & \cdots & g_2^{(\beta)} & \ddots & g_0^{(\beta)} & g_{-1}^{(\beta)} & g_{-2}^{(\beta)} \\ g_{M-3}^{(\beta)} & g_{M-4}^{(\beta)} & \cdots & g_2^{(\beta)} & g_1^{(\beta)} & g_0^{(\beta)} & g_{-1}^{(\beta)} \\ g_{M-2}^{(\beta)} & g_{M-3}^{(\beta)} & g_{M-4}^{(\beta)} & \cdots & g_2^{(\beta)} & g_1^{(\beta)} & g_0^{(\beta)} \end{pmatrix}.$$

Application of the Laplace transform on both sides of the system(3)with respect to t yields

$$\widetilde{U}(t_k)=(sI-\mu s^{1-\alpha}A^{-1}B)^{-1}U(t_0).$$

Note that matrices A and B are both symmetric positive define, so $A^{-1}B$ is also a symmetric positive define matrix and there exists a nonsingular matrix Q, such that

$$A^{-1}B=Q\Lambda Q^{-1},$$

where $\Lambda=(\lambda_1,\lambda_2,\ldots,\lambda_{N-1})$, λ_j being the eigenvalues of $A^{-1}B$.

Hence, we have

$$\widetilde{U}(t_k)=Q(sI-\mu S^{1-\alpha}\Lambda)^{-1}Q^{-1}U(t_0).$$

Using the well known inverse Laplace transform

$$E_\alpha(-\omega t^\alpha)=L^{-1}\left\{\frac{s^{\alpha-1}}{s^\alpha+\omega}\right\}$$

to the above equation and then one has

$$\widetilde{U}(t_k)=QL^{-1}\{(sI-\mu s^{1-\alpha}\Lambda^{-1}\}Q^{-1}U(t_0)=QE_\alpha(\mu t_k^\alpha\Lambda)Q^{-1}U(t_0).$$

where $E_\alpha(z)$ is the Mittag—Leffler function

$$E_\alpha(z)=\sum_{k=0}^{\infty}\frac{z^k}{\Gamma(\alpha k+1)},\alpha>0.$$

For Mittag—Leffler function $E_\alpha(z)$, we utilize the following global Padé approximation with degree 2 [8]

$$E_\alpha(z)\ \frac{1-\dfrac{1}{\Gamma(1-\alpha)q_0}z}{1-\dfrac{q_1}{1_0}z+\dfrac{1}{q_0}z^2}+\vartheta(z^2),$$

where

$$q_0=\frac{\dfrac{\Gamma(1-\alpha)}{\Gamma(1+\alpha)}-\dfrac{\Gamma(1+\alpha)\Gamma(1-\alpha)}{\Gamma(1-2\alpha)}}{\Gamma(1+\alpha)\Gamma(1-\alpha)-1},q_1=\frac{\Gamma(1+\alpha)-\dfrac{\Gamma(1+\alpha)}{\Gamma(1-2\alpha)}}{\Gamma(1+\alpha)\Gamma(1-\alpha)-1}$$

Substituting (5) into (4) yields

$$U(t_k)=Q\left[I-\frac{1}{\Gamma(1-\alpha)q_0}(\mu(k\tau)^{\alpha}\Lambda)\right]$$
$$\times\left[I-\frac{q_1}{q_0}(\mu(k\tau)^{\alpha}\Lambda)+\frac{1}{q_0}(\mu(k\tau)^{\alpha}\Lambda)^2\right]^{-1}Q^{-1}U(t_0)+\vartheta(\tau^2),$$

neglecting the high—order term $\vartheta(\tau^2)$ and denoting U^k as the numerical solution of $U(t_k)$, then one obtain numerical algorithm for Eq. (1)as follows,

$$U^k=Q\left[I-\frac{1}{\Gamma(1-\alpha)q_0}(\mu(k\tau)^{\alpha}\Lambda)\right]$$
$$\times\left[I-\frac{q_1}{q_0}(\mu(k\tau)^{\alpha}\Lambda)+\frac{1}{q_0}(\mu(k\tau)^{\alpha}\Lambda)^2\right]^{-1}Q^{-1}U^0.$$

Obviously, we easily know that the convergence order(local truncation error)of method(6)is $\vartheta(\tau^2+h^4)$ by using the above analysis.

4. Numerical example

In this section, we present two examples to show that our numerical method can be used to solve this type of fractional differential equation.

Example 1. Consider the following space—time Riesz—Caputo fractional diffusion equation

$${}_{C}D^{\alpha}_{0,t}u(x,t)=\frac{\partial^2 u(x,t)}{\partial|x|^{\beta}},0<x<\pi,0\leqslant t\leqslant 1,$$

subject to the corresponding initial and boundary value conditions

$$u(x,0)=x^4(\pi-x)^4,\quad 0\leqslant x\leqslant\pi,$$
$$u(0,t)=u(\pi,t)=0,\quad 0\leqslant t\leqslant T.$$

Using the method in [5], one can get the analytical solution of the above equation as

$$u(x,t)=\sum_{k=1}^{\infty}\frac{48(1-(-1)^k(\pi^4k^4-180\pi^2k^2+1680))}{\pi k^9}E_{\alpha}(-k^{\beta}t^{\alpha})\sin(kx).$$

We give the comparison of the numerical solution with the analytical solution at different cases by Figs. 4. 1and 4. 2. From these figures, we see that the numerical solution is good in line with the exact solution.

Example 2([5]). Consider the space—time Riesz—Caputo fractional diffusion equation. The analytical solution is

$$u(x,t)=\sum_{k=1}^{\infty}\frac{8(-1)^{k+1}-4}{k^3}E_{\alpha}(-k^{\beta}t^{\alpha})\sin(kx).$$

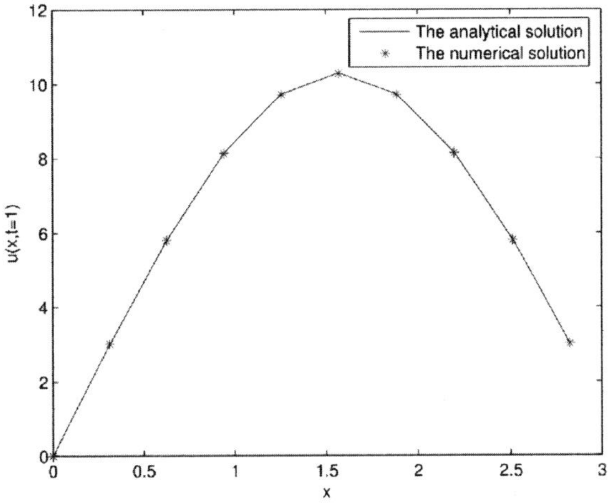

Fig. 4.1. The comparison of numerical solution with analytical solution for $\alpha = 0.895$, $\beta = 1.95$ ($\tau = \frac{1}{10}$, $h = \frac{\pi}{10}$).

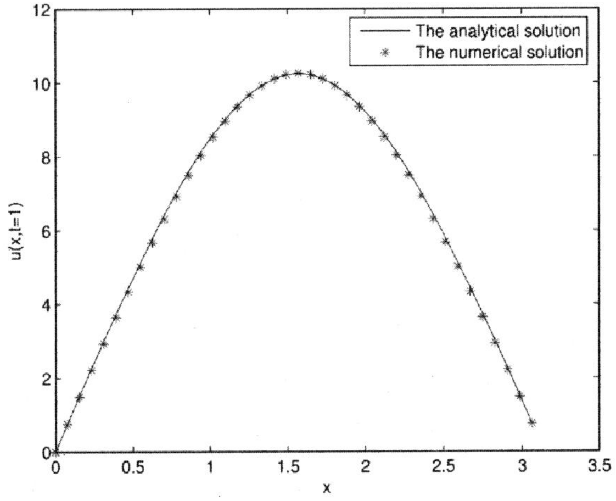

Fig. 4.2. The comparison of numerical solution with analytical solution for $\alpha = 0.905$, $\beta = 1.95$ ($\tau = \frac{1}{40}$, $h = \frac{\pi}{40}$).

Figs. 4. 3—4. 6 displaythe numerical solution and analytical solution surfaces for different cases, respectively. From these figures, it can be seen that our numerical method is again in good agreement with the analytical solution. Furthermore, we give the error surfaces by Figs. 4. 7and 4. 8.

5. Conclusion

In this paper, we propose a numerical method for solving space—time Riesz—Caputo fractional diffusion equation based on Laplace and inverse Laplace transforms and the global Padé approximation methods. The method can be ex-

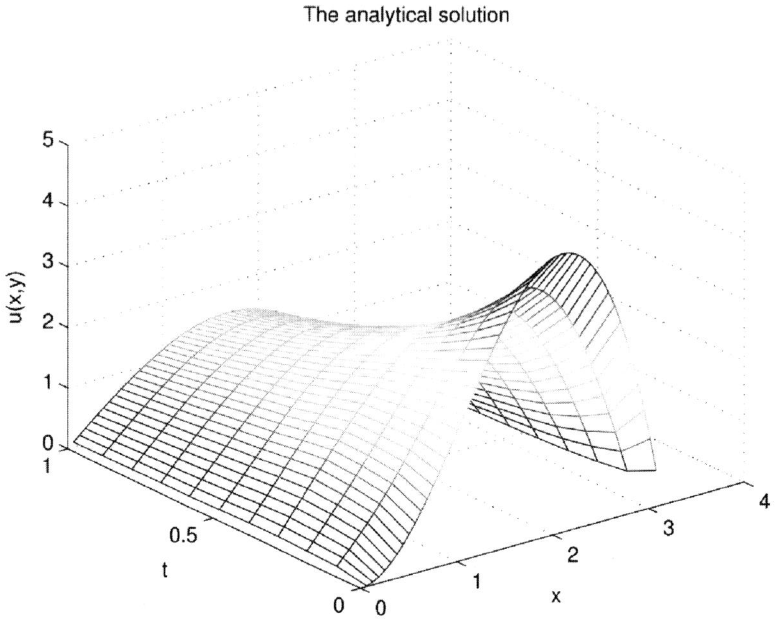

Fig. 4.3. The analytical solution surface for $\alpha = 0.9,\ \beta = 1.96\,(\tau = \frac{1}{10},\ h = \frac{\pi}{50})$.

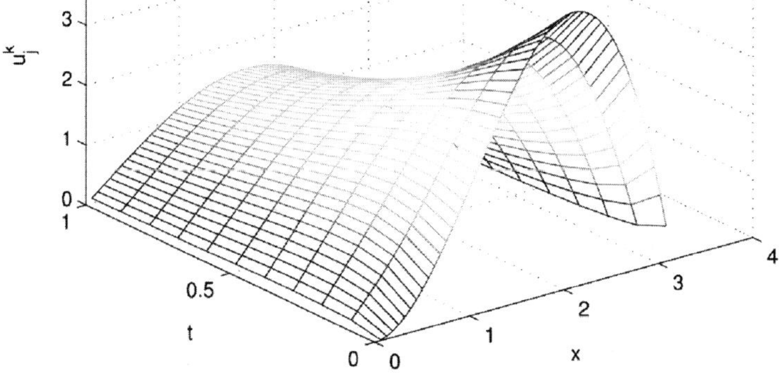

Fig. 4.4. The numerical solution surface for $\alpha = 0.9,\ \beta = 1.96\,(\tau = \frac{1}{10},\ h = \frac{\pi}{50})$.

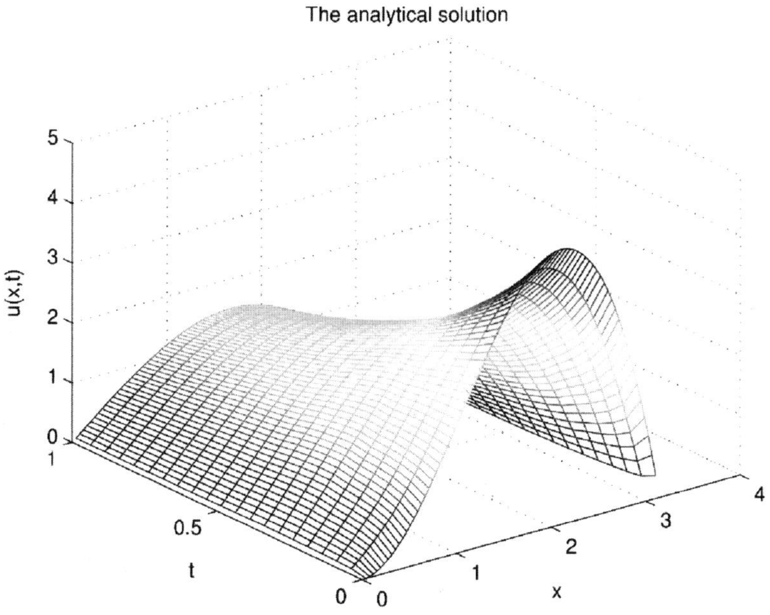

Fig. 4.5. The analytical solution surface for $\alpha = 0.906,\ \beta = 1.938\ (\tau = \frac{1}{20},\ h = \frac{\pi}{80})$.

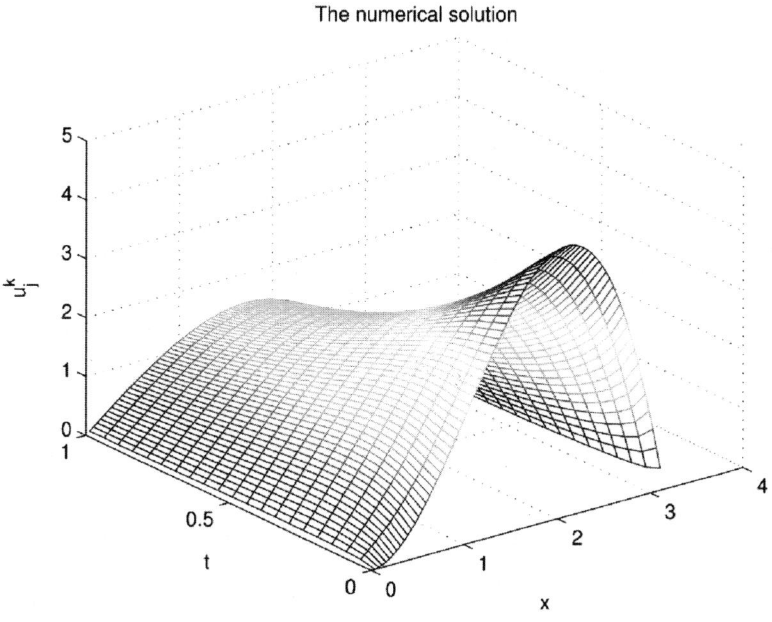

Fig. 4.6. The numerical solution surface for $\alpha = 0.906,\ \beta = 1.938\ (\tau = \frac{1}{20},\ h = \frac{\pi}{80})$.

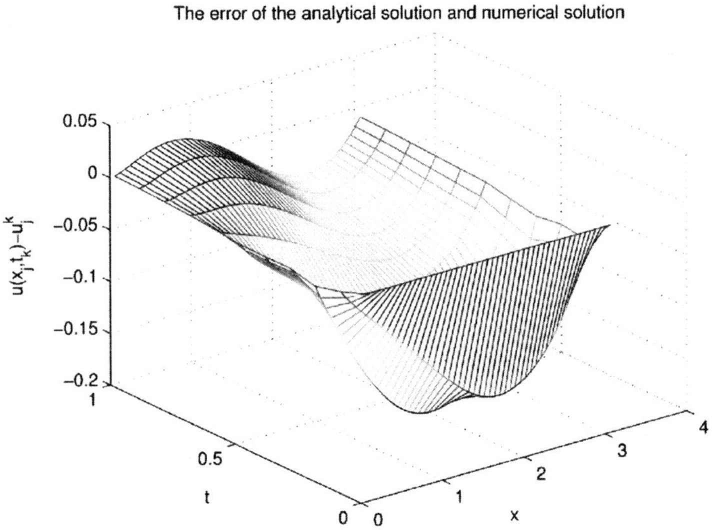

Fig. 4.7. The error of the analytical solution and numerical solution for $\alpha = 0.9$, $\beta = 1.96$ ($\tau = \frac{1}{10}$, $h = \frac{\pi}{50}$).

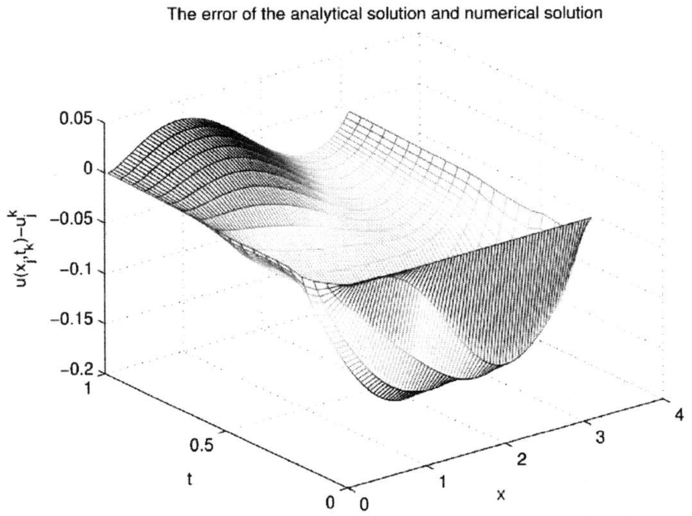

Fig. 4.8. The error of the analytical solution and numerical solution for $\alpha = 0.906$, $\beta = 1.938$ ($\tau = \frac{1}{20}$, $h = \frac{\pi}{80}$).

tended to these types of fractional differential equations. Finally, numerical results are march to the theory analysis.

References

[1]K. B. Oldham, J. Spanier, The Fractional Calculus, Theory and Applications of Differ-

entiation and Integration to Arbitrary Order, Academic Press, New York, 1974.

[2]I. Podlubny, Fractional Differential Equations, Academic Press, San Diego, 1999.

[3] M. M. Meerschaert, C. Tadjeran, Finite difference approximations for two－sided space－fractional partial differential equations, Appl. Numer. Math. 56(2006)80－90.

[4]S. B. Yuste, L. Acedo, An explicit finite difference method and a new von Neumann－type stability analysis for fractional diffusion equations, SIAM J. Numer. Anal. 42(5)(2005) 1862－1874.

[5]Q. Yang, I. Turner, F. Liu, Analytical and numerical solutions for the time and space －symmetric fractional diffusion equation, ANZIAM J. 50(C)(2008)C800－C814.

[6]C. P. Li, F. H. Zeng, Numerical Methods for Fractional Calculus, Chapman and Hall/ CRC, Boca Raton, USA, 2015.

[7] H. F. Ding, C. P. Li, Y. Q. Chen, High－order algorithms for Riesz derivative and their applications(II), J. Comput. Phys. 293(2015)218－237.

[8]C. B. Zeng, Y. Q. Chen, Global Padé approximations of the generalized Mittag－Leffler function and its inverse, arXiv:1310. 5592v2[math. CA].

（本文发表于 2016 年《Journal of Computational and Applied Mathematics》299 卷）

On the probabilistic Hausdorff distance and a class of probabilistic decomposable measures

Yonghong Shen *

摘要:本文进一步讨论了概率 Hausdorff 距离的一些有用性质.特别地,对于 Menger 概率度量空间的重要性质我们给出了一个直接的证明.随后,我们提出了基于三角模的概率可分解测度,其中测度值通过一个概率分布函数来刻画.同时,为更好地解释和理解相关的定义,构造了几个适当的例子.此外,如果给定一个 Menger 概率度量空间,借助概率 Hausdoff 距离可以诱导出一类概率可分解测度.同时,这类测度对于最强的三角模是 $\sigma-\perp-$次可分解概率测度.进一步,所有可测集类构成了一个代数.最后,借助一类概率可分解测度和三角模可以诱导出一个外概率测度.借助这类测度,在非严格连续 Archimedean 三角模条件下可以获得一个 Menger 概率伪度量空间.

In this paper, some useful properties associated with the probabilistic Hausdorff distance are further derived. Especially, we provide a direct proof for an existing important result. Afterwards, the t−norm−based probabilistic decomposable measure is presented, in which the value of measure is characterized by a probability distribution function. Meantime, several examples are constructed to illustrate different notions, and then further properties are examined. Moreover, for a given Menger PM−space, a probabilistic decomposable measure can be induced by means of the resulting probabilistic Hausdorff distance. We prove that this type of measure is $\sigma-\perp-$probabilistic subdecomposable measure for the strongest t−norm. Furthermore, we also prove that the class of all measurable

* 作者简介:沈永红(1982—),男,甘肃西和人,天水师范学院数学与统计学院副教授、博士,主要从事经典与模糊微分方程稳定性方面的研究。

sets forms an algebra. Finally, an outer probabilistic measure is induced by a class of probabilistic decomposable measures and the t-norm. Based on this kind of measure, a Menger probabilistic pseudometric space can be obtained for a non-strict continuous Archimedean t-norm.

1. INTRODUCTION

The classical measure is one of the most important concepts in mathematics, in which the additivity is the main characteristic. From a theoretical point of view, this property is very effective and convenient, but often too inflexible or too rigid in dealing with practical problems. For this reason, the fuzzy measure proposed by Sugeno in[26]can be regarded as an extension of classical measure in which the additivity is replaced by a weaker condition, i. e., monotonicity. It should be noted that the notion of fuzziness in Sugeno's fuzzy measure is different from that introduced by Zadeh. So far, there have been many different fuzzy measures, such as the belief measure, the possibility measure, the plausibility measure, and the decomposable measure, etc. The non-additivity is the main feature of the fuzzy measure, so it is also called a non-additive measure[30]. A specialized and systematic research of the non-additive measure can be found in Denneberg's monograph[4].

Among the fuzzy measures mentioned above, the decomposable measure, independently introduced by Dubois and Prade[5], and Weber[31], is very important, because some fuzzy measures, such as the k-additive measure, the probability measure and the possibility measure, are only special classes of the decomposable measure[8]. In the last two decades, the theory of decomposable measure has been widely and in-depth studied by various authors[13, 19, 20, 29, 30, 32]. More importantly, decomposable measures and the corresponding integrals have provided a basis for the pseudo-analysis with important applications in nonlinear(differential) equations, decision theory, aggregation analysis, optimization, optimal control and fuzzy systems[6-8, 15, 16, 21, 23, 24, 27, 29]. On the theoretical side, the relationship between the decomposable measure and the probabilistic metric was discussed in[9, 22]. More specifically, a Menger probabilistic metric space can be induced by a given decomposable measure of(NSA)-type and another separable metric space. Furthermore, a complete Menger probabilistic metric space can also be obtained if the separable metric space is complete. Recently, the subdecomposable and supdecomposable measures, which are a generalization of the decomposable measure, are further studied by Cavallo et al. [1] and Qiu et al. [17,18], respectively. It should be pointed out that Qiu and Zhang [17] considered the relationship between the decomposable measure and the classical Hausdorff distance, and then they proved that for a given normalized compact metric space, a type of superdecomposable measure can be induced by constructing a classical Hausdorff distance. Afterwards, Qiu et al. [18] further concluded that every non-strict continuous Archimedean t-norm-based decomposable measure can be extended to a σ-algebra with the help of the fuzzy pseudometric. All the abovementioned works show that it is deserved to do further researches on the relationship between the measure and different metrics.

As we all know, the probabilistic (statistic) metric can be regarded as an appropriate extension of the classical metric. Within the framework of probabilistic metric theory, to characterize the uncertainty or randomness of the distance, the probability distribution function is used, instead of a determinate nonnegative real number. In this case, the distance between two points is viewed as a random variable. However, many practical problems associated with the measure theory also account uncertainty or randomness. For instance, if we measure the efficiency of a worker by recording its productivity per hour, we can find that the recording number is not a single value usually, but a series of values of statistical data. Normally, the average value is chosen to measure the productivity of the worker. However, it is more appropriate to regard the study subject as a statistical rather than a deterministical one in many similar situations. For modeling these problems in which we only have a probabilistic information about the measure of a set, Hutník and Mesiar [12] introduced the notion of σ_T-submeasure based on the classical numerical submeasure. Specifically, the value of measure was replaced by a distance distribution function and the triangular norm served as an important tool in the generalization of the subadditivity condition. In the sequel, Halčinová et al. [10] considered several possible generalizations of σ_T-submeasure (meant to be a probabilistic submeasure) in connection with probabilistic metric spaces related to various triangle functions. Especially, Menger probabilistic pseudometrics and several pseudo-metrics and metrics on a ring induced by such probabilistic submeasures were constructed for different types of triangular functions. Soon after, they further discussed different probabilistic extensions of the classical submeasures related to aggregation functions in[11], such as triangular norm — based probabilistic submeasures and more general semicopula — based probabilistic submeasures. As a natural generalization, this paper will deal with a class of decomposable measures from a probabilistic point of view. Similar to the notion of a probabilistic submeasure, this extended measure will be called the probabilistic decomposable measure. Moreover, from a different perspective, we shall consider the relationship between the probabilistic decomposable mea— sure and the metric in a Menger probabilistic metric space.

The rest of the paper is organized as follows. In Section 2, some concepts and related results associated with t—norms and probabilistic metric spaces are introduced. In Section 3, the probabilistic Hausdorff distance is introduced and some related properties are examined. Section 4 proposes a concept of probabilistic decomposable measure based on the Menger probabilistic metric space and inspired by the t—conorm—based decomposable measure. Meantime, several examples are provided in order to interpret the related definitions. Then further properties associated with the proposed probabilistic decomposable measure are derived. Afterwards, a class of probabilistic decomposable measures is constructed using the probabilistic Hausdorff distance, and some related results are established. In Section 5, based on the probabilistic decomposable measure and the t—norm, an outer probabilistic measure is constructed and some related properties are examined. Finally, some concluding remarks are given in Section 6.

2. PRELIMINARIES

Definition 2.1(Klement et al. [14]). A triangular norm(t—norm for short)is a binary operation on the unit interval[0,1], i. e. , a function: $[0,1]^2 \to [0,1]$ such that for all $x, y, z \in [0,1]$ the following four axioms are satisfied:

(TN—1) $xTy = yTx$ (commutativity);

(TN—2) $xT(yTz) = (xTy)Tz$ (associativity);

(TN—3) $xTy \leqslant xTz$ whenever $y \leqslant z$ (monotonicity);

(TN—4) $xT1 = x$ (boundary condition).

In particular, a t—norm T is said to be continuous if it is a continuous function in$[0,1]^2$. A t—norm T is said to be strictly monotone if $xTy<xTz$ whenever $x>0$and $y<z$. A t—norm is strict if it is continuous and strictly monotone.

The following are the four basic t—norms:

[1](Minimum t—norm)$x\,T_M y=min\{x,y\}$(the strongest t—norm);

[2](Product t—norm)$x\,T_P y=x\cdot y$;

[3](Łukasiewicz t—norm)$x\,T_L y=max\{x+y-1,0\}$;

[4](Drastic t—norm)(the weakest t—norm)

$$xT_D y=\begin{cases}\min(x,y) & if\ \max\{x,y\}=1,\\ 0 & otherwise.\end{cases}$$

Obviously,T_M,T_P,T_L are continuous t—norms exceptT_D.

Definition 2.2 (Klementet al. [14]). A triangular conorm(t—conorm for short) is a binary operation on the unit interval[0,1],i. e. ,a function $\perp:[0,1]^2\to[0,1]$such that for all $x,y,z\in[0,1]$ satisfies(TN—1)—(TN—3)and(SN—4) $x\perp 0=x$(boundary condition).

In fact,the associativity of t—norm implies that a t—norm can be extended to an n—ary operation in the usual way by induction,defining for each n—tuple $(x_1,x_2,...,x_n)\in[0,1]^n$,

$$T_{i-1}^{n}x_i=T_{i-1}^{n-1}x_iT\,x_n$$

In addition,since each t—norm T is weaker than T_M,it is possible to extend a t—norm to a countably infinitary operation,i. e. ,for each sequence $\{x_i\}_{i\in\mathbb{N}}\in[0,1]^{\mathbb{N}}$

$$T_{i-1}^{\infty}x_i=\lim_{n\to\infty}T_{i-1}^{n}x_i$$

Especially,if $x_1=x_2=\cdots=x_n=x$, then $T_{i-1}^{n}x_i$ can be briefly written as $x_T^{(n)}=xTxT\cdots Tx$.

A t—norm is called Archimedean if there exists an integer $n\in\mathbb{N}$such that $x_T^{(n)}<$y for all $(x,y)\in(0,1)^2$. If T is a continuous Archimedean t—norm,then $xTx<x$ for all $x\in(0,1)$.

Theorem 2.1. [Weber[31]]A function T $:[0,1]^2\in[0,1]$ is a continuous Arichimedean t—norm iff there exists a continuous and strictly decreasing function $f:[0,1]\to[0,\infty]$ with f(1)=0 such that

$$xTy=f^{(-1)}(f(x)+f(y))$$

where $f^{(-1)}$ denotes the pseudoinverse of f defined by

$$f^{(-1)}(y)=\begin{cases} f^{(-1)}(y) & if y\in[0,f(0)], \\ 0 & if y\in[f(0),\infty]. \end{cases}$$

Moreover, T is strict iff $f(0)=\infty$, T is non-strict iff $f(0)<\infty$.

Note that the pseudoinverse satisfies the following identities:

$$f^{(-1)}(f(x))=x, x\in[0,1],$$

$$f(f^{(-1)}(y))=\min\{y,f(0)\}, y\in[0,\infty]$$

Remark 1(Weber[31]). The function f from Theorem 2.1 is called an additive generator of T. It is unique up to multiplication by positive numbers. In the nonstrict case, the additive generator with $f(0)=1$ is called the normed generator.

Moreover, for a continuous Archimedean t-norm T, we have that

$$T_{i-1}^{\infty} x_i = f^{(-1)}\left(\sum_{i-1}^{M} f(x_i)\right) \text{ where } M\in \mathbb{N}\cup\{\infty\}$$

For a t-conorm an analogous statement still holds.

Definition 2.3. [Hadžic′ and Pap [9]] A distribution function on $[-\infty,\infty]$(the extended real line) is a function $F:[-\infty,\infty]\to[0,1]$ which is left-continuous on $\mathbb{R}$, non-decreasing and satisfies $F(-\infty)=0, F(\infty)=1$.

The Dirac distribution function $H_a:[-\infty,\infty]\to[0,1]$ is defined for $a\in[-\infty,\infty]$ by

$$H_a(u)=\begin{cases} 0 & if\ u\in[-\infty,a], \\ 1 & if\ u\in[a,\infty]. \end{cases}$$

And for $a=\infty$ by

$$H_a(u)=\begin{cases} 0 & if\ u\in[-\infty,\infty], \\ 1 & if\ u=\infty. \end{cases}$$

Let Δ denote the set of all distribution functions on $[-\infty,\infty]$. For two distribution functions F and G define $F\leqslant G$ if and only if $F(x)\leqslant G(x)$ for every $x\in[-\infty,\infty]$. Then it can easily be seen that the ordered pair $(\Delta,\leqslant)$ is a complete lattice with the maximal element $H_{-\infty}$ and the minimal element H_{∞}.

Definition 2.4(Hadžic′ and Pap[9]). A distance distribution function $F:[-\infty,\infty]\to[0,1]$ is a distribution function with support contained in $[0,\infty]$. The family of all distance distribution functions will be denoted by Δ^+. Simultaneously, we write

$$D^+=\{F|F\in\Delta^+, \lim_{t\to\infty}F(t)=0\}.$$

Obviously, for every $F \in \Delta^+$, since F is equal to zero on $[-\infty, 0]$, the notation Δ^+ usually stands for the set of nondecreasing functions defined on $[0, \infty]$ which satisfy $F(0)=0$ and $F(\infty)=1$. Similarly, D^+ consists of all nondecreasing functions F defined on $[0,\infty)$ which satisfy $F(0)=0$ and $\lim_{n\to\infty} F(t)=1$. Observe that the ordered pair $(\Delta^+, \leqslant)$ is also a complete lattice with the maximal element H_0 and the minimal element H_∞.

Definition 2.5 (Schweizer and Sklar[25]). A probabilistic metric space (briefly, PM-space) is an ordered pair (X, F), where X is a nonempty set, F is a mapping from $X \times X$ to Δ^+ (F(p,q) is denoted by $F_{p,q}$ for every $(p,q) \in X \times X$), satisfying the following conditions:

(MPM-1) $F_{p,q}=H_0$ iff $p=q$;

(MPM-2) $F_{p,q}=F_{q,p}$ for every $p, q \in X$;

(MPM-3) if $F_{p,q}(t)=1$, $F_{q,r}(s)=1$, then $F_{q,r}(t+s)=1$ for all $p, q, r \in X$ and $t, s > 0$.

If the condition (MPM-1) is replaced by the condition (MPM-1') $F_{p,p}=H_0$ for every $p \in X$, then F is called a probabilistic pseudometric on X.

Definition 2.6 (Schweizer and Sklar[25]). A Menger probabilistic metric space (briefly, Menger PM-space) is a triple (X, F, T), where (X, F) is a PM-space, T is a t-norm and the following condition is satisfied:

(MPM-3') $F_{p,r}(t+s) \geqslant F_{p,q}(t) * F_{q,r}(s)$ for all $p, q, r \in X$ and $t, s \geqslant 0$.

Remark 2. If the conditions (MPM-1'), (MPM-2) and (MPM-3') are satisfied, then the triple (X, F, T) is called a Menger probabilistic pseudometric space.

Lemma 2.2. [Tan[28]] Let (X, F, T) be a Menger PM-space. Then for a given $y \in X$ the function $F_{x,y}(t): X \to [0,1]$ is lower semi-continuous for every $x \in X$ and $t \in [0, \infty)$.

Definition 2.7 (Hadžić and Pap[9]). Let (X, F) be a probabilistic metric space and A a be a nonempty subset of X. The probabilistic diameter of A is a mapping $D_A(t): [0,\infty) \to [0,1]$ defined by

$$D_A(t)=\sup_{u<t} \inf_{p,q \in A} F_{p,q}(u)$$

The set A is said to be probabilistic bounded iff $\sup_t D_A(t)=1$, which means that $D_A \in D^+$. In this paper, for a nonempty set X, we denote by P(X), PB(X) and PBC(X) the power set of X, the collection of all probabilistic bounded subsets of

X and the collection of all probabilistic bounded and closed subsets of X, respectively.

3. PROBABILISTIC HAUSDORDORFF DISTANCE

In 1974, Constantin and Bocsan[3] extended the classical Hausdorff distance to the probabilistic setting and introduced the notion of probabilistic Hausdorff distance. For completeness, we recall here this concept.

(i) Given $p \in X, B \in P(X)$, the "probabilistic distance" from p to B is defined as

$$d_{P,B}(t)=d_{B,P}(t)=\begin{cases} 0 & t=0, \\ \sup_{s<t}\sup_{q\in B} F_{p,q}(s) & t\in(0,\infty), \end{cases} \tag{1}$$

with the convention $d_{p,\varnothing}=1-H_0$,

(ii) Given $A, B \in P(X)$, the "probabilistic distance" from A to B is defined as

$$d_{A,B}(t)=\begin{cases} 0 & t=0, \\ \sup_{s<t}\inf_{p\in A}\sup_{q\in B} F_{p,q}(s) & t\in(0,\infty). \end{cases} \tag{2}$$

with the convention $d_{\varnothing,B}=H_0$, For convenience, we write $\widetilde{d_{A,B}}(s)=\inf_{P\in A}\sup_{q\in B} F_{p,q}(s)$, Then

$$d_{A,B}(t)=\sup_{s<t}\widetilde{d_{A,B}}(s).$$

Base on the above formulas 1 and 2, we can obtain the following definition.

Definition3. 1. Let (X,F,T) be a Menger PM-space and let $A, B \in P(X)$. The probabilistic Hausdorff distance between A and B is a mapping $H_{A,B}:[0,\infty)\to[0,1]$ defined by

$$\begin{aligned} H_{A,B}(t) &= \sup_{s<t}\min\{\widetilde{d_{A,B}}(s),\widetilde{d_{B,A}}(s)\} \\ &= \sup_{s<t}\min\{\inf_{p\in A}\sup_{q\in B} F_{p,q}(s), \inf_{q\in A}\sup_{p\in B} F_{p,q}(s)\}, t\in(0,\infty) \end{aligned} \tag{3}$$

where $H_{A,B}(0)=0$.

According to the above definition, some results related to the probabilistic Hausdorff distance can be obtained.

Theorem 3. 1. Let (X,F,T) be a Menger PM-space. For a given $B \in PB(X)$, $t \geq 0$, the function $d_{P,B}(t): X \to [0,1]$ is lower semi-continuous.

Proof. The result is an immediate consequence of Lemma 2. 2 and the formula(1).

Theorem 3. 2. Let(X,F,T) be a Menger PM−space. Given p ∈ X and B ∈ P(X), the function $d_{P,B}(t):[0,\infty)\to[0,1]$ is left−continuous.

Proof. If B=∅, the result is immediate, since $d_{P,B}(t)=1-H_0(t)$ for every t > 0. Then we can suppose that B is nonempty. For every p ∈ B and $s<t$ the distribution function $F_{P,B}(s):[0,\infty)\to[0,1]$ is left−continuous, so is the function $sup_{q\in B}F_{P,B}(s)$, and hence, $d_{P,B}(t)$ is left−continuous.

Corollary 3. 3. Let(X,F,T) be a Menger PM−space. Given two sets $A,B\in PB(X)$, the function $d_{A,B}(t):[0,\infty)\to[0,1]$ is left−continuous.

Theorem 3. 4. Let(X,F,T) be a Menger PM−space. If $\{A_n\}$ is a decreasing sequence in PB(X)such that $\bigcap_{n=1}^{\infty}A_n=A$, then $\lim_{n\to\infty}d_{P,A_n}(t)=d_{P,A}(t)$, for every $p\in X$ and t ⩾ 0.

Proof. For t=0, the result is obvious. Let t > 0. Since $A\subseteq A_n$ for every $n\in\mathbb{N}$, it follows from the formula(1)that

$$\lim_{n\to\infty}d_{P,A_n}(t)\geqslant d_{P,A}(t)$$

Suppose that $\lim_{n\to\infty}d_{P,A_n}(t)=\alpha(t)>\beta(t)=d_{P,A}(t)$ for any t > 0. Since the sequence $\{d_{P,A_n}(t)\}$ is nonincreasing with respect to n, we know that $F_{P,q}(t)>\alpha(t)$ for all $q\in A_n, n\in\mathbb{N}$. On the other hand, by the formula(1), $\beta(t)=d_{P,A}(t)\sup_{s<t}\sup_{q\in A}F_{P,q}(s)$, for any s<t, which implies that $F_{P,q}(t)<\beta(t)$, for all q∈ A. If q∈ A, then q∈ A_n for every n∈ ℕ, since $\bigcap_{n=1}^{\infty}A_n=A$, Thus, we obtain that $F_{P,q}(t)>\alpha(t)$ for all q∈ A which leads to a contradiction. So we conclude that $\lim_{n\to\infty}d_{P,A_n}(t)=d_{P,A}(t)$, for every $p\in X$and $t\geqslant 0$.

Theorem 3. 5. Let(X,F,T) be a Menger PM−space and let X be a compact set. If {An} is a decreasing sequence in PB(X)such that $\bigcap_{n=1}^{\infty}A_n=A$, then $\lim_{n\to\infty}H_{A_n,X}(t)=H_{A,X}(t)$, for every t ⩾ 0.

Proof. Since $A_n, A\subseteq X$, according to Definition 3. 1, we have

$$H_{A_n,X}(t)=\sup_{s<t}\inf_{q\in X}\sup_{q\in A_n}F_{p,q}(s)=d_{X,A_n}(t)$$

$$H_{A,X}(t)=\sup_{s<t}\inf_{q\in X}\sup_{q\in A_n}F_{p,q}(s)=d_{X,,A}(t)$$

Therefore, it suffices to show that

$$\lim_{n\to\infty}d_{X,A_n}(t)=d_{X,A}(t)$$

In particular, if $t=0$, the result is trivial. So we can assume that $t>0$. Since $A_n\subseteq A$ for every $n\in N$, it follows from the formula (2) that

$$\lim_{n\to\infty} d_{X,A_n}(t) \geqslant d_{X,A}(t).$$

Suppose that $\lim_{n\to\infty} d_{X,A_n}(t) = \alpha(t) > \beta(t) = d_{X,A}(t)$, Since X is compact, for any $s < t$ there exists $q_0 \in X$ such that

$d_{q_0,A}(t) = \sup_{s<t} \sup_{P\in A} F_{P,q_0}(s) = \sup_{s<t} \inf_{q\in X} \sup_{p\in A} F_{P,q}(s) = d_{X,A}(t)$, Therefore, we obtain that

$$d_{q_0,A_n}(t) \geqslant d_{X,A_n}(t) \geqslant \alpha(t),$$

Letting $n \to \infty$, we get

$$\lim_{n\to\infty} d_{q_0,A_n}(t) \geqslant \alpha(t) > \beta(t) = d_{q_0,A}(t)$$

By Theorem 3. 4, this contradicts the fact that

$$\lim_{n\to\infty} d_{q_0,A_n}(t) = d_{q_0,A}(t)$$

Consequently, we conclude then $\lim_{n\to\infty} H_{A_{n,X}}(t) = H_{A,X}(t)$.

Theorem 3. 6. Let (X, F, T) be a Menger PM—space under a continuous t—norm T. Then $(PBC(X), H, T)$ is also a Menger PM—space.

Proof. The conditions (MPM－1') and (MPM－2) are obvious. Suppose that there exist A, B∈PBC(X) with A ≠ B such that $H_{A,B}(t) = H_0(t)$ for every $t \geqslant 0$. Without loss of generality, we can assume that there exists $p_0 \in A$ such that $P_0 \notin B$. From Definition 3. 1 we have that

$$\sup_{s<t} \inf_{p\in A} \sup_{q\in B} F_{p,q}(s) = 1.$$

Therefore, we have that

$$\sup_{s<t} \sup_{q\in B} F_{p_0,q}(s) = \sup_{s<t} d_{p_0,B}(s) = 1.$$

From (1) we have $d_{p_0,B}(t) = 1$. Since B is closed, there exists $q_0 \in B$ such that $F_{P_0,q_0}(t) = 1$, which leads to a contradiction. Consequently, the condition (MPM－1) holds.

Next, we have to show that the condition (MPM－3') is satisfied.

Let A, B, C⊂PBC(X). If at least one of these three sets is empty, by Definition 3. 1 it can easily be verified that the inequality is true. Moreover, if s=0 or t=0, the inequality is also obvious. Thus, we assume that these three sets are non－empty and t > 0, s > 0.

Set $u < s$, $v < t$. For every $p \in A$, we may assume that $\sup_{q\in B} F_{p,q}(u) > 0$. Then for each $\varepsilon \in (0, \sup_{q\in B} F_{p,q}(u))$ there exists $q_p \in B$ such that

$$\sup_{q\in B} F_{p,q}(u) - \varepsilon \leqslant F_{p,q_p}(u)$$

Moreover, since F is a Menger probabilistic metric, it follows that

$$F_{p,q}(u)T\ \sup_{r\in C}F_{p,r}(v)\leqslant\sup_{r\in C}F_{p,r}(u+v)$$

for every $q\in B$.

Thus we can obtain

$$(\sup_{q\in B}F_{p,q}(u)-\varepsilon)T\ \inf_{q\in B}\sup_{r\in C}F_{q,r}(v)\leqslant F_{p,q_p}(u)T\ \sup_{r\in C}F_{q_p,r}(v)\leqslant\sup_{r\in C}F_{p,r}(u+v)$$

By the arbitrariness of ε and the continuity of T, we have

$$\sup_{q\in B}F_{p,q}(u)T\ \inf_{q\in B}\sup_{r\in C}F_{q,r}(v)\leqslant\sup_{r\in C}F_{p,r}(u+v)\tag{4}$$

Then we have that

$$\inf_{p\in A}\sup_{q\in B}F_{p,q}(u)T\ \inf_{q\in B}\sup_{r\in C}F_{q,r}(v)\leqslant\inf_{p\in A}\sup_{r\in C}F_{p,r}(u+v)$$

which implies that

$$\tilde{d}_{A,B}(u)T\ \tilde{d}_{B,C}(v)\leqslant\tilde{d}_{A,C}(u+v)\tag{5}$$

In addition, if $\sup_{q\in B}F_{p,q}(u)=0$ for every $p\in A$, then the inequality(4) still holds. So the inequality(5) is also true in this case. Analogously, we also obtain that

$$\tilde{d}_{B,A}(u)T\ \tilde{d}_{C,B}(v)\leqslant\tilde{d}_{C,A}(u+v)$$

Therefore, we have

$$\begin{aligned}&\min\{\tilde{d}_{A,B}(u),\tilde{d}_{B,A}(u)\}T\ \min\{\tilde{d}_{B,C}(v),\tilde{d}_{C,B}(v)\}\\ &\leqslant\min\{\tilde{d}_{A,B}(u)t\ \tilde{d}_{B,C}(v),\tilde{d}_{B,A}(u)T\ \tilde{d}_{C,B}(v)\}\\ &\leqslant\min\{\tilde{d}_{A,C}(u+v),\tilde{d}_{C,A}(u+v)\}\end{aligned}$$

Furthermore, we can get that

$$\begin{aligned}&\sup_{u<s}\min\{\tilde{d}_{A,B}(u),\tilde{d}_{B,A}(u)\}T\ \sup_{v<t}\min\{\tilde{d}_{B,C}(v),\tilde{d}_{C,B}(v)\}\\ &\leqslant\sup_{u<s,v<t}\min\{\tilde{d}_{A,C}(u+v),\tilde{d}_{C,A}(u+v)\}\\ &\leqslant\sup_{u+v<s+t}\min\{\tilde{d}_{A,C}(u+v),\tilde{d}_{C,A}(u+v)\}\end{aligned}$$

Then, it follows from the above inequality that

$$H_{A,B}(s)T\ H_{B,C}(t)\leqslant H_{A,C}(s+t)$$

Hence, we conclude that $(PBC(X),H,T)$ is a Menger PM−space. This completes the proof.

Remark 3. Here, a direct proof for Theorem 3.6 is given, which also can be obtained as a corollary of Theorems 4.2.6−4.2.12 in[2].

As a direct consequence of Theorem 3.6 we have the following result.

Corollary 3.7. Let (X,F,T) be a Menger PM−space under a continuous t−

norm T. Then $(PB(X), H, T)$ is a Menger probabilistic pseudometric space.

4. PROBABILISTIC DECOMPOSABLE MEASURES

In this section, we shall introduce several notions: a probabilistic decomposable measure based on the classical measure, a probability measure, a probabilistic submeasure and a decomposable measure. Meantime, a class of special probabilistic decomposable measures is constructed by means of the probabilistic Hausdorff distance. Finally, some related properties are discussed.

In the sequel, we will recall some basic notions associated with different measures which come from[9,14].

Let X be a nonempty set and A be a $\sigma-$algebra of subsets of X. A classical measure is a set function m defined on A and with values in $[0,\infty]$ such that $m(\varnothing)=0$ and

$$m(\bigcup_{n=1}^{\infty} A_n)=\sum_{n=1}^{\infty} m(A_n)$$

for every sequence $\{A_n\}$ of pairwise disjoint elements from A. Especially, when the range of m is the interval $[0,1]$ with $m(X)=1$, then m is called a probability measure.

Notice that each measure is finitely additive in the sense that the sequence $\{A_n\}$ consists of finite number of nonempty sets. Moreover, if the set function m is continuous, the opposite statement is true.

As a generalization of the classical notion of finite additivity for a set function $m: A \to [0,1]$, a function $F: [0,1]^2 \to [0,1]$

(i. e. , a functional dependence) of $m(A\cup B)$ on $m(A)$ and $m(B)$ is introduced to satisfy the following condition:

$$m(A\cup B)=F(m(A), m(B)) \tag{6}$$

whenever $A, B \in A$ with $A\cap B=\varnothing$. In fact, the formula(6) can also be extended to countable additivity. Considering that the properties of F meets the formula(6), F can be replaced e. g. with t-conorms. In this case, the measure is called a t-conorm-based decomposable measure. The following are some important decomposable measures from[1,31]. Let $\perp$ be a t-conorm. A set function $m: A\to [0,1]$ with $m(\varnothing)=0$ and $m(X)=1$ will be called

(i) a $\perp$-decomposable measure iff $m(A\cup B)=m(A)\perp m(B)$ for every pair of disjoint elements $A, B \in A$;

(ii) a $\sigma-\perp-$decomposable measure iff $m(\bigcup_{n=1}^{\infty} A_n)=\perp_{n=1}^{\infty} m(A_n)$ for every sequence $\{A_n\}$ of pairwise disjoint elements of A;

(iii) a $\perp-$subdecomposable measure iff $m(A\cup B)\leqslant m(A)\perp m(B)$;

(iv) a $\sigma-\perp-$subdecomposable measure iff $m(\bigcup_{n=1}^{\infty} A_n)\leqslant \perp_{n=1}^{\infty} m(A_n)$;

(5) continuous from below(above) iff $\lim_{n\to\infty} m(A_n)=m(A)$ for each increasing(decreasing) sequence $\{A_n\}$ from A with $\lim_{n\to\infty} A_n=A$.

As it is well known, a distance distribution function is adopted to characterize the distance between two points. This metric includes the classical metric as a special case. Based on this idea, Hutník and Mesiar[12] extended a numerical submeasure to a probabilistic submeasure using t-norms. It should be noted that a distance distribution function is employed to characterize a measure of a set. Naturally, it can be viewed as a probabilistic generalization of classical submeasure. For more details, we refer to [12]. Here, we shall consider the corresponding probabilistic generalizations of several decomposable measures in a different way. Similarly, a t-norm will act as a case of functional dependence. Based on the above statement, we formulate the following notions.

Definition 4.1. Let T be a t-norm. A mapping $\mathfrak{M}: A\to\Delta^{+}$ *with* $\mathfrak{M}_{\varnothing}=H_0$ is called

(i) a $T-$probabilistic decomposable measure iff

$$\mathfrak{M}_{A\cup B}(t)=\mathfrak{M}_A(t)T\,\mathfrak{M}_B(t),$$

for every pair of disjoint elements A and B of A and for each $t\geqslant 0$;

(ii) a $\sigma-T-$probabilistic decomposable measure iff

$$\mathfrak{M}_{\bigcup_{n=1}^{\infty} A_n}(t)=T_{n=1}^{\infty}\mathfrak{W}_{A_n}(t),$$

for every sequence $\{A_n\}$ of pairwise disjoint elements of A and for each t $\geqslant 0$;

(iii) a $T-$probabilistic subdecomposable measure iff

$$\mathfrak{W}_{A\cup B}(t)\leqslant\mathfrak{W}_A(t)T\,\mathfrak{W}_B(t),$$

for $A, B\in A$ and $t\geqslant 0$;

(iv) a $\sigma-T-$probabilistic subdecomposable measure iff

$$\mathfrak{M}_{\bigcup_{n=1}^{\infty} A_n}(t)\leqslant T_{n=1}^{\infty}\mathfrak{W}_{A_n}(t),$$

for every sequence $\{A_n\}$ of elements of A and $t\geqslant 0$;

(v) a $T-$probabilistic sudecomposable measure iff

$$\mathfrak{W}_{A\cup B}(t)\geqslant\mathfrak{W}_A(t)T\,\mathfrak{W}_B(t)$$

for $A, B \in \mathcal{A}$ and $t \geqslant 0$;

(vi) a $\sigma - T -$ probabilistic supdecomposable measure iff

$$\mathfrak{M}_{\bigcup_{n=1}^{\infty} A_n}(t) \geqslant T_{n=1}^{\infty} \mathfrak{W}_{A_n}(t),$$

for every sequence $\{A_n\}$ of elements of $\mathcal{A}$ and $t \geqslant 0$;

(vii) continuous from below (above) iff $\lim_{n\to\infty} \mathfrak{W}_{A_n}(t) = \mathfrak{W}_A(t)$ for each increasing (decreasing) sequence $\{A_n\}$ of sets from $\mathcal{A}$ with $\lim_{n\to\infty} A_n = A$ and for every $t \geqslant 0$.

Remark 4. In Definition 4.1, the notion of a $T-$ probabilistic supdecomposable measure corresponds to a $s_{\max,T}-$submeasure in the context of non-Archimedean Menger spaces, see([10], Definition 2.3).

Let M be a $T-$probabilistic decomposable measure and $A, B \in \mathcal{A}$ with $A \subseteq B$. Then from Definition 4.1 it follows that

$$\begin{aligned} \mathfrak{W}_B(t) = \mathfrak{W}_{A\cup(B-A)} &= \mathfrak{W}_A(t) T\, \mathfrak{W}_{(B-A)}(t), \\ &\leqslant \mathfrak{W}_A(t) T 1, \\ &= \mathfrak{W}_A(t), \end{aligned}$$

For every $t \geqslant 0$. Thus, we can say that every t-norm-based probabilistic decomposable measure is antimonotone. Obviously, this property is different from those of the classical measure, probability measure and t-conorm-based decomposable measure. For this issue, we will make a reasonable explanation from a probabilistic point of view. In fact, if $\mathfrak{W}_A(t)$ is interpreted as the probability that the "classical measure of the set A is less than t", then, for $A \subseteq B$, it is obvious that the probability that the "classical measure of the set A is less than t" is greater than or equal to the probability that the "classical measure of the set B is less than t".

Example 1. Let $m: \mathcal{A} \to [0, \infty)$ be a classical finite measure. For every $t \geqslant 0$, define

$$\mathfrak{W}_A(t) = \begin{cases} 1 & t > m(A), \\ 0 & t \leqslant m(A). \end{cases}$$

It can easily be verified that M is a $T-$probabilistic subdecomposable measure for any t-norm T. This example shows that every classical monotone measure can be regarded as a probabilistic subdecomposable measure.

Example 2. Let $A \in P(X)$. Define

$$\mathfrak{W}_A(t)=\begin{cases} H_0(t) & A=\varnothing, \\ 0 & A\neq\varnothing, t=0, \\ a & A\neq\varnothing, 0<t\leqslant r, \\ b & A\neq\varnothing, r<t\leqslant 2r, \\ 1 & A\neq\varnothing, t>2r, \end{cases}$$

Since $\lim_{N\to\infty}\bigcup_{n=1}^{N}E_n=E$, by Theorem 3. 5, we have

$$\lim_{N\to\infty}H_{\bigcap_{n=1}^{N}E_{n,X}^c}(t)=H_{E^c,X}(t)$$

Thus, we conclude that

$$\begin{aligned}\overline{\mathfrak{M}}_E^H(t) &= \lim_{N\to\infty}\overline{\mathfrak{M}}_{\bigcup_{n=1}^{N}E_n}^H(t) \\ &= \lim_{N\to\infty}\min\{\overline{\mathfrak{M}}_{E_1}^H(t), \overline{\mathfrak{M}}_{E_2}^H(t), \cdots \overline{\mathfrak{M}}_{E_N}^H(t)\} \\ &= \inf_{n\in\mathbb{N}}\{\overline{\mathfrak{M}}_{E_n}^H(t)\}\end{aligned}$$

which means that $\overline{\mathfrak{M}}^H$ is a $\sigma-T-$probabilistic decomposable measure for the t$-$norm T_M.

5. AN OUTER PROBABILISTIC MEASURE INDUCED BY THE PROBABILISTIC DECOMPOSABLE MEASURE

In this section, we shall construct a kind of outer probabilistic measure using a given probabilistic decomposable measure. Meantime, the corresponding properties are discussed.

Definition 5. 1. Let $\mathfrak{M}$ be a $\sigma-T-$probabilistic decomposable measure on a σ $-$algebra $A\subseteq P(X)$. The mapping $\mathfrak{M}_A^*: P(X)\to\Delta^+$ defined by

$$\mathfrak{M}_A^*(t)=sup\{T_{n=1}^{\infty}\mathfrak{M}_{A_n}(t)\mid A_n\in\mathbf{A}, A\subseteq\bigcup_{n=1}^{\infty}A_n\}, A\in P(X), t\geqslant 0\} \quad (7)$$

is called an outer probabilistic measure induced by the probabilistic decomposable measure $\mathfrak{M}$.

Theorem 5. 1. Let $\mathfrak{M}^*$ be an outer probabilistic measure induced by a $\sigma-T-$probabilistic decomposable measure $\mathfrak{M}$. Then

(i) $\mathfrak{M}_A^*(t)=\mathfrak{M}_A(t)$ for every $A\in\mathbf{A}$ and $t\geqslant 0$;

(ii) $\mathfrak{M}_{\varnothing}^*(t)=H_0$;

(iii) $\mathfrak{M}^*$ *is antimonotone*, *i. e.*, $\mathfrak{M}_A^*\geqslant\mathfrak{M}_B^*$ whenever $A\subseteq B$.

Proof.

(i)For each $A \in \mathbf{A}$, we can write $A = A \cup \varnothing \cup \varnothing \cup \varnothing \cup \cdots$. By Definition 5.1, for every $t \geqslant 0$, it follows that

$$\begin{aligned}\mathfrak{M}_A^*(t) &\geqslant \mathfrak{M}_A(t) T \mathfrak{M}_{\varnothing}(t) T \mathfrak{M}_{\varnothing}(t) \cdots \\ &= \mathfrak{M}_A(t) T H_0(t) T H_0(t) \cdots \\ &= \mathfrak{M}_A(t)\end{aligned}$$

On the other hand, by (ii) and (iv) of Theorem 4.1, for every $A \in \mathbf{A}; A \subseteq \bigcup_{n=1}^{\infty} A_n$ with $A_n \in \mathbf{A}$, we can obtain that

$$\mathfrak{M}_A(t) \geqslant \mathfrak{M}_{\bigcup_{n=1}^{\infty} A_n}(t) \geqslant T_{n=1}^{\infty} \mathfrak{M}_{A_n}(t)$$

Then we have $\mathfrak{M}_A(t) \geqslant \sup\{T_{n=1}^{\infty} \mathfrak{M}_{A_n}(t)\} = \mathfrak{M}_A^*(t)$. Thus we conclude that $\mathfrak{M}_A^*(t) = \mathfrak{M}_A(t)$.

(ii)From(i), we know that $\mathfrak{M}_{\varnothing}^* = \mathfrak{M}_{\varnothing} = H_0$.

(iii)Since $A \subseteq B$, it is obvious that if the sequence $\{A_n\}$ is a cover of B in A, then it must be a cover of A. Hence, we have

$$\mathfrak{M}_A^* \geqslant \mathfrak{M}_B^* \text{ for every } t \geqslant 0.$$

Theorem 5.2. Let $\mathfrak{M}^*$ be an outer probabilistic measure induced by the $\sigma - T$ $-$probabilistic decomposable measure $\mathfrak{M}$, where T is a non$-$strict continuous Archimedean t$-$norm. Then

(i)$\mathfrak{M}^*$ is *a* $\sigma - T -$probabilistic supdecomposable measure;

(ii)$\mathfrak{M}^*$ is a $aT-$probabilistic supdecomposable measure.

Proof.

(i)Let $A \in P(X)$ and $\{A_n\}$ be a sequence of sets from P(X) such that $A \subseteq \bigcup_{n=1}^{\infty} A_n$. For every $n \in \mathbb{N}$, suppose that the sequence $\{A_{n,m}\}_{m \in \mathbb{N}}$ is a cover of A_n in A. Set

$$B_{n,1} = A_{n,1}, B_{n,m} = A_{n,m} - (A_{n,1} \cup A_{n,2} \cup \cdots \cup A_{n,m-1})\ (m \geqslant 2).$$

It can easily be seen that the sequence $\{B_{n,m}\}_{m \in \mathbb{N}}$ is a disjoint sequence of sets from A such that $B_{n,m} \subseteq A_{n,m}$ and $\bigcup_{m=1}^{\infty} B_{n,m} = \bigcup_{m=1}^{\infty} A_{n,m}$. Then, for every $n \in \mathbb{N}$ and $t \geqslant 0$, we have

$$A_n \subseteq \bigcup_{m=1}^{\infty} B_{n,m}, \qquad T_{m=1}^{\infty} \mathfrak{M}_{B_{n,m}}(t) \geqslant \mathfrak{M}_{A_n}^*(t) - \frac{1}{k}$$

for any fixed $k \in \mathbb{N}$.

The sequence $\{B_{n,m}\}_{n,m \in \mathbb{N}}$ constitutes a countable class of disjoint elements from A that covers A. If f is a normed additive generator of T, then by Theorem

4.2, we can obtain that

$$\mathfrak{M}^*_{\bigcup_{n=1} A_n}(t) \geqslant T^{\infty}_{m=1}\,\mathfrak{M}_{\bigcup_{n=1} B_{n,m}}(t) = T^{\infty}_{n=1}\,T^{\infty}_{m=1}\,\mathfrak{M}_{B_{n,m}}(t)$$

$$= f^{(-1)}\Big(\sum_{n=1}^{\infty}\sum_{m=1}^{\infty} f(\mathfrak{M}_{B_{n,m}}(t))\Big)$$

$$= f^{(-1)}\Big(\sum_{n=1}^{\infty}\Big(\sum_{m=1}^{\infty} f(\mathfrak{M}_{B_{n,m}}(t))\Big)\Big)$$

$$= f^{(-1)}\Big(\sum_{n=1}^{\infty} f^{\circ} f^{(-1)}\Big(\sum_{m=1}^{\infty} f(\mathfrak{M}_{B_{n,m}}(t))\Big)\Big)$$

$$\Big(\sum_{m=1}^{\infty} f(\mathfrak{M}_{B_{n,m}}(t)) \leqslant 1\Big)$$

$$= f^{(-1)}\Big(\Big(\sum_{n=1}^{\infty} f(T^{\infty}_{m=1}\,\mathfrak{M}_{B_{n,m}}(t))\Big)\Big)$$

$$= f^{(-1)}\left(\sum_{n=1}^{\infty} f\left(\max\left\{\mathfrak{M}^*_{A_n}(t) - \frac{1}{k}, 0\right\}\right)\right)$$

for every $t \geqslant 0$. Letting $k \to \infty$, the continuity of f and $f^{(-1)}$ implies that

$$\mathfrak{M}^*_{\bigcup_{n=1} A_n}(t) \geqslant f^{(-1)}\Big(\sum_{n=1}^{\infty} f(\mathfrak{M}^*_{A_n}(t))\Big) = T^{\infty}_{n=1}\,\mathfrak{M}^*_{A_n}(t)$$

Thus, we conclude that $\mathfrak{M}^*$ is *a* $\sigma - T -$probabilistic supdecomposable measure.

(ii) It is a direct consequence of Theorem 5.1(ii) and the above conclusion (i).

Theorem 5.3. Let $\mathfrak{M}^*$ be an outer probabilistic measure induced by a $\sigma - T -$ probabilistic decomposable measure $\mathfrak{M}$, where T is a non-strict continuous Archimedean t-norm. For any $A, B, C \in P(X)$ we have

$$\mathfrak{M}^*_{A\Delta C}(t) \geqslant \mathfrak{M}^*_{A\Delta B}(t)\, T\, \mathfrak{M}^*_{B\Delta C}(t)\ (t \geqslant 0)$$

where $A\Delta B$ denotes the symmetric difference of sets A and B.

Proof. Since $A\Delta C \subseteq (A\Delta B) \cup (B\Delta C)$, by Theorems 5.1 and 5.2, we can get that

$$\mathfrak{M}^*_{A\Delta C}(t) \geqslant \mathfrak{M}^*_{(A\Delta B)\cup(B\Delta C)}(t) \geqslant \mathfrak{M}^*_{A\Delta B}(t)\Delta\ \mathfrak{M}^*_{B\Delta C}(t).$$

Theorem 5.4. Let $\mathfrak{M}^*$ be an outer probabilistic measure induced by a $\sigma - T -$ probabilistic decomposable measure $\mathfrak{M}$, where T is a non-strict continuous Archimedean t-norm. Define the mapping $H^{\mathfrak{M}^*}: P(X) \times P(X) \to \Delta^+$ by

$$H^{\mathfrak{M}^*}_{A,B} = \mathfrak{M}^*_{A\Delta B}$$

Then $(P(X), H^{\mathfrak{M}^*}, T)$ is a Menger probabilistic pseudometric space.

Proof. Since $H^{\mathfrak{M}^*}_{A,A} = \mathfrak{M}^*_{\varnothing} = H_0$ and $\mathfrak{M}^*_{A\Delta B} = \mathfrak{M}^*_{B\Delta A}$, the conditions(MPM－1') and(MPM－2)are obvious. Now,it suffices to verify that the condition(MPM－3')holds. For any A,B,C∈P(X)and t P≥0,by Theorem 5. 3,it follows that

$$H^{\mathfrak{M}^*}_{A,C}(t) = \mathfrak{M}^*_{A\Delta C}(t) \geqslant \mathfrak{M}^*_{A\Delta B}(t) T\, \mathfrak{M}^*_{B\Delta C}(t) = H^{\mathfrak{M}^*}_{A,B}(t) T\, H^{\mathfrak{M}^*}_{B,C}(t).$$

Moreover,observe that $H^{\mathfrak{M}^*}_{A,C}(t)$, $H^{\mathfrak{M}^*}_{A,B}(t)$, $H^{\mathfrak{M}^*}_{B,C}(t)$ and $H^{M}_{B;C}$ðtÞ are non－decreasing with respect to t. Then,for any $s,t \geqslant 0$,we can obtain that

$$H^{\mathfrak{M}^*}_{A,C}(s+t) \geqslant H^{\mathfrak{M}^*}_{A,B}(s+t) T\, H^{\mathfrak{M}^*}_{B,C}(s+t) \geqslant H^{\mathfrak{M}^*}_{A,B}(t) T\, H^{\mathfrak{M}^*}_{B,C}(s).$$

Thus,we conclude that $H^{\mathfrak{M}^*}$ is a Menger probabilistic pseudometric on $P(X)$,that is,$(P(X), H^{\mathfrak{M}^*}, T)$ is a Menger probabilistic pseudometric space.

6. CONCLUSIONS

In this paper,we have further examined some related properties of the probabilistic Hausdorff distance,and have shown that,for a given a Menger PM－space under a continuous t－norm,this type of distance is exactly a probabilistic metric on the collection of all probabilistic bounded and closed subsets. More generally,it is a probabilistic pseudometric on the collection of all probabilistic bounded subsets. In addition,we have introduced the concepts of probabilistic decomposable measure, probabilistic subdecomposable measure and probabilistic supdecomposable measure based on t－norms,and then we have discussed several intrinsic relationships among them. Furthermore,for a given probabilistic Hausdorff distance in a Menger PM－space,we have proved that it induces a probabilistic decomposable measure for the t－norm $>_M$. Moreover,we have proved that the class of all measurable sets forms an algebra on its power set. At last,we have introduced an outer probabilistic measure based on a σ－T－probabilistic decomposable measure. Interestingly,for a non－strict continuous t－norm,this type of probabilisitic decomposable measure induces a probabilistic pseudometric on the power set of a given nonempty set. Based on the statement mentioned above,when some specific conditions are not considered,one can find the main line of this paper is that for a given Menger PM－space we can induce a probabilistic Hausdorff distance,and then obtain a class of probabilistic decomposable measures using this distance and a t－norm. Moreover,a kind of outer probabilistic measure can be constructed based on the probabilistic decomposable measure with respect to a t－norm. Finally, a Menger probabilistic pseudometric space can be induced by means of the outer probabilistic measure. The main results and methods presented in this paper extend and generalize some known results of the classical measure and the decomposable measure. Furthermore, from the probabilistic point of view, this paper provides a general theoretical framework to deal with some practical problems based on the measure theory.

Acknowledgements

This work was supported by "Qing Lan" Talent Engineering Funds by Tianshui Normal University and the Humanity and Social Science Youth Foundation of Ministry of Education of China(No. 13YJC630012). The author would like to express his sincere thanks to the editor and referees for their valuable suggestions. In particular, the author is very grateful to the referees for the Refs. [2, 10－12, 22] they have kindly provided.

References

[1]B. Cavallo, L. D'Apizzo, M. Squillante, Independence and convergence in non－additive settings, Fuzzy Optim. Decis. Mat. 8(2009)29－43.

[2]S. S. Chang, Y. J. Cho, S. M. Kang, Probabilistic Metric Spaces and Nonlinear Operator Theory, Sichuan Univ. Press, 1994.

[3]G. Constantin, G. Bocsan, On some measures of noncompactness in probabilistic metric spaces, Proc. 5th Conf. Probab. Theory(1974)163－168.

[4]D. Denneberg, Non－Additive Measure and Integral, Kluwer Academic Publishers, Dordrecht, 1994.

[5]D. Dubois, M. Prade, A class of fuzzy measures based on triangular norms, Int. J. Gen. Syst. 8(1982)43－61.

[6]D. Dubois, H. Prade, Aggregation of decomposable measures with application to utility theory, Theory Decis. 41(1996)59－95.

[7]A. Dvořák, M. Holčapek, Fuzzy measures and integrals defined on algebras of fuzzy subsets over complete residuated lattices, Inform. Sci. 185(2012)205－229.

[8]M. Grabisch, T. Murofushi, M. Sugeno(Eds.), Fuzzy Measures and Integrals: Theory and Applications, Physica－Verlag, Heidelberg, 2000.

[9]O. Hadžić, E. Pap, Fixed Point Theory in Probabilistic Metric Spaces, Kluwer Academic Publishers, Dordrecht, 2001.

[10]L. Halcˇinová, O. Hutník, R. Mesiar, On some classes of distance distribution function—valued submeasures, Nonlinear Anal. 74(2011)1545—1554.

[11]L. Halcˇinová, O. Hutník, R. Mesiar, On distance distribution functions—valued submeasures related to aggregation functions, Fuzzy Sets Syst. 194(2012)15—30.

[12]O. Hutník, R. Mesiar, On a certain class of submeasures based on triangular norms, Internat. J. Uncertain. Fuzziness Knowl. —Based Syst. 17(2009)297—316.

[13]E. P. Klement, S. Weber, Generalized measures, Fuzzy Sets Syst. 40(1991)375—394.

[14]E. P. Klement, R. Mesiar, E. Pap, Triangular Norms, Kluwer Academic Publishers, Dordrecht, 2000.

[15]E. P. Klement, R. Mesiar, E. Pap, Measure—based aggregation operators, Fuzzy Sets Syst. 142(2004)3—14.

[16]A. Kolesárová, A. Stupnˇanová, J. Beganová, Aggregation—based extensions of fuzzy measures, Fuzzy Sets Syst. 194(2012)1—14.

[17]D. Qiu, W. Q. Zhang, On decomposable measures induced by metrics, J. Appl. Math. vol. 2012, Article ID 701206, 8 pp.

[18]D. Qiu, W. Q. Zhang, C. Li, Extension of a class of decomposable measures using fuzzy pseudometrics, Fuzzy Sets Syst. 222(2013)33—44.

[19]E. Pap, Lebesgue and Saks decompositions of ⊥—decomposable measures, Fuzzy Sets Syst. 38(1990)345—353.

[20]E. Pap, Extension of the continuous t—conorm decomposable measure, Univ. Nov. Sadu. Zb. Rad. Prirod. —Mat. Fak. Ser. Mat. 20(1990)121—130.

[21]E. Pap, Decomposable measures and applications on differential equations, Rend. Mat. Palermo Set II 28(1992)387—403.

[22]E. Pap, O. Hadžic′, R. Mesiar, A fixed point theorem in probabilistic metric spaces and an application, J. Math. Anal. Appl. 202(1996)433—449.

[23]E. Pap, Decomposable measures and nonlinear equations, Fuzzy Sets Syst. 92(1997)205—221.

[24]E. Pap, Applications of decomposable measures on nonlinear differential equations, Novi Sad J. Math. 31(2001)89—98.

[25]B. Schweizer, A. Sklar, Statistical metric spaces, Pacific. J. Math. 10(1960)313—334.

[26]M. Sugeno, Theory of Fuzzy Integrals and Its Applications, Ph. D. thesis, Tokyo Institute of Technology, Tokyo, 1974.

[27]A. Stupnˇanová, Special fuzzy measures on infinite countable sets and related aggregation functions, Fuzzy Sets Syst. 167(2011)57—64.

[28]D. H. Tan, On the probabilistic Hausdorff distance and fixed point theorems for multivalued contractions, Acta Math. Vietnam 15(1990)61—68.

[29]Z. Y. Wang, G. J. Klir, Fuzzy Measure Theory, Plenum Press, New York, 1992.

[30]Z. Y. Wang, G. J. Klir, Generalized Measure Theory, Springer Press, New York, 2009.

[31]S. Weber, ⊥ − Decomposable measures and integral for Archimedean t − conorms ⊥, J. Math. Anal. Appl. 101(1984)114−138.

[32]D. L. Zhang, C. M. Guo, Integrals of set − valued functions for ⊥ − decomposable measures, Fuzzy Sets Syst. 78(1996)341−346.

(本文发表于 2014 年《Information Sciences》263 卷第 4 期)

Pullback attroctors for a nonautonomons damped wave equation with infinite delays in weighted space

Yanping Ran Qihong Shi *

摘要:利用能量方法,证明了带无穷时滞阻尼波方程在加权空间 $C_{\gamma,H_0^1(\Omega),L^2(\Omega)}$ 和 $C_{\gamma,D(A),H_0^1(\Omega)}$ and 中解的存在性,并利用直接简单的紧性方法获得了由该方程产生的过程的拉回吸引子的存在性.

1 Introduction

In the present paper, we consider the following functional wave equation:

$$\begin{cases} \dfrac{\partial^2 u}{\partial t^2}+\dfrac{\partial u}{\partial t}+\Delta u = f(t,u(t-\rho(t))) & \\ +\displaystyle\int_{-\infty}^{0} G(t,s,u(t+s,x))ds+\mathrm{g}(\mathrm{t},\mathrm{x}) & \text{in } [\tau,+\infty]\times\Omega, \\ u\,|_{\partial\Omega} = 0 & \text{on } [\tau,+\infty]\times\partial\Omega, \\ u(t,x) = \varphi(t-\tau,x), & t\in(-\infty,\tau], x\in\Omega, \\ \dfrac{\partial u}{\partial t} = \dfrac{\partial\varphi(t-\tau,x)}{\partial t}, & t\in(-\infty,\tau], x\in\Omega, \end{cases} \tag{1.1}$$

where$\Omega\subseteq\mathbb{R}^n$ is a smooth bounded domain with smooth boundary, and$\tau\in\mathbb{R}$ is initial time, $\varphi$is the initial data in the interval of time $(-\infty,\tau]$, f, G are nonlinear terms containing some hereditary characteristic and $g$denotes a non-delay external force.

This equation models the propagation of sound and vibration of the overhead transmission line and the thin films with the internal delays and the excitation of

---

* 作者简介:冉延平(1976—),男,甘肃靖远人,天水师范学院数学与统计学院副教授、博士,主要从事微分方程理论方面的研究。

the external force. It is also employed to describe the variation from the configuration at the rest of a homogeneous and isotropic linear viscoelastic solid with memory[9,23]. The nonlinear terms of(1. 1) suggest that the source intensity of the system may depend on the whole history and the non—autonomous force.

Actually, partial differential equations with variable or distributed delay have attracted the attention [7,12,13]. Retarded differential equation or partial differential equations become nonautonomous in essence. Particularly, the study on the asymptotic behaviorof nonautonomous dynamical systems has an important significance in predicting theevolution of the system in modern physics and many engineer problems. As the mostappropriate concept, pullback attractors for non-autonomous dynamical systems have beenproposed and developed by many many researchers [15,23—25].

Formally, (1. 1) may be regarded as a generalization of the classical wave equation, andits asymptotic behavior has been investigated in many articles by several authors [4,10]. Recently, Caraballo, Kloeden and Real [12] considered the pullback and forward attractorsof a damped wave equation with finite variable delays by the theory of two—parametersemigroups. Wang [13] studied the pullback attractors for the multi—valued damped waveequation with finite delays. In-addition, we also note that pullback attractors for reaction—diffusion equations in an infinite delay case were investigated in [15]. Motivated by theseresults, our goal in this paper is to study the existence of solutions, in particular, thepullback attractor for the wave equation (1. 1) in weighted energy space. To our knowledge, there is no rigorous result for this damped wave equation in the literature. Generally, for the wave equation with delays, these methods [12,13] can not be directlyapplicable. Moreover, due to the presence of mixture delays, it is much difficult in theweighted space to obtain the existence of the solutions and the pullback attractors for (1. 1). In this paper, by splitting the term related to the delay in the initial datum and the evolution of the solution and using Galerkin approximation [13,15], we obtain the existence andsome properties of the solutions for (1. 1) in$C_{\gamma,H_0^1(\Omega),L^2(\Omega)}$ and $C_{\gamma,D(A),H_0^1(\Omega)}$, respectively. Furthermore, we employ the idea of [15,30] treating Navier—Stokes equations to deal withthe infinite delay term. We use the decompose of the processes to obtain the existence ofpullback attractors in higher regularity space for (1. 1) under some weaker hypotheses.

The structure of this paper is following. In this section, we introduce some functionspaces useful for the establishment of the abstract variational formulation of the problem, and some assumptions on the delays terms. In Section 2, we prove the existenceand uniqueness of the solutions with respect to initial data. In Section 3, 4, we are devoted to prove the existence of pullback attractors in$C_{\gamma,H_0^1(\Omega)}$ and $C_{\gamma,D(A),H_0^1(\Omega)}$ for thenonautonomous infinite dimensional dynamical systems associated with the damped waveequations with mixture delays.

To start with, we consider some usual abstract spaces and notation. Let $(X, \|\cdot\|_X)$, $(Y, \|\cdot\|_Y)$ be two Banach spaces such that the injection $X\subseteq Y$ is continuous. Consider two Banach spaces

$$C_{\gamma,X} = \{\varphi \in C^0((-\infty,0],X): \exists \lim_{s\to-\infty} e^{\gamma s}\varphi(s) \in X\},$$

and

$$C_{\gamma,X,Y} = \{\varphi \in C^0((-\infty,0];X) \cap C^1((-\infty,0];Y): \\ \exists \lim_{s\to-\infty} e^{\gamma s}\varphi(s) \in X, \lim_{s\to-\infty} e^{\gamma s}\varphi'(s) \in Y\}$$

which have the norm

$$\|\varphi\|_{\gamma,X}^2 = \sup_{s\in(-\infty,0]} e^{2\gamma s}\|\varphi(s)\|_X^2 < \infty, \text{and } \|\varphi\|_{\gamma,X,Y}^2 \\ = \|\varphi\|_{\gamma,X}^2 + \|\varphi'\|_{\gamma,Y}^2 < \infty$$

respectively, which allow us to deal with infinite delays, for$\gamma > 0$.

Hereafter, we will use these phase space (with a suitable $\gamma > 0$) for our problem.

Let$Au = -\Delta u$ for any$u \in D(A)$, where$D(A) = \{u \in H_0^1(\Omega): Au = L^2(\Omega)\} = H_0^1(\Omega) \cap H^2(\Omega)$. Given function$u:(-\infty,T] \to X$, for each $t \le T$, we denote by u_t the function defined on $(-\infty,0]$ by the relation$u_t(s) = u(t+s)$, $s \in (-\infty, 0]$. The inner products of $L^2(\Omega)$ and $H_0^1(\Omega)$ are denoted by $(\cdot,\cdot)$ and $((\cdot,\cdot))$, respectively. The notation $B_X(a,r)$ will be used to denote the open ball of center a and radius r in X.

In order to state our problem in correct framework, let us make the following hypotheses on the function$f$and$g$.

$(H1)$ $f \in C(\mathbb{R}\times\mathbb{R},\mathbb{R})$ satisfies

$$|f(t,v)|^2 \le k_1^2 + k_2^2 e^{-2\gamma\rho(t)}|v|^2, \forall t,v \in \mathbb{R}, \quad (1.2)$$

where$k_1, k_2 > 0$ and $\rho(t) \in C(\mathbb{R},\mathbb{R}^+)$.

(H2) $G \in C(\mathbb{R} \times \mathbb{R}^- \times \mathbb{R}, \mathbb{R})$ satisfies

$|G(t,s,v)| \le m_0(s) + m_1(s)|v|, \forall t,v \in \mathbb{R}, s \in \mathbb{R}^-$, (1.3)

where$m_0(\cdot), m_1(\cdot) \in C(\mathbb{R}_-, \mathbb{R}^+)$ and $m_1(\cdot)e^{-\gamma\cdot} \in L^1(\mathbb{R}_-)$. We will denote$m_0 = \int_{-\infty}^{0} m(s)ds, m_1 = \int_{-\infty}^{0} m_1(s)e^{-\gamma s}ds$.

(H3) The non-autonomous term$g \in L^2_{loc}(\mathbb{R}; L^2(\Omega))$ satisfies that

$$\sup_{\tau \le t} e^{-\delta\tau}\int_{-\infty}^{\tau} e^{\delta s}\|g(s)\|_X^2 ds < \infty, \forall t \in \mathbb{R}, \quad (1.4)$$

for each$\delta = \alpha$, where α will be given in Theorem 2.3, the local 2-power integral is the Bochner integral.

(H4) There exist two positive function$L_f, L_G: \mathbb{R}_- \to \mathbb{R}_+$ such that, for any $t \in [\tau, T]$,

$$s \in (-\infty, 0] \text{ and all } u, v \in \mathbb{R},$$

$$|f(t,u) - f(t,v)| \le L_f(-\rho(t))|u - v|,$$

$$|G(t,s,u) - G(t,s,v)| \le L_G(s)|u - v|,$$

where $L_f(\cdot)e^{\gamma\cdot}$ and $L_g(\cdot)e^{-\gamma\cdot}$ are both in$L^2(\mathbb{R}_-)$. Now we will gives the definitions of solution for problem (1.1).

Definition 1.1 A weak solution (1.1) in the interval $(-\infty, T]$, with initial datum $\varphi \in C_{\gamma, H_0^1(\Omega), L^2(\Omega)}$, is a function $u \in C^0((-\infty, T]; H_0^1(\Omega)) \cap C^1((-\infty, T]; L^2(\Omega))$ with $u_\tau = \varphi$, such that, for all$v \in H_0^1(\Omega)$ and $t \in (\tau, T)$,

$$\frac{d^2}{dt^2}(u,v) + \frac{d}{dt}(u,v) + ((\Delta u, v)) = (f(t, u(t-\rho(t)))$$

$$+ \int_{-\infty}^{0} G(t,s,u(t+s,x))ds + g(t,x), v), \quad (1.7)$$

where the equation (1.1) must be understood in the sense of $D'(\tau, T)$.

Remark 1.2 if u is a weak solution of (1.1) in the sense given above, then u satisfies an energy equality. Namely,

$$\begin{aligned}
\|\varphi\|^2_{\gamma, H_0^1(\Omega), L^2(\Omega)} &= \|\varphi' + \varepsilon\varphi - \varepsilon\varphi\|^2_{\gamma, L^2(\Omega)} + \|\nabla\varphi\|^2_{\gamma, L^2(\Omega)} \\
&\le 2\|\varphi' + \varepsilon\varphi\|^2_{\gamma, L^2(\Omega)} + 2\varepsilon^2\|\varphi\|^2_{\gamma, L^2(\Omega)} + \|\nabla\varphi\|^2_{\gamma, L^2(\Omega)} \\
&\le 2\|\varphi' + \varepsilon\varphi\|^2_{\gamma, L^2(\Omega)} + 2\varepsilon^2\lambda_1^{-1}\|\nabla\varphi\|^2_{\gamma, L^2(\Omega)} + \|\nabla\varphi\|^2_{\gamma, L^2(\Omega)} \\
&\le 2\|\varphi' + \varepsilon\varphi\|^2_{\gamma, L^2(\Omega)} + (1 + 2\varepsilon^2\lambda_1^{-1})\|\nabla\varphi\|^2_{\gamma, L^2(\Omega)} \\
&\le C(\varepsilon)\|\varphi\|^2_{\gamma, \varepsilon}.
\end{aligned}$$

$$\|u'\|^2_{L^2(\Omega)} + \|\nabla u\|^2_{L^2(\Omega)} + \int_s^t \|u'(r)\|^2_{L^2(\Omega)}dr = \int_s^t \{(f(r, u(r-\rho(r))) + \int_{-\infty}^{0} G(r,$$

$z,u(r+z,x))dz+g(r),u'(r))\}dr,\forall s,t\in[\tau,T]$. (1.8)

For convenience, in the sequel C denotes an arbitrary positive constant, which may be different from line to line and even in the same line.

2 Existence of solutions

In this section, we establish the existence of weak solutions for (1.1) by a compactnessmethod using a Faedo—Galerkin scheme. Let us denote

$$\lambda_1=\inf_{v\in H_0^1(\Omega)?\{0\}}\frac{\|\nabla v\|_{L^2(\Omega)}^2}{\|v\|_{L^2(\Omega)}^2}>0.$$

This implies $\|\nabla u\|_{L^2(\Omega)}^2\geq\lambda_1\|v\|_{L^2(\Omega)}^2$.

Proposition 2.1 $\forall\varepsilon\in\mathbb{R}$, $\forall\varphi\in C_{\gamma,H_0^1(\Omega),L^2(\Omega)}$, the norm $\|\varphi\|_{\gamma,\varepsilon}^2=\|\varphi'+\varepsilon\varphi\|_{\gamma,L^2(\Omega)}^2+\|\nabla\varphi\|_{\gamma,L^2(\Omega)}^2$ is equivalent the norm $\|\varphi\|_{\gamma,H_0^1(\Omega),L^2(\Omega)}^2$

i. e. $\exists C(\varepsilon)=\max\{2,1+2\varepsilon^2\lambda_1^{-1}\}$ such that

$$\frac{1}{C(\varepsilon)}\|\varphi\|_{\gamma,\varepsilon}^2\leq\|\varphi\|_{\gamma,H_0^1(\Omega),L^2(\Omega)}^2\leq C(\varepsilon)\|\varphi\|_{\gamma,\varepsilon}^2. \quad (2.1)$$

Proof. Let $C(\varepsilon)=\max\{2,1+2\varepsilon^2\lambda_1^{-1}\}$. It follows that

$$\begin{aligned}\|\varphi\|_{\gamma,H_0^1(\Omega),L^2(\Omega)}^2&=\|\varphi'+\varepsilon\varphi-\varepsilon\varphi\|_{\gamma,L^2(\Omega)}^2+\|\nabla\varphi\|_{\gamma,L^2(\Omega)}^2\\&\leq 2\|\varphi'+\varepsilon\varphi\|_{\gamma,L^2(\Omega)}^2+2\varepsilon^2\lambda_1^{-1}\|\nabla\varphi\|_{\gamma,L^2(\Omega)}^2+\|\nabla\varphi\|_{\gamma,L^2(\Omega)}^2\\&\leq C(\varepsilon)\|\varphi\|_{\gamma,\varepsilon}^2.\end{aligned}$$

On the other hand

$$\begin{aligned}\|\varphi\|_{\gamma,\varepsilon}^2&=\|\varphi'+\varepsilon\varphi\|_{\gamma,L^2(\Omega)}^2+\|\nabla\varphi\|_{\gamma,L^2(\Omega)}^2\\&\leq 2\|\varphi'\|_{\gamma,L^2(\Omega)}^2+2\varepsilon^2\|\varphi\|_{\gamma,L^2(\Omega)}^2+\|\nabla\varphi\|_{\gamma,L^2(\Omega)}^2\\&\leq 2\|\varphi'\|_{\gamma,L^2(\Omega)}^2+2\varepsilon^2\lambda_1^{-1}\|\nabla\varphi\|_{\gamma,L^2(\Omega)}^2+\|\nabla\varphi\|_{\gamma,L^2(\Omega)}^2\\&\leq 2\|\varphi'\|_{\gamma,L^2(\Omega)}^2+(2\varepsilon^2\lambda_1^{-1}+1)\|\nabla\varphi\|_{\gamma,L^2(\Omega)}^2\\&\leq C(\varepsilon)\|\varphi\|_{\gamma,H_0^1(\Omega),L^2(\Omega)}^2.\end{aligned}$$

Proposition 2.2 $u\in C^0((-\infty,T];D(A))\cap C^1((-\infty,T];H_0^1(\Omega))$, $\forall T>\tau$. $\forall\varepsilon\in\mathbb{R}$, $\forall\varphi\in C_{\gamma,D(A),H_0^1(\Omega)}$, the norm $\|\varphi\|_{\gamma,\varepsilon}^2=\|\varphi'+\varepsilon\varphi\|_{\gamma,H_0^1(\Omega)}^2+\|\varphi\|_{\gamma,D(A)}^2$ is equivalent the norm $\|\varphi\|_{C_{\gamma,D(A),H_0^1(\Omega)}}^2$ i. e. $\exists C(\varepsilon)=\max\{2,1+2\varepsilon^2\lambda_1^{-1}\}$, suchthat

$$\frac{1}{C(\varepsilon)}\|\varphi\|_{\gamma,\varepsilon}^2\leq\|\varphi\|_{C_{\gamma,D(A),H_0^1(\Omega)}}^2\leq C(\varepsilon)\|\varphi\|_{\gamma,\varepsilon}^2. \quad (2.2)$$

Theorem 2.3(1)Take γ,α such that

$$\frac{m_1}{2\lambda_1}-m_1+\frac{\sqrt{(2\lambda_1+1)^2m_1^2+16k_2^2\lambda_1}}{2\lambda_1}<\alpha<$$

$$\min\{2\gamma,\frac{\sqrt{2\lambda_1}}{4}-2m_1,\frac{1}{3+4\lambda_1}-2m_1\}. \quad (2.3)$$

If hypotheses (H1)—(H4) holds, $\varphi\in C_{\gamma,H_0^1(\Omega),L^2(\Omega)}$ and for each$\tau\in\mathbb{R}$, then there exists aunique solution u(t) of problem (1.1),andu(t) satisfies

$$u\in C^0((-\infty,T];H_0^1(\Omega))\cap C^1((-\infty,T];L^2(\Omega)),\forall T>\tau..$$

(2) If in addition $\nabla g\in L^2_{loc}(\mathbb{R};\Omega)$, $f\in C^1(\mathbb{R}\times\mathbb{R},\mathbb{R})$ and there exists $L_f>0$ such that

$$|f'_v(t,v)|\le\tilde{L}_f',\forall t,v\in\mathbb{R}, \quad (2.4)$$

$$|G'_v(t,s,v)|\le L_G'(s),\forall t,v\in\mathbb{R},s\in\mathbb{R}_- \quad (2.5)$$

where$L_G':\mathbb{R}_-\to\mathbb{R}_+$, $L_G'(s)e^{-\iota\gamma s}\in L^2(\mathbb{R}_-)$, $\iota>0$. Then for $\varphi\in C_{\gamma,C_{\gamma,D(A)},H_0^1(\Omega)}$ and for each$\tau\in\mathbb{R}$, then there exists a unique solution u(t) of problem (1.1),and u(t) satisfies

$$u\in C^0((-\infty,T];D(A))\cap C^1((-\infty,T];H_0^1(\Omega)),\forall T>\tau.$$

Proof. For the existence,we split the proof into several steps.

Step 1:Let us consider $\{v_j\}$, the orthonormal basis of $L^2(\Omega)$ of all the eigenfunctions of the operator$A$ inΩ with homogeneous Dirichlet boundary conditions. Denote$V_m=span\{v_1,\cdots,v_m\}$, and consider the projector $P_m u=\sum_{j=1}^{m}(u,v_j)v_j,u\in L^2(\Omega)$.

Let$u_m(t)=\sum_{j=1}^{m}\alpha_{m,j}(t)v_j$ and the coefficients $\alpha_{m,j}$ are required to satisfy the following ordinary function differential equations:

$$\frac{d^2}{dt^2}(u_m(t),v_j)+\frac{d}{dt}(u_m(t),v_j)+((\Delta u_m(t),v_j))=(f(t,u_m(t-\rho(t))),v_j)$$

$$+(\int_{-\infty}^{0}G(t,s,u_m(t+s,x))ds+g(t,x),v_j),1\le j\le m, \quad (2.6)$$

$$u_m(t)=P_m\varphi(t-\tau),\frac{du_m(t)}{dt}=\frac{dP_m\varphi(t-\tau)}{dt},t\in(-\infty,\tau].$$

By[[18],Th. 1.1,p. 36],we know the equations (2.6) with infinite delay havea uniqueness solution for all time and in particular in $[\tau,T]$.

The functionu_m,u_m' are in $C^0((-\infty,T];H_0^1(\Omega))$, u_m', u"_m is in $L^2((\tau,T];L^2(\Omega))$.

Step 2:A priori estimates.

According to the existence of(2.6),let $u_m(t)$ be a solution of the ordinary functional differential system

$$\frac{d^2 u_m(t)}{dt^2}+\frac{du_m(t)}{dt}+P_m\Delta u_m(t)=P_m(f(t,u_m(t-\rho(t)))$$
$$+\int_{-\infty}^{0}P_mG(t,s,u_m(t+s,x))ds+P_mg(t,x)),t\in[\tau,T], \tag{2.7}$$

$$u_m(t)=P_m\varphi(t-\tau),\frac{du_m(t)}{dt}=\frac{dP_m\varphi(t-\tau)}{dt},\qquad t\in(-\infty,\tau].$$

Let $\beta>0$ be chosen later on, and let $v_m=u'_m+\beta u_m$. Multiplying (2.7) by v_m, we obtain

$$\frac{1}{2}\frac{d}{dt}(\|v_m(t)\|^2_{L^2(\Omega)}+\|\nabla u_m(t)\|^2_{L^2(\Omega)})+\beta\|\nabla u_m(t)\|^2_{L^2(\Omega)}$$
$$+(1-\beta)\|v_m(t)\|^2_{L^2(\Omega)} \tag{2.8}$$
$$-\beta(1-\beta)(u_m(t),v_m(t))=(f(t,u_m(t-\rho(t)))$$
$$+\int_{-\infty}^{0}G(t,s,u_m(t+s,x))ds+g(t,x),v_m(t)).$$

Choosing $\beta>0$ small enough such that $1-\beta-\frac{\beta}{2\lambda_1}>\frac{\beta}{2}$, so we have

$$\beta\|\nabla u_m(t)\|^2_{L^2(\Omega)}+(1-\beta)\|v_m(t)\|^2_{L^2(\Omega)}-\beta(1-\beta)(u_m(t),v_m(t))$$
$$\geq\beta\|\nabla u_m(t)\|^2_{L^2(\Omega)}+(1-\beta)\|v_m(t)\|^2_{L^2(\Omega)}$$
$$-\frac{\beta(1-\beta)}{\sqrt{\lambda_1}}\|\nabla u_m(t)\|^2_{L^2(\Omega)}\|v_m(t)\|^2_{L^2(\Omega)}$$
$$\geq\beta\|\nabla u_m(t)\|^2_{L^2(\Omega)}+(1-\beta)\|v_m(t)\|^2_{L^2(\Omega)}$$
$$-\frac{\beta}{\sqrt{\lambda_1}}\|\nabla u_m(t)\|^2_{L^2(\Omega)}\|v_m(t)\|^2_{L^2(\Omega)} \tag{2.9}$$
$$\geq\beta\|\nabla u_m(t)\|^2_{L^2(\Omega)}+(1-\beta)\|v_m(t)\|^2_{L^2(\Omega)}$$
$$-\frac{\beta}{2}\|\nabla u_m(t)\|^2_{L^2(\Omega)}-\frac{\beta}{2\lambda_1}\|v_m(t)\|^2_{L^2(\Omega)}$$
$$\geq\frac{\beta}{2}(\|\nabla u_m(t)\|^2_{L^2(\Omega)}+\|v_m(t)\|^2_{L^2(\Omega)}).$$

Using (H1)−(H3) and Young's inequality, we have

$$|(f(t,u_m(t-\rho(t))),v_m)|\leq\int_\Omega|f(t,u_m(t-\rho(t)))v_m|dx$$
$$\leq\frac{\beta}{8}\|v_m(t)\|^2_{L^2(\Omega)}+\frac{2k_1^2}{\beta}|\Omega|+\frac{2k_2^2}{\beta}\|u_{mt}\|^2_{\gamma,L^2(\Omega)}$$
$$\leq\frac{\beta}{8}\|v_m(t)\|^2_{L^2(\Omega)}+\frac{2k_1^2}{\beta}|\Omega|+\frac{2k_2^2}{\lambda_1\beta}\|\nabla u_{mt}\|^2_{\gamma,L^2(\Omega)}$$

and

$$|(g(t,x),v_m(t))| \leq \frac{\beta}{16}\|v_m(t)\|_{L^2(\Omega)}^2 + \frac{4}{\beta}\|g(t)\|_{L^2(\Omega)}^2,$$

And

$$\left|\int_\Omega \int_{-\infty}^0 G(t,s,u_m(t+s,x))v_m(t)dsdx\right|$$

$$\leq \int_\Omega \int_{-\infty}^0 (|m_0(s)| + m_1(s)|u(t+s)|)v_m(t)dsdx$$

$$\leq m_0\int_\Omega |v_m(t)|dx + \int_\Omega \int_{-\infty}^0 m_1(s)e^{-\gamma s}e^{\gamma s}|u_m(t+s)||v_m(t)|dsdx$$

$$\leq \frac{m_0^2}{\beta}|\Omega| + \frac{\beta}{8}\|v_m(t)\|_{L^2(\Omega)}^2 + \int_\Omega \left(\frac{e^{2\gamma s}|u_m(t+s)|^2}{4}\right.$$

$$\left. + |v_m(t)|^2\right)dx\int_{-\infty}^0 m_1(s)e^{-\gamma s}ds$$

$$\leq \frac{8m_0^2}{\beta}|\Omega| + \frac{\beta}{16}\|v_m(t)\|_{L^2(\Omega)}^2 + \frac{m_1\|u_{mt}\|_{L^2(\Omega)}^2}{4} + m_1\|v_m(t)\|_{L^2(\Omega)}^2.$$

Hence,

$$\frac{d}{dt}(\|v_m(t)\|_{L^2(\Omega)}^2 + \|\nabla u_m(t)\|_{L^2(\Omega)}^2)$$

$$+(\frac{\beta}{2} - 2m_1)(\|v_m(t)\|_{L^2(\Omega)}^2 + \|\nabla u_m(t)\|_{L^2(\Omega)}^2)$$

$$\leq \frac{4k_1^2}{\beta}|\Omega| + \frac{8m_0^2}{\beta}|\Omega| + (\frac{4k_2^2}{\lambda_1\beta} + \frac{m_1}{2\lambda_1})\|\nabla u_{mt}\|_{\gamma,L^2(\Omega)}^2 + \frac{8}{\beta}\|g(t)\|_{L^2(\Omega)}^2. \tag{2.10}$$

Thus,

$$\frac{d}{dt}e^{\alpha t}(\|v_m(t)\|_{L^2(\Omega)}^2 + \|\nabla u_m(t)\|_{L^2(\Omega)}^2)$$

$$= \alpha e^{\alpha t}(\|v_m(t)\|_{L^2(\Omega)}^2 + \|\nabla u_m(t)\|_{L^2(\Omega)}^2)$$

$$+ e^{\alpha t}\frac{d}{dt}(\|v_m(t)\|_{L^2(\Omega)}^2 + \|\nabla u_m(t)\|_{L^2(\Omega)}^2)$$

$$\leq \alpha e^{\alpha t}(\|v_m(t)\|_{L^2(\Omega)}^2 + \|\nabla u_m(t)\|_{L^2(\Omega)}^2)$$

$$-(\frac{\beta}{2} - 2m_1)e^{\alpha t}(\|v_m(t)\|_{L^2(\Omega)}^2 + \|\nabla u_m(t)\|_{L^2(\Omega)}^2)$$

$$+(\frac{4k_1^2}{\beta} + \frac{8m_0^2}{\beta})e^{\alpha t}|\Omega| + (\frac{4k_2^2}{\lambda_1\beta} + \frac{m_1}{2\lambda_1})e^{\alpha t}\|\nabla u_{mt}\|_{\gamma,L^2(\Omega)}^2 + \frac{8}{\beta}e^{\alpha t}\|g(t)\|_{L^2(\Omega)}^2$$

$$\leq (\alpha - (\frac{\beta}{2} - 2m_1))e^{\alpha t}(\|v_m(t)\|_{L^2(\Omega)}^2$$

$$+\|\nabla u_m(t)\|^2_{L^2(\Omega)})+(\frac{4k_1^2}{\beta}+\frac{8m_0^2}{\beta})|\Omega|e^{\alpha t}$$

$$+(\frac{4k_2^2}{\lambda_1\beta}+\frac{m_1}{2\lambda_1})\|\nabla u_{mt}\|^2_{\gamma,L^2(\Omega)}e^{\alpha t}+\frac{8}{\beta}\|g(t)\|^2_{L^2(\Omega)}e^{\alpha t}. \tag{2.11}$$

Integrating from τ to$t$, we have

$$e^{\alpha t}(\|v_m(t)\|^2_{L^2(\Omega)}+\|\nabla u_m(t)\|^2_{L^2(\Omega)})$$

$$\leq e^{\alpha\tau}(\|v_m(\tau)\|^2_{L^2(\Omega)}+\|\nabla u_m(\tau)\|^2_{L^2(\Omega)})+(\frac{4k_1^2}{\beta}+\frac{8m_0^2}{\beta})|\Omega|\int_\tau^t e^{\alpha s}ds$$

$$+(\frac{4k_2^2}{\lambda_1\beta}+\frac{m_1}{2\lambda_1})\int_\tau^t e^{\alpha s}\|\nabla u_{ms}\|^2_{\gamma,L^2(\Omega)}ds+\frac{8}{\beta}\int_\tau^t e^{\alpha s}\|g(s)\|^2_{L^2(\Omega)}ds$$

$$+(\alpha-(\frac{\beta}{2}-2m_1))\int_\tau^t e^{\alpha s}(\|v_m(s)\|^2_{L^2(\Omega)}+\|\nabla u_m(s)\|^2_{L^2(\Omega)})ds. \tag{2.12}$$

Let $\beta=2\alpha+4m_1$. Replacing t by $t+\theta$ and multiplying (2.12) by $e^{-\alpha(t+\theta)}e^{2\gamma\theta}$, we have

$$e^{2\gamma\theta}(\|v_m(t+\theta)\|^2_{L^2(\Omega)}+\|\nabla u_m(t+\theta)\|^2_{L^2(\Omega)})$$

$$\leq e^{-\alpha(t-\tau)}e^{(2\gamma-\alpha)\theta}(\|v_m(\tau)\|^2_{L^2(\Omega)}+\|\nabla u_m(\tau)\|^2_{L^2(\Omega)})$$

$$+(\frac{4k_1^2}{\beta}+\frac{8m_0^2}{\beta})|\Omega|e^{-\alpha t}e^{(2\gamma-\alpha)\theta}\int_\tau^{t+\theta}e^{\alpha s}ds$$

$$+(\frac{4k_2^2}{\lambda_1\beta}+\frac{m_1}{2\lambda_1})e^{-\alpha t}e^{(2\gamma-\alpha)\theta}\int_\tau^{t+\theta}e^{\alpha s}\|\nabla u_{ms}\|^2_{\gamma,L^2(\Omega)}ds$$

$$+\frac{8}{\beta}e^{-\alpha t}e^{(2\gamma-\alpha)\theta}\int_\tau^{t+\theta}e^{\alpha s}\|g(s)\|^2_{L^2(\Omega)}ds. \tag{2.13}$$

By(2.1),we have

$$e^{2\gamma\theta}(\|u'_m(t+\theta)\|^2_{L^2(\Omega)}+\|\nabla u_m(t+\theta)\|^2_{L^2(\Omega)})$$

$$\leq C^2(\beta)e^{-\alpha(t-\tau)}e^{(2\gamma-\alpha)\theta}(\|u'_m(\tau)\|^2_{L^2(\Omega)}$$

$$+\|\nabla u_m(\tau)\|^2_{L^2(\Omega)})+(\frac{4k_1^2}{\alpha\beta}+\frac{8m_0^2}{\alpha\beta})|\Omega|C(\beta)$$

$$+(\frac{4k_2^2}{\lambda_1\beta}+\frac{m_1}{2\lambda_1})C(\beta)e^{-\alpha t}\int_\tau^t e^{\alpha s}(\|u'_{ms}\|^2_{\gamma,L^2(\Omega)}+\|\nabla u_{ms}\|^2_{\gamma,L^2(\Omega)})ds.$$

Further,

$$\|u_{mt}\|^2_{C_{\gamma,H_0^1(\Omega),L^2(\Omega)}}\leq\max\{\sup_{\theta\in(-\infty,\tau-t]}e^{2\gamma\theta}\|\varphi(\theta+t-\tau)\|^2_{H_0^1(\Omega),L^2(\Omega)},$$

$$\sup_{\theta\in[\tau-t,t]}[C^2(\beta)e^{-\alpha(t-\tau)}e^{(2\gamma-\alpha)\theta}(\|u'_m(\tau)\|^2_{L^2(\Omega)}+\|\nabla u_m(\tau)\|^2_{L^2(\Omega)})$$

$$+(\frac{4k_1^2}{\alpha\beta}+\frac{8m_0^2}{\alpha\beta})|\Omega|C(\beta)+(\frac{4k_2^2}{\lambda_1\beta}+\frac{m_1}{2\lambda_1})C(\beta)e^{-\alpha t}\int_\tau^t e^{\alpha s}(\|u'_{ms}\|^2_{\gamma,L^2(\Omega)}$$

$$+\|\nabla u_{ms}\|^2_{\gamma,L^2(\Omega)})ds+\frac{8}{\beta}C(\beta)e^{-\alpha t}\int_\tau^t e^{\alpha s}\|g(s)\|^2_{L^2(\Omega)}ds]\}.$$

Assuming that $\alpha < 2\gamma$, we have

$$e^{-\alpha(t-\tau)}e^{(2\gamma-\alpha)\theta}(\| u'_m(\tau) \|^2_{L^2(\Omega)} + \| \nabla u_m(\tau) \|^2_{L^2(\Omega)}) \le e^{-\alpha(t-\tau)} \| u(\tau) \|^2_{C_{\gamma,H_0^1(\Omega),L^2(\Omega)}}.$$

And

$$\begin{aligned} \varepsilon > 0 & \sup_{\theta\in(-\infty,\tau-t]} e^{2\gamma\theta} \| \varphi(\theta+t-\tau) \|^2_{H_0^1(\Omega),L^2(\Omega)} \\ &= \sup_{\theta\in(-\infty,0]} e^{-2\gamma(\theta-(t-\tau))} e^{-2\gamma(t-\tau)} \| \varphi(\theta) \|^2_{H_0^1(\Omega),L^2(\Omega)} \\ &= \sup_{\theta\in(-\infty,0]} e^{-2\gamma(t-\tau)} \| \varphi(\theta) \|^2_{H_0^1(\Omega),L^2(\Omega)} \\ &\le \sup_{\theta\in(-\infty,0]} e^{-\alpha(t-\tau)} \| \varphi(\theta) \|^2_{H_0^1(\Omega),L^2(\Omega)}. \end{aligned}$$

Collecting these inequalities, we deduce that

$$\| u_{mt} \|^2_{C_{\gamma,H_0^1(\Omega),L^2(\Omega)}} \le (2+C^2(\beta))e^{-\alpha(t-\tau)} \| \varphi \|^2_{C_{\gamma,H_0^1(\Omega),L^2(\Omega)}} + (\frac{4k_1^2}{\alpha\beta} + \frac{8m_0^2}{\alpha\beta}) \mid \Omega \mid C(\beta)$$

$$+ (\frac{4k_2^2}{\lambda_1\beta} + \frac{m_1}{2\lambda_1})C(\beta)e^{-\alpha t}\int_\tau^t e^{\alpha s} \| u_{ms} \|^2_{C_{\gamma,H_0^1(\Omega),L^2(\Omega)}} ds + \frac{8}{\beta}C(\beta)e^{-\alpha t}\int_\tau^t e^{\alpha s} \| g(s) \|^2_{L^2(\Omega)} ds$$

By the Gronwall lemma, we have

$$\| u_{mt} \|^2_{C_{\gamma,H_0^1(\Omega),L^2(\Omega)}} \le (2+C^2(\beta))e^{-\alpha(t-\tau)} \| \varphi \|^2_{C_{\gamma,H_0^1(\Omega),L^2(\Omega)}} + (\frac{4k_1^2}{\alpha\beta} + \frac{8m_0^2}{\alpha\beta}) \mid \Omega \mid C(\beta)$$

$$+ \frac{8}{\beta}C(\beta)e^{-\alpha t}\int_\tau^t e^{\alpha s} \| g(s) \|^2_{L^2(\Omega)} ds + (2+C^2(\beta))e^{-(\alpha-(\frac{4k_2^2}{\lambda_1\beta}+\frac{m_1}{2\lambda_1})C(\beta))(t-\tau)} \| \varphi \|^2_{C_{\gamma,H_0^1(\Omega),L^2(\Omega)}}$$

$$+ \frac{2}{\alpha}e^{-(\alpha-(\frac{4k_2^2}{\lambda_1\beta}+\frac{m_1}{2\lambda_1})C(\beta))t}\int_\tau^t e^{(\alpha-(\frac{4k_2^2}{\lambda_1\beta}+\frac{m_1}{2\lambda_1})C(\beta))s} \| g(s) \|^2_{L^2(\Omega)} ds.$$

Then, using (H3) and(2.3), we obtain the estimate

$$\| u_{mt} \|^2_{C_{\gamma,H_0^1(\Omega),L^2(\Omega)}} \le C(\tau,T,R),\ \forall t \in [\tau,T],\ \| \varphi \|_{C_{\gamma,H_0^1(\Omega),L^2(\Omega)}} \le R, m \ge 1, \quad (2.14)$$

where constant C, depending on someconstants of the problem (namely λ_1, g, f, G), and on α, τ, T and $R > 0$. This implies that

$$\{u_{mt}, u'_{mt}\} \text{ is bounded in } L^\infty(\tau,T;H_0^1(\Omega)\times L^2(\Omega)). \quad (2.15)$$

Step 3: Approximation $C_{\gamma,H_0^1(\Omega),L^2(\Omega)}$ of the initial datum.

For the initial datum $\varphi \in C_{\gamma,H_0^1(\Omega),L^2(\Omega)}$, we have used the projections in the Galerkin method in step 1. Let us check that

$$P_m\varphi \to \varphi \text{ in } C_{\gamma,H_0^1(\Omega),L^2(\Omega)}. \quad (2.16)$$

Indeed, if not, there exists $\varepsilon > 0$ and a subsequence, that we relabel the same, such that

$$e^{2\gamma\theta_m}(\| P_m\varphi(\theta_m) - \varphi(\theta_m) \|^2_{H_0^1(\Omega)} + \| P_m\varphi'(\theta_m) - \varphi'(\theta_m) \|^2_{L^2(\Omega)}) > \varepsilon. \quad (2.17)$$

We can assume that $\theta_m \to -\infty$, otherwise, if $\theta_m \to \theta$, then $P_m\varphi(\theta_m) \to \varphi(\theta)$, $P_m\varphi'(\theta_m) \to \varphi'(\theta)$ in $C_{\gamma,H_0^1(\Omega),L^2(\Omega)}$,

Since

$$e^{2\gamma\theta_m}(\|P_m\varphi(\theta_m)-\varphi(\theta_m)\|^2_{H_0^1(\Omega)}+\|P_m\varphi'(\theta_m)-\varphi'(\theta_m)\|^2_{L^2(\Omega)})$$
$$\le e^{2\gamma\theta_m}\|P_m\varphi(\theta_m)-\varphi_m(\theta)\|^2_{H_0^1(\Omega)}+e^{2\gamma\theta_m}\|P_m\varphi(\theta)-\varphi(\theta)\|^2_{H_0^1(\Omega)}$$
$$+e^{2\gamma\theta_m}\|P_m\varphi'(\theta_m)-\varphi'(\theta_m)\|^2_{L^2(\Omega)}+e^{2\gamma\theta_m}\|P_m\varphi'(\theta)$$
$$-\varphi'(\theta)\|^2_{L^2(\Omega)}\to 0 \text{ as } m\to+\infty.$$

Let us denote $x_1=\lim\limits_{m\to\infty}\lim\limits_{\theta_m\to-\infty}e^{\gamma\theta_m}\varphi(\theta_m)$, $x_2=\lim\limits_{m\to\infty}\lim\limits_{\theta_m\to-\infty}e^{\gamma\theta_m}\varphi'(\theta_m)$.

$$e^{2\gamma\theta_m}(\|P_m\varphi(\theta_m)-\varphi(\theta_m)\|^2_{H_0^1(\Omega)}+\|P_m\varphi'(\theta_m)-\varphi'(\theta_m)\|^2_{L^2(\Omega)})$$
$$=\|P_m(e^{\gamma\theta_m}\varphi(\theta_m)-e^{\gamma\theta_m}\varphi(\theta_m))\|^2_{H_0^1(\Omega)}+\|P_me^{\gamma\theta_m}\varphi'(\theta_m)-e^{\gamma\theta_m}\varphi'(\theta_m)\|^2_{L^2(\Omega)}$$
$$\le\|P_m(e^{\gamma\theta_m}\varphi(\theta_m))-P_mx_1\|^2_{H_0^1(\Omega)}+\|P_mx_1$$
$$-x_1\|^2_{H_0^1(\Omega)}+\|x_1-e^{\gamma\theta_m}\varphi(\theta_m)\|^2_{H_0^1(\Omega)}$$
$$+\|P_m(e^{\gamma\theta_m}\varphi'(\theta_m))-P_mx_2\|^2_{L^2(\Omega)}+\|P_mx_2$$
$$-x_2\|^2_{L^2(\Omega)}+\|x_2-e^{\gamma\theta_m}\varphi'(\theta_m)\|^2_{L^2(\Omega)}$$
$$\to 0 \text{ as } m\to+\infty.$$

$$e^{2\gamma\theta_m}(\|P_m\varphi(\theta_m)-\varphi(\theta_m)\|^2_{H_0^1(\Omega)}+\|P_m\varphi'(\theta_m)-\varphi'(\theta_m)\|^2_{L^2(\Omega)})$$
$$=\|P_m(e^{\gamma\theta_m}\varphi(\theta_m)-e^{\gamma\theta_m}\varphi(\theta_m))\|^2_{H_0^1(\Omega)}+\|P_me^{\gamma\theta_m}\varphi'(\theta_m)-e^{\gamma\theta_m}\varphi'(\theta_m)\|^2_{L^2(\Omega)}$$
$$\le\|P_m(e^{\gamma\theta_m}\varphi(\theta_m))-P_mx_1\|^2_{H_0^1(\Omega)}$$
$$+\|P_mx_1-x_1\|^2_{H_0^1(\Omega)}+\|x_1-e^{\gamma\theta_m}\varphi(\theta_m)\|^2_{H_0^1(\Omega)}$$
$$+\|P_m(e^{\gamma\theta_m}\varphi'(\theta_m))-P_mx_2\|^2_{L^2(\Omega)}+\|P_mx_2-x_2\|^2_{L^2(\Omega)}$$
$$+\|x_2-e^{\gamma\theta_m}\varphi'(\theta_m)\|^2_{L^2(\Omega)}\to 0 \text{ as } m\to+\infty.$$

This is a contradiction with(2.17),so (2.16) holds.

Step 4:Energy estimates and compactness results. Now we apply some well—known compactness

results with energy method to pass to the limit in a subsequence of $\{u_m\}$ to obtain a solution of (1.1).

Thanks to(2.15),(H1) and (H2),we can extract a subsequence,still denoted m,such that

$$\begin{cases} u_m \xrightarrow{*} u \quad \text{weakly star in } L^\infty(-\infty,T;H_0^1(\Omega)), \\ u'_m \xrightarrow{*} u' \quad \text{weakly star in } L^\infty(-\infty,T;L^2(\Omega)), \\ f(\cdot,u_m(\cdot-\rho(\cdot))\rightharpoonup\zeta \quad \text{weakly in } L^2(\tau,T;L^2(\Omega)), \\ \int_{-\infty}^0 G(\cdot,s,u_{m.}(\cdot+s))ds\rightharpoonup\xi \quad \text{weakly in } L^2(\tau,T;L^2(\Omega)),\text{as } m\to\infty. \end{cases} \tag{2.18}$$

It is easy to pass to the limit in(2. 6) and (2. 7)by the classical compactness theorem [27 theorem 2. 3. chapter Ⅲ] and we find u is a solution of (1. 1), $\zeta,\xi\in L^2(\tau,T;L^2(\Omega))$ such that

$$u\in L^{\infty}(-\infty,T;H_0^1(\Omega)),\quad u'\in L^{\infty}(-\infty,T;L^2(\Omega)). \tag{2.19}$$

By(2. 14) and (2. 18), we can assume that

$$u_m(t_m)\rightharpoonup u(t_0)\text{ weakly in }H_0^1(\Omega),u'_m(t_m)\rightharpoonup u'(t_0)\text{ weakly in }L^2(\Omega),$$
$$a.e.\,t\in[\tau,T].$$

$$u_m\to u\quad\text{in}H_0^1(\Omega),\quad u'_m\to u'\quad\text{in}L^2(\Omega),a.e.\,t\in[\tau,T]. \tag{2.20}$$

Since$H_0^1(\Omega)\to\to L^2(\Omega)\to\to H^{-1}(\Omega)$ and $\{u''_m\}\in L^{\infty}(\tau,T;H^{-1}(\Omega))$, $\{u_m\}$ and $\{u'_m\}$ are equicontinuous on $[\tau,T]$ with value in $L^2(\Omega)$ and $H^{-1}(\Omega)$.

we find that

$$u_m\to u\quad\text{in }C([\tau,T];L^2(\Omega)),u'_m\to u'$$
$$\text{in}C([\tau,T];H^{-1}(\Omega)),a.e.\,t\in[\tau,T]. \tag{2.21}$$

Combining with(2. 14) and (2. 21), we can claim that, for any sequence $\{t_m\}\subset[\tau,T]$, with $t_m\to t_0\in[\tau,T]$, one have

$$u_m(t_m)\rightharpoonup u(t_0)\text{ weakly in }H_0^1(\Omega),u'_m(t_m)\rightharpoonup u'(t_0)\text{ weakly in }L^2(\Omega),$$
$$a.e.\,t\in[\tau,T]. \tag{2.22}$$

Let denote$u$ is $\varphi\in(-\infty,\tau)$ in $[\tau,T]$ and is the above limit in $[\tau,T]$.

Now, we will prove the fact that

$$u_m\to u\text{in }C([\tau,T];H_0^1(\Omega)),u'_m\to u'\text{in }C([\tau,T];L^2(\Omega)).$$
$$f(\cdot,u_m(\cdot-\rho(\cdot)))\to f(\cdot,u(\cdot-\rho(\cdot))),\text{in }L^2(\tau,T;L^2(\Omega)), \tag{2.23}$$
$$\int_{-\infty}^0 G(\cdot,s,u_m(\cdot+s))ds\to\int_{-\infty}^0 G(\cdot,s,u(\cdot+s))ds,\text{in }L^2(\tau,T;L^2(\Omega)),$$
$$a.e.\,t\in[\tau,T].$$

Let us prove these results above by contradiction.

If this were not so, then, taking into account that $u\in C(-\infty,T;H_0^1(\Omega))$ and $u'\in C(-\infty,T;L^2(\Omega))$, there would exist $\varepsilon>0$ a value $t_0\in[\tau,T]$ and subsequences (relabelled the same) $\{u_m\}$, $\{u'_m\}$ and $\{t_m\}\subset[\tau,T]$ with $\lim_{m\to+\infty}t_m=t_0$ such that

$$\|u_m(t_m)-u(t_0)\|^2_{H_0^1(\Omega)}\geq\varepsilon,\ \|u'_m(t_m)-u'(t_0)\|^2_{L^2(\Omega)}\geq\varepsilon,\ \forall m. \tag{2.24}$$

It will be absurd by the energy method.

By multiplying(2.6) by u'_{jm} and these relations for$j = 1, \cdots, m$, we obtain (2.24)

$$\frac{d}{2dt}(\| u_m'(t) \|^2_{L^2(\Omega)} + \| \nabla u_m(t) \|^2_{L^2(\Omega)}) + \| u'_m(t) \|^2_{L^2(\Omega)} \tag{2.25}$$

$$= (f(t, u_m(t - \rho(t))) + \int_{-\infty}^{0} G(t, s, u_m(t + s, x))ds$$

$$+ g(t), u_m'(t)), \forall t \in [\tau, T].$$

Thus, $\forall t, s \in [\tau, T]$, we have the following energy equality,

$$\frac{1}{2}(\| u_m'(t) \|^2_{L^2(\Omega)} + \| \nabla u_m(t) \|^2_{L^2(\Omega)}) + \int_s^t \| u'_m(s) \|^2_{L^2(\Omega)} ds$$

$$= \frac{1}{2}(\| u_m'(s) \|^2_{L^2(\Omega)} + \| \nabla u_m(s) \|^2_{L^2(\Omega)}) + \int_s^t (f(s, u_m(s - \rho(s)))ds$$

$$+ \int_{-\infty}^{0} G(r, z, u_m(r + z, x))dz + g(r), u_m'(r))dr. \tag{2.26}$$

By (H1),(H2) and(2.14),we have for $\tau \leq s \leq t \leq T$

$$\int_s^t (f(s, u_m(s - \rho(s)))u_m'(r))ds$$

$$\leq \int_s^t \int_\Omega \left| f(s, u_m(s - \rho(s)))u_m'(r) \right| dxdr$$

$$\leq \int_s^t \frac{k_1^2 | \Omega | + k_2^2 \| u_{mr} \|^2 C_{\gamma, L^2(\Omega)}}{2} dr + \int_s^t \frac{\| u_m'(r) \|^2}{2} dr$$

$$\leq C(t - s),$$

And

$$\int_s^t \int_\Omega \int_{-\infty}^{0} G(r, z, u_m(r + z, x))u_m'(r))dzdxds$$

$$\leq \int_s^t \int_\Omega \int_{-\infty}^{0} (| m_0(z) | + m_1(z) | u(r + s) |)u'_m(r)dzdxdr$$

$$\leq \int_s^t m_0 \int_\Omega \left| u'_m(r) \right| dxdr + \int_s^t \int_\Omega \int_{-\infty}^{0} m_1(r)e^{-\gamma z}e^{\gamma z} | u_m(r + z) | | u'_m(r) | dzdxdr$$

$$\leq \int_s^t \int_\Omega (\frac{e^{2\gamma r} | u_m(r + z) |^2}{2} + \frac{| u'_m(r) |^2}{2})dx \int_{-\infty}^{0} m_1(z)e^{-\gamma z}dzdr$$

$$+ \int_s^t (\frac{m_0^2}{2} | \Omega | + \frac{1}{2} \| u'_m(r) \|^2_{L^2(\Omega)})dr$$

$$\leq \frac{1}{2} \int_s^t (m_0^2 | \Omega | + \| u'_m(r) \|^2_{L^2(\Omega)})dr$$

$$+ \frac{m_1}{2} \int_s^t (\| u_{mr} \|^2_{L^2(\Omega)} + \| u'_m(r) \|^2_{L^2(\Omega)})dr$$

$$\leq C(t - s). \tag{2.28}$$

Combining with(2.26)—(2.28) that the following energy inequality holds for all u_m :

$$\frac{1}{2}(\|u_m'(t)\|_{L^2(\Omega)}^2+\|\nabla u_m(t)\|_{L^2(\Omega)}^2)+\int_s^t\|u'_m(s)\|_{L^2(\Omega)}^2ds \tag{2.29}$$

$$\leq\frac{1}{2}(\|u_m'(s)\|_{L^2(\Omega)}^2+\|\nabla u_m(s)\|_{L^2(\Omega)}^2)$$

$$+\int_s^t\int_\Omega g(r)u_m'(r)dxdr+C(t-s),\forall t,s\in[\tau,T].$$

On the other hand by(2.18), passing to limit in (2.8), we have

$u\in C^0((-\infty,0];H_0^1(\Omega))\cap C^1((-\infty,0];L^2(\Omega))$ is a solutionof similar problem to (1.1), namely

$$\begin{cases}\dfrac{d^2}{dt^2}(u,v)+\dfrac{d}{dt}(u,v)+((\Delta u,v))=(f(t,u(t-\rho(t)))\\ \qquad+\displaystyle\int_{-\infty}^0 G(t,s,u(t+s,x))ds+g(t,x),v),\forall v\in H_0^1(\Omega),\\ u(\tau)=\varphi(0).\end{cases}$$

Replace $f(t,u(t-\rho(t)))$ and $\int_{-\infty}^0 G(t,s,u(t+s,x))ds$ by $\zeta(t)$ and $\xi(t)$, respectively.

Therefore, it satisfies the energy equality

$$\frac{1}{2}(\|u'(t)\|_{L^2(\Omega)}^2+\|\nabla u(t)\|_{L^2(\Omega)}^2)+\int_s^t\|u'(s)\|_{L^2(\Omega)}^2ds$$

$$=\frac{1}{2}(\|u'(s)\|_{L^2(\Omega)}^2+\|\nabla u(s)\|_{L^2(\Omega)}^2)$$

$$+\int_s^t\int_\Omega(\zeta(r)+\xi(r)+g(r)))u'(r)dxdr,\forall t,s\in[\tau,T].$$

Also, by the last convergence in(2.18), $\forall t,s\in[\tau,T]$, we deduce that

$$\int_s^t\|\zeta(r)\|_{L^2(\Omega)}^2dr\leq\liminf_{m\to+\infty}\int_s^t\int_\Omega\left|f(r,u(r-\rho(r)))\right|^2dxdr\leq C(t-s),$$

$$\int_s^t\|\xi(r)\|_{L^2(\Omega)}^2dr\leq\liminf_{m\to+\infty}\int_s^t\int_\Omega\int_{-\infty}^0\left|G(r,z,u(r+z))\right|dzdxdr\leq C(t-s).$$

So, $u$also satisfies the inequality (2.28) with the same constant C.

Now, we define the function $J_m, J:[\tau,T]\to\mathbb{R}$ from (2.28) and the analogous inequality for u,

$$J_m(t)=\frac{1}{2}(\|u_m'(t)\|_{L^2(\Omega)}^2+\|\nabla u_m(t)\|_{L^2(\Omega)}^2)-\int_s^t\int_\Omega g(r)u_m'(r)dxdr-Ct,$$

$$J(t)=\frac{1}{2}(\|u'(t)\|^2_{L^2(\Omega)}+\|\nabla u(t)\|^2_{L^2(\Omega)})-\int_s^t\int_\Omega g(r)u'(r)dxdr-Ct,$$

it is clear that J_m and J are non-increasingfunction and continuous and

$$J_m(t)\rightarrow J(t)a.e.t\in[\tau,T]. \tag{2.30}$$

Now we are ready to prove that

$$u_m(t_m)\rightarrow u(t_0) \text{ in } H_0^1(\Omega), u'_m(t_m)\rightarrow u'(t_0) \text{ in } L^2(\Omega). \tag{2.31}$$

Firstly, from the(2.21), we have

$$\|u(t_0)\|_{H_0^1(\Omega)}\leq\liminf_{m\rightarrow+\infty}\|u_m(t_m)\|_{H_0^1(\Omega)},$$
$$\|u'(t_0)\|_{L^2(\Omega)}\leq\liminf_{m\rightarrow+\infty}\|u'_m(t_m)\|_{L^2(\Omega)}. \tag{2.32}$$

Therefore, we need only show that

$$\limsup_{m\rightarrow+\infty}\|u_m(t_m)\|_{H_0^1(\Omega)}\leq\|u(t_0)\|_{H_0^1(\Omega)},$$
$$\limsup_{m\rightarrow+\infty}\|u'_m(t_m)\|_{L^2(\Omega)}\leq\|u'(t_0)\|_{L^2(\Omega)}, \tag{2.33}$$

which jointly with(2.32) implies (2.31).

Now, we discuss(2.33) in two case which $t_0=\tau$ and $t_0>\tau$. From step 3 and (2.29), the case $t_0=\tau$ follows directly. For the case $t_0>\tau$, we choose sequence $J_m(t_m)-J(t_0)\leq|J_m(\tilde{t}_{k_\varepsilon})-J(\tilde{t}_{k_\varepsilon})|+|J(\tilde{t}_{k_\varepsilon})-J(t_0)|<\varepsilon.$, $\{\tilde{t}_k \lim_{k\rightarrow+\infty}\tilde{t}_k\nearrow t_0\}$, which satisfies (2.30). Since $u(\cdot)$ is continuous and J_m is non-increasing and for all $\{\tilde{t}_k: \lim_{k\rightarrow+\infty}\tilde{t}_k\nearrow t_0\}$ the convergence (2.30), for any $\varepsilon>0$ and $m\geq m(k_\varepsilon)>0$ large enough, one has that

$$J_m(t_m)-J(t_0)\leq|J_m(\tilde{t}_{k_\varepsilon})-J(\tilde{t}_{k_\varepsilon})|+|J(\tilde{t}_{k_\varepsilon})-J(t_0)|<\varepsilon. \tag{2.34}$$

By(2.34) and (2.18), we conclude that

$$\begin{aligned}
&\frac{1}{2}(\|u'_m(t_m)\|^2_{L^2(\Omega)}+\|\nabla u_m(t_m)\|^2_{L^2(\Omega)})\\
&\leq\frac{1}{2}(\|u'(t_0)\|^2_{L^2(\Omega)}+\|\nabla u(t_0)\|^2_{L^2(\Omega)})+(Ct_m-Ct_0)\\
&+(\int_s^{t_m}\int_\Omega|g(r)u'_m(r)|dxdr-\int_s^{t_0}\int_\Omega|g(r)u'(r)|dxdr)\\
&\leq\frac{1}{2}(\|u'(t_0)\|^2_{L^2(\Omega)}+\|\nabla u(t_0)\|^2_{L^2(\Omega)})+\varepsilon
\end{aligned}\tag{2.35}$$

which implies (2.33) holds. So (2.31) holds and the assumption (2.24) is false. Thus, the first result of (2.23) is true.

Thanks to step 3 and(2.23), we have,

$$u_{mt}\rightarrow u_t \text{in} C_{\gamma,H_0^1(\Omega),L^2(\Omega)}, \forall t\leq T.$$

Indeed,

$$\sup_{\theta\le 0} e^{2\gamma\theta} \| u_m(t+\theta)-u(t+\theta) \|^2_{H_0^1(\Omega),L^2(\Omega)}$$

$$= \max\{ \sup_{\theta\in(-\infty,\tau-t]} e^{2\gamma\theta} \| P_m\varphi(\theta+t-\tau)-\varphi(\theta+t-\tau) \|^2_{H_0^1(\Omega),L^2(\Omega)},$$

$$\sup_{\theta\in[\tau-t,0]} e^{2\gamma\theta} \| u_m(t+\theta)-u(t+\theta) \|^2_{H_0^1(\Omega),L^2(\Omega)} \}$$

$$\le \max\{ e^{2\gamma(\tau-t)} \| P_m\varphi-\varphi \|^2_{C_{\gamma,H_0^1(\Omega),L^2(\Omega)}}, \max_{\theta\in[\tau,t]}(\| u'_m(\theta)-u'(\theta) \|^2_{L^2(\Omega)}$$

$$+ \| \nabla u_m(\theta)-\nabla u(\theta) \|^2_{H_0^1(\Omega)})\} \to 0.$$

Therefore, from the above convergence, (H1), (H2), the weak limit ζ and ξ from (2. 18), we can identity the remaining results of (2. 23). Thus, we can finally pass to the limit in (2. 26), concluding that u solves (1. 1).

The uniqueness of solution in the following way. Consider two solutions, u and v of (1. 1) with the same initial data, and let $w=u-v$. The function $w\in C^0((-\infty,T];H_0^1(\Omega))\cap C^1((-\infty,T];L^2(\Omega))$, $\forall T>\tau$ satisfies

$$\begin{cases} \dfrac{\partial^2 w}{\partial t^2}+\dfrac{\partial w}{\partial t}+\Delta w = f(t,u(t-\rho(t)))-f(t,v(t-\rho(t))) \\ +\displaystyle\int_{-\infty}^{0}(G(t,s,u(t+s,x))-G(t,s,v(t+s,x)))ds, \text{in } [\tau,+\infty]\times\Omega, \\ w\mid\partial\Omega=0, \quad \text{on } [\tau,+\infty]\times\partial\Omega, \\ w(0,x)=0, \quad t\in(-\infty,\tau], x\in\Omega, \\ \dfrac{\partial w}{\partial t}=0, \quad t\in(-\infty,\tau], x\in\Omega. \end{cases} \tag{2. 36}$$

Multiplying(2. 36) by w', we obtain that

$$\frac{d}{2dt}(\| w'(t) \|^2_{L^2(\Omega)} + \| \nabla w(t) \|^2_{L^2(\Omega)}) + \| w'(t) \|^2_{L^2(\Omega)}$$

$$=(f(t,u(t-\rho(t)))-f(t,v(t-\rho(t)))$$

$$+\int_{-\infty}^{0}(G(t,s,u(t+s,x))-G(t,s,v(t+s,x)))ds, w'(t)).$$

Integrating the equation and by (H3), (H4), we have for all $t\in[\tau,T]$ that

$$(\| w'(t) \|^2_{L^2(\Omega)} + \| \nabla w(t) \|^2_{L^2(\Omega)}) + 2\int_{\tau}^{t} \| w'(s) \|^2_{L^2(\Omega)} ds$$

$$\le 2\left(\frac{\tilde{L}_f+\tilde{L}_G}{\lambda_1}+\tilde{L}_f+\tilde{L}_G\right)\int_{\tau}^{t} \| w_s \|^2_{C_{\gamma,H_0^1(\Omega),L^2(\Omega)}} ds.$$

Observe that $w(\theta)=0$, if $\theta\le\tau$. Therefore,

$$\| w_s \|^2_{C_{\gamma,H_0^1(\Omega),L^2(\Omega)}} = \sup_{\theta\le 0} e^{2\gamma\theta} (\| w'(s+\theta) \|^2_{L^2(\Omega)} + \| \nabla w(s+\theta) \|^2_{H_0^1(\Omega)})$$
$$= \sup_{\theta\in[\tau-s,0]} (\| w'(s+\theta) \|^2_{L^2(\Omega)} + \| \nabla w(s+\theta) \|^2_{H_0^1(\Omega)})$$
$$= \sup_{r\in[\tau,s]} (\| w'(r) \|^2_{L^2(\Omega)} + \| \nabla w(r) \|^2_{H_0^1(\Omega)}),$$

where $\tilde{L}_f, \tilde{L}_G$ are constant.

So, this yields that

$$\sup_{r\in[\tau,s]} (\| w'(r) \|^2_{L^2(\Omega)} + \| \nabla w(r) \|^2_{L^2(\Omega)})$$
$$\le 2(\frac{\tilde{L}_f + \tilde{L}_G}{\lambda_1} + \tilde{L}_f + \tilde{L}_G) \int_\tau^t \sup_{r\in[\tau,s]} (\| w'(r) \|^2_{L^2(\Omega)} + \| \nabla w(r) \|^2_{H_0^1(\Omega)}) ds.$$

Hence, the Gronwall lemma finishes the proof of uniqueness.

Proposition 2.4 (Continuity of Solutions w. r. t. Initial Data)

(Continuity of Solutions w. r. t. Initial Data)

Under the assumptions of Theorem 2.3, the solutions obtained for (1.1) are continuous with respect to the initial continuous with respect to the initial condition. Namely denoting u_i for $i=1,2$, the corresponding solution to initial data

$\varphi \in C_{\gamma,H_0^1(\Omega),L^2(\Omega)}$, the following estimates hold:

$$\max_{r\in[\tau,t]} (\| u'_1(r) - u'_2(r) \|^2_{L^2(\Omega)} + \| \nabla u_1(r) - \nabla u_2(r) \|^2_{L^2(\Omega)})$$
$$\le (\| \varphi'_1(0) - \varphi'_2(0) \|^2_{L^2(\Omega)} + \| \nabla \varphi_1(0) - \nabla \varphi_2(0) \|^2_{L^2(\Omega)} \quad (2.37)$$
$$+ (\frac{\lambda_1 + 1}{\gamma\lambda_1})(\tilde{L}_f + \tilde{L}_G) \| \varphi_1 - \varphi_2 \|^2_{C_{\gamma,H_0^1(\Omega),L^2(\Omega)}}) e^{2(\frac{\lambda_1+1}{\lambda_1})(\tilde{L}_f+\tilde{L}_G)(t-\tau)},$$

$$\| u_{1t} - u_{2t} \|^2_{C_{\gamma,H_0^1(\Omega),L^2(\Omega)}} \quad (2.38)$$
$$\le (1 + \frac{\lambda_1 + 1}{\gamma\lambda_1})(\tilde{L}_f + \tilde{L}_G) \| \varphi_1 - \varphi_2 \|^2_{C_{\gamma,H_0^1(\Omega),L^2(\Omega)}} e^{2(\frac{\lambda_1+1}{\lambda_1})(\tilde{L}_f+\tilde{L}_G)(t-\tau)}.$$

Proof. Consider the equation (1.1) satisfied by $u_i, i=1,2$, acting on the element $u'_1 - u'_2$ and take the difference, This gives

$$\frac{d}{2dt} (\| u'_1(t) - u'_2(t) \|^2_{L^2(\Omega)} + \| \nabla(u_1(t) - u_2(t)) \|^2_{L^2(\Omega)})$$
$$+ \| (u'_1(t) - u'_2(t)) \|^2_{L^2(\Omega)}$$
$$= (f(t,u_1(t-\rho(t))) - f(t,u_2(t-\rho(t)))$$
$$+ \int_{-\infty}^0 (G(t,s,u_1(t+s,x)) - G(t,s,u_2(t+s,x))) ds, u'_1(t) - u'_2(t)).$$

Arguing as the uniqueness proof of Theorem 2.3, we have

$$\| u'_1(t) - u'_2(t) \|^2_{L^2(\Omega)} + \| \nabla u_1(t) - \nabla u_2(t) \|^2_{L^2(\Omega)}$$

$$\leq (\| u'_1(\tau) - u'_2(\tau) \|^2_{L^2(\Omega)} + \| \nabla u_1(\tau) - \nabla u_2(\tau) \|^2_{L^2(\Omega)})$$
$$+ 2(1 + \frac{1}{\lambda_1})(\tilde{L}_f + \tilde{L}_G) \int_\tau^t \| u_{1s} - u_{2s} \|^2_{C_{\gamma,H_0^1(\Omega),L^2(\Omega)}} ds.$$

Since the fact that, for $s \in [\tau, t]$, one has

$$\| u_{1s} - u_{2s} \|^2_{C_{\gamma,H_0^1(\Omega),L^2(\Omega)}} = \sup_{\theta \leq 0} e^{2\gamma\theta} (\| u'_1(s+\theta) - u'_2(s+\theta) \|^2_{L^2(\Omega)}$$
$$+ \| \nabla u_1(s+\theta) - \nabla u_2(s+\theta) \|^2_{H_0^1(\Omega)})$$
$$= \max\{ \sup_{\theta \in [-\infty, \tau-s]} e^{2\gamma\theta} \| \varphi_1(s-\tau+\theta) - \varphi_2(s-\tau+\theta) \|^2,$$
$$\sup_{\theta \in [\tau-s, 0]} (\| u'_1(s+\theta) - u'_2(s+\theta) \|^2_{L^2(\Omega)}$$
$$+ \| \nabla u_1(s+\theta) - \nabla u_2(s+\theta) \|^2_{H_0^1(\Omega)})\} \tag{2.39}$$
$$\leq \max\{ e^{2\gamma(\tau-s)} \| \varphi_1 - \varphi_2 \|^2_{C_{\gamma,H_0^1(\Omega),L^2(\Omega)}},$$
$$\max_{\theta \in [\tau, s]} (\| u'_1(\theta) - u'_2(\theta) \|^2_{L^2(\Omega)} + \| \nabla u_1(\theta) - \nabla u_2(\theta) \|^2_{H_0^1(\Omega)})\},$$

we conclude that, for all $t \in [\tau, T]$,

$$\| u'_1(t) - u'_2(t) \|^2_{L^2(\Omega)} + \| \nabla u_1(t) - \nabla u_2(t) \|^2_{L^2(\Omega)}$$
$$\leq \| \varphi'_1(0) - \varphi'_2(0) \|^2_{L^2(\Omega)} + \| \nabla \varphi_1(0) - \nabla \varphi_2(0) \|^2_{L^2(\Omega)}$$
$$+ 2(1 + \frac{1}{\lambda_1})(\tilde{L}_f + \tilde{L}_G) \int_\tau^t \| u_{1s} - u_{2s} \|^2_{C_{\gamma,H_0^1(\Omega),L^2(\Omega)}} ds$$
$$\leq \| \varphi'_1(0) - \varphi'_2(0) \|^2_{L^2(\Omega)} + \| \nabla \varphi_1(0) - \nabla \varphi_2(0) \|^2_{L^2(\Omega)}$$
$$+ 2(1 + \frac{1}{\lambda_1})(\tilde{L}_f + \tilde{L}_G) \| \varphi_1 - \varphi_2 \|^2_{C_{\gamma,H_0^1(\Omega),L^2(\Omega)}} \int_\tau^t e^{2\gamma(\tau-s)} ds$$
$$+ 2(1 + \frac{1}{\lambda_1})(\tilde{L}_f + \tilde{L}_G) \int_\tau^t \max_{\theta \in [\tau, s]} (\| u'_1(\theta) - u'_2(\theta) \|^2_{L^2(\Omega)}$$
$$+ \| \nabla u_1(\theta) - \nabla u_2(\theta) \|^2_{H_0^1(\Omega)}) ds$$

Substitutingt by $r \in [\tau, t]$, and considering the maximum when varying this r, from the above we can conclude that

$$\max_{r \in [\tau, t]} (\| u'_1(r) - u'_2(r) \|^2_{L^2(\Omega)} + \| \nabla u_1(r) - \nabla u_2(r) \|^2_{L^2(\Omega)})$$
$$\leq (\| \varphi'_1(0) - \varphi'_2(0) \|^2_{L^2(\Omega)} + \| \nabla \varphi_1(0) - \nabla \varphi_2(0) \|^2_{L^2(\Omega)})$$
$$+ (\frac{\lambda_1 + 1}{\gamma \lambda_1})(\tilde{L}_f + \tilde{L}_G) \| \varphi_1 - \varphi_2 \|^2_{C_{\gamma,H_0^1(\Omega),L^2(\Omega)}}$$
$$+ 2(\frac{\lambda_1 + 1}{\lambda_1})(\tilde{L}_f + \tilde{L}_G) \int_\tau^t \max_{r \in [\tau, t]} (\| u'_1(r) - u'_2(r) \|^2_{L^2(\Omega)}$$
$$+ \| \nabla u_1(r) - \nabla u_2(r) \|^2_{H_0^1(\Omega)}) ds$$

Hence, by the Gronwall lemma, we obtain (2.37), (2.38) follows from (2.37) and (2.39) easily.

proposition 2.5 (Continuity of Solutions w. r. t. Initial Time)

Under the same assumption as proposition(2.4), the solutions obtain for (1.1) are continuous with respect to the initial time; i.e., let us denote

$u(\cdot;s,\varphi)$ the solution of (1.1) with initial times. Then for each

$t\in[\tau,T]$ and $\varphi\in C_{\gamma,H_0^1(\Omega),L^2(\Omega)}$ fixed, the mapping $[\tau,T]\Rightarrow\exists s\rightarrow u_t(\cdot;s,\varphi)\in C_{\gamma,H_0^1(\Omega),L^2(\Omega)}$ is continuous.

proof. The principle of the proof the same as Proposition 7 [15]. We only show the sketch. We will prove the continuity of mapping from left and right of fixed $s_0\in(\tau,T)$ (the extremal cases $s_0=\tau$ or T is analogous) in two steps.

Step 1: Assume that$s_0<s$, and let $t\geq s$ and $\varphi\in C_{\gamma,H_0^1(\Omega),L^2(\Omega)}$ be fixed. Then, by(2.38), we have

$$\|u_t(\cdot;s_0,\varphi)-u_t(\cdot;s,\varphi)\|^2_{C_{\gamma,H_0^1(\Omega),L^2(\Omega)}}\leq(1+\frac{\lambda_1+1}{\gamma\lambda_1})(\tilde{L}_f+\tilde{L}_G)$$

$$\|u_s(\cdot;s_0,\varphi)-\varphi\|^2_{C_{\gamma,H_0^1(\Omega),L^2(\Omega)}}e^{2(\frac{\lambda+1}{\lambda})(\tilde{L}_f+\tilde{L}_G)(t-s)}\tag{2.40}$$

We note that for an$\varepsilon>0$ y, there exists $\tilde{T}>0$

$$\|u_s(\cdot;s_0,\varphi)-\varphi\|^2_{C_{\gamma,H_0^1(\Omega),L^2(\Omega)}}=\sup_{\theta\leq0}e^{2\gamma\theta}(\|u(s+\theta;s_0,\varphi)-\varphi(\theta)\|^2_{C_{\gamma,H_0^1(\Omega),L^2(\Omega)}}$$

$$=\max\{\sup_{\theta\in[-\infty,-\tilde{T}]}e^{2\gamma\theta}\|u(s+\theta;s_0,\varphi)-\varphi(\theta)\|^2_{C_{\gamma,H_0^1(\Omega),L^2(\Omega)}}$$

$$\sup_{\theta\in[-\tilde{T},0]}e^{2\gamma\theta}\|u(s+\theta;s_0,\varphi)-\varphi(\theta)\|^2_{C_{\gamma,H_0^1(\Omega),L^2(\Omega)}}\}\xrightarrow{s\searrow s_0}0.\tag{2.41}$$

Step 2: Let us now prove the continuity of the map from the left. Assume that$s<s_0$, and $t\geq s_0$ and $\varphi\in C_{\gamma,H_0^1(\Omega),L^2(\Omega)}$ be fixed. Again, by 2.38, we have

$$\|u_t(\cdot;s_0,\varphi)-u_t(\cdot;s,\varphi)\|^2_{C_{\gamma,H_0^1(\Omega),L^2(\Omega)}}$$

$$\leq(1+\frac{\lambda_1+1}{\gamma\lambda_1})(\tilde{L}_f+\tilde{L}_G)\|\varphi-u_{s_0}(\cdot;s,\varphi)\|^2_{C_{\gamma,H_0^1(\Omega),L^2(\Omega)}}e^{2(\frac{\lambda+1}{\lambda})(\tilde{L}_f+\tilde{L}_G)(t-s_0)}.\tag{2.42}$$

We will prove that

$$\|\varphi-u_{s_0}(\cdot;s,\varphi)\|^2_{C_{\gamma,H_0^1(\Omega),L^2(\Omega)}}\xrightarrow{s\nearrow s_0}0.\tag{2.43}$$

If not, we can obtain $\exists\varepsilon>0$ and a subsequenceu_{s_n}, $s_n\nearrow s_0$, such that

$$\|u_{s_0}(\cdot;s_n,\varphi)-\varphi\|^2_{C_{\gamma,H_0^1(\Omega),L^2(\Omega)}}\geq\varepsilon,\ \forall n\in\mathbb{N}.\tag{2.44}$$

Let$v^s(\cdot)=u(\cdot+s-s_0)$ is a weak solution with initial data $v^s_{s_0}=\varphi$ and $v(s_0)=u(s_0)=\varphi(s_0-s_0)$ and with $f(r)$ and $\int_{-\infty}^0G(r,z,u(r+z,x))ds$ and $g(r)$ replaced by $f^s(r)=f(r+s-s_0)$ and

$$\int_{-\infty}^0G^s(r+s-s_0,z,u(r+s-s_0+z,x))dz=\int_{-\infty}^0G(r,z,u(r+z,x))dz$$

and $g^s(r) = g(r+s-s_0)$, respectively. As $s\nearrow s_0$ for any $T > s_0$, we have the fact $g^s \to g$ in $L^2(s_0,T;H^{-1}(\Omega))$. Arguing as in the proof of Theorem 2.3, we have $v^{s_n} \to v \in C([s_0,T],H_0^1(\Omega))$ and $(v^{s_n})' \to v' \in C^1([s_0,T],L^2(\Omega))$, as $n\to\infty$.

On the one hand, the initial data $v(s_0) = \varphi(0)$ implies that

$$\sup_{\theta\in[-\tilde{T},0]} e^{2\gamma\theta}(\| u(s_0+\theta;s_n,\varphi)-\varphi(\theta)\|^2_{H_0^1(\Omega),L^2(\Omega)} \to 0 \text{ as } n\to\infty, \tag{2.45}$$

for all $\tilde{T} > 0$.

On the other hand, for any $\tilde{T} > 0$,

$$\begin{aligned}\| u_{s_0}(\cdot;s_n,\varphi)-\varphi\|^2_{C_{\gamma,H_0^1(\Omega),L^2(\Omega)}} &= \max\{\sup_{\theta\in(-\infty,-\tilde{T}]} e^{2\gamma\theta}\|\varphi(\theta+s_0-s_n)\\ &-\varphi(\theta+t-\tau)\|^2_{H_0^1(\Omega),L^2(\Omega)},\ \sup_{\theta\in[-\tilde{T},0]} e^{2\gamma\theta}\|u(s_0+\theta;s_n,\varphi)\\ &-\varphi(\theta)\|^2_{H_0^1(\Omega),L^2(\Omega)}\},\end{aligned}$$

where for any $\varepsilon > 0$, for some fixed $\bar{s} \geq s_0 - s_n$, there exists $\tilde{T} > 0$ depending on ε, such that

$$\sup_{\theta\in(-\infty,-\tilde{T}]} e^{2\gamma\theta}\|\varphi(\theta+s_0-s_n)-\varphi(\theta+t-\tau)\|^2_{H_0^1(\Omega),L^2(\Omega)} \leq \varepsilon \quad \text{if} \quad \theta \leq -\tilde{T}-\tilde{s} = -\tilde{T}. \tag{2.46}$$

Thus, using (2.45) and (2.46), we deduce that contradiction of (2.43).

Therefore, we obtain

$$\| u_t(\cdot;s_0,\varphi)-u_t(\cdot;s,\varphi)\|^2_{C_{\gamma,H_0^1(\Omega),L^2(\Omega)}} \to 0 \quad \text{as} \quad s\nearrow s_0.$$

3 Existence of pullback attractors in $C_{\gamma,H_0^1(\Omega),L^2(\Omega)}$

Let X be a complete metric space with metric $d_X(\cdot,\cdot)$.

Denote by $H_X^*(\cdot,\cdot)$ the Hausdorff semi-distance between two nonempty subsets of a complete metric space X, which are defined by

$$H_X^*(A,B) = \sup_{a\in A}\inf_{b\in B} d_X(a,b).$$

Definition 3.1 A biparametric family of mapping $U(t,\tau):X\to X, \infty > t \geq \tau > -\infty$, is called to be a process if it satisfies:

(1) $U(\tau,\tau)x = x, \forall\tau\in\mathbb{R}, x\in X$;

(2) $U(t,s)U(s,\tau)x = U(t,\tau)x$, for all $-\infty < \tau \leq s \leq t < \infty, x\in X$.

Let D be a nonempty class of parameterized sets $D = \{D(t)\}_{t\in\mathbb{R}}$, with $D(t) \subset X$ for all $t\in\mathbb{R}$.

Definition 3.2 Let $\{U(t,\tau)\}$ is a process on X.

(1) It is said that $Q = \{Q(t)\}_{t\in\mathbb{R}} \in D$ is pullback D − absorbing for the process U if for any $t \in \mathbb{R}$ and any $? = \{B(t)\}_{t\in\mathbb{R}} \in D$, there exists $t_0(t,?) \in \mathbb{R}^+$ such that

$$U(t,t-s)B(t-s) \subset Q(t), \forall s \geq t_0(t,?).$$

(2) $\{U(t,\tau)\}$ is said that to be pullback D −asymptotically compact in X, if for any $t \in \mathbb{R}$ and any $? = \{B(t)\}_{t\in\mathbb{R}} \in D$, any sequence $0 \leq s_n \to -\infty$, any sequence $x_n \in B(t-s_n)$, the sequence $\{U(t,t-s_n)x_n\}$ is relatively compact in X.

Definition 3.3 A family of nonempty compact subsets $A = \{A(t)\}_{t\in\mathbb{R}}$ of X is called to be a pullback attractor for the process $\{U(t,\tau)\}$, if

(1) $A = \{A(t)\}_{t\in\mathbb{R}}$ is invariant, i. e. ,

$$U(t,\tau)A(t) = A(t), \forall t \geq \tau, \tau \in \mathbb{R};$$

(2) A is pullback attracting, i. e. , for every bounded set B of X and any fixed $t \in \mathbb{R}$,

$$\lim_{s\to+\infty} H_X^*(U(t,t-s)B, A(t)) = 0.$$

The general existence of pullback attractors has been given as follows (see [[29], Th. 18] and also [[30], Th. 7]).

Theorem 3.4 Let $\{U(t,\tau)\}$ be a process on Banach space X. Assume that $U(t,\tau)$ has a pullback D −absorbing set $Q = \{Q(t)\}_{t\in\mathbb{R}} \in D$ and $\{U(t,\tau)\}$ is D − pullback asymptotically compact. Then family $A = \{A_D(t)\}_{t\in\mathbb{R}}$ is D −pullback attractors, defined by

$$A_D(t) = \bigcap_{T\in\mathbb{R}^-} \overline{\bigcup_{s\geq T} U(t,t-s)Q(t-s)}^X \subset Q(t), t \in \mathbb{R},$$

which is minimal in the sense that if $C = \{C(t): t \in \mathbb{R}\} \subset (X)$ is a family of closed sets such that

$$\lim_{s\to+\infty} H_X^*(U(t,t-s)B(t-s), C(t)) = 0,$$

then $A_D(t) \subset C(t)$.

The following result will be useful in order to prove the process $\{U(t,\tau)\}$ on Banach space X is D −pullback asymptotically compact (see[[13], Th. 5]).

Theorem 3.5 Let $\{U(t,\tau)\}$ be a process on Banach space X, with a pullback D −absorbing set $Q = \{Q(t)\}_{t\in\mathbb{R}} \in D$. Suppose that U can be written as

$U(t,\tau) = U_1(t,\tau) + U_2(t,\tau)$, $\forall t \geq \tau$, and for any fixed $t \in \mathbb{R}$,

(1) $\lim_{s\to\infty} \| U_2(t,t-s)Q(t-s) \|_X = 0$;

(2) for any fixed $s > 0$, every sequence $\{y_n\} \subset U_1(t,t-s)Q(t-s)$ is a Cauchy

sequence in X .

Lemma 3. 6 Under the assumption of Theorem2. 3, the following estimates hold for a solution to (1. 1) for all $t \geq \tau$:

$$\| u_t \|^2_{C_{\gamma, H_0^1(\Omega), L^2(\Omega)}} \leq 6e^{-\alpha(t-\tau)} \| \varphi \|^2_{C_{\gamma, H_0^1(\Omega), L^2(\Omega)}}$$

$$+ \frac{4k_1^2 + 8m_0^2}{(\alpha + 2m_1)\alpha} | \Omega | + \frac{8}{(\alpha + 2m_1)} e^{-\alpha t} \int_{-\infty}^{t} e^{\alpha s} \| g(s) \|^2_{L^2(\Omega)} ds$$

$$+ 6e^{-(\alpha - (\frac{4k_2^2}{(\alpha+2m_1)\lambda_1} + \frac{m_1}{\lambda_1}))(t-\tau)} \| \varphi \|^2_{C_{\gamma, H_0^1(\Omega), L^2(\Omega)}}$$

$$+ \frac{2}{(\alpha + 2m_1)} e^{-(\alpha - (\frac{4k_2^2}{(\alpha+2m_1)\lambda_1} + \frac{m_1}{\lambda_1}))t} \int_{-\infty}^{t} e^{(\alpha - (\frac{4k_2^2}{(\alpha+2m_1)\lambda_1} + \frac{m_1}{\lambda_1}))s} \| g(s) \|^2_{L^2(\Omega)} ds. \tag{3. 1}$$

Moreover, from (H3), the family ? $= \{B(t) : t \in \mathbb{R}\}$, with $B_t = B_{C_{\gamma, H_0^1(\Omega), L^2(\Omega)}}(0, \tilde{\rho}(t))$, where

$$\tilde{\rho}^2(t) = C + \frac{8}{(\alpha + 2m_1)} e^{-\alpha t} \int_{-\infty}^{t} e^{\alpha s} \| g(s) \|^2_{L^2(\Omega)} ds$$

$$+ \frac{2}{(\alpha + 2m_1)} e^{-(\alpha - (\frac{4k_2^2}{(\alpha+2m_1)\lambda_1} + \frac{m_1}{\lambda_1}))t} \int_{-\infty}^{t} e^{(\alpha - (\frac{4k_2^2}{(\alpha+2m_1)\lambda_1} + \frac{m_1}{\lambda_1}))s} \| g(s) \|^2_{L^2(\Omega)} ds, \tag{3. 2}$$

is pullback absorbing for bounded sets for the process $\{U(t, \tau)\}$.

Proof

Since the uniform estimates for the solution of (1. 1) associating with the process $\{U(t, \tau)\}$ are analogous to the proof of Theorem2. 3, we only sketch the main ideas.

Multiplying (1. 1) by $v = u' + \beta u$, by the Young inequality and Proposition 3, and using that $\beta = 2(\alpha + 2m_1)$, $C(\beta) = 2$, we arrive at

$$e^{2\gamma\theta} (\| u'(t+\theta) \|^2_{L^2(\Omega)} + \| \nabla u(t+\theta) \|^2_{L^2(\Omega)})$$

$$\leq 6e^{-\alpha(t-\tau)} (\| u'(\tau) \|^2_{L^2(\Omega)} + \| \nabla u(\tau) \|^2_{L^2(\Omega)}) + \frac{4k_1^2 + 8m_0^2}{(\alpha + 2m_1)\alpha} | \Omega | e^{(2\gamma - \alpha)\theta}$$

$$+ (\frac{4k_2^2}{(\alpha + 2m_1)\lambda_1} + \frac{m_1}{\lambda_1}) e^{-\alpha t} e^{(2\gamma - \alpha)\theta} \int_{-\infty}^{t} e^{\alpha s} \| u_s \|^2_{\gamma, H_0^1(\Omega), L^2(\Omega)} ds$$

$$+ \frac{8}{(\alpha + 2m_1)} e^{-\alpha t} e^{(2\gamma - \alpha)\theta} \int_{-\infty}^{t} e^{\alpha s} \| g(s) \|^2_{L^2(\Omega)} ds. \tag{3. 3}$$

Splitting the term related to the delay in the initial data and the evolution of solution for $t \geq \tau$, and by (2. 3), we have

$$\| u_t \|^2_{C_{\gamma, H_0^1(\Omega), L^2(\Omega)}} \leq 6e^{-\alpha(t-\tau)} \| \varphi \|^2_{C_{\gamma, H_0^1(\Omega), L^2(\Omega)}}$$

$$+ \frac{4k_1^2 + 8m_0^2}{(\alpha + 2m_1)\alpha} | \Omega | + (\frac{4k_2^2}{(\alpha + 2m_1)\lambda_1} + \frac{m_1}{\lambda_1}) e^{-\alpha t} \int_{-\infty}^{t} e^{\alpha s} \| u_s \|^2_{C_{\gamma, H_0^1(\Omega), L^2(\Omega)}} ds$$

$$+\frac{8}{(\alpha+2m_1)}e^{-\alpha t}\int_{-\infty}^{t}e^{\alpha s}\parallel g(s)\parallel_{L^2(\Omega)}^2 ds.$$

By the Gronwall lemma, we conclude that(3. 1). From the assumption (H3) and(3. 1), (3. 2) is obviously.

Now, we state and prove the main result in this section.

Theorem 3. 7 Suppose that (H1), (H2), (H3) and(3. 2) hold true. Then the process $\{U(t,\tau)\}$ on $C_{\gamma,H_0^1(\Omega),L^2(\Omega)}$ possesses a unique pullback attractor $\{A_{D,C_{\gamma,H_0^1(\Omega),L^2(\Omega)}}(t)\}_{t\in\mathbb{R}}$ in $D_{C_{\gamma,H_0^1(\Omega),L^2(\Omega)}}$. Moreover, every $A_{D,C_{\gamma,H_0^1(\Omega),L^2(\Omega)}}(t)$ is connected in $C_{\gamma,H_0^1(\Omega),L^2(\Omega)}$.

Proof

By condition(2. 3) and(3. 1), we can conclude that

$$e^{2\gamma\theta}(\parallel u'(t+\theta)\parallel_{L^2(\Omega)}^2+\parallel \nabla u(t+\theta)\parallel_{L^2(\Omega)}^2)$$

$$\leq 6e^{-\alpha(t-\tau)}e^{(2\gamma-\alpha)\theta}\parallel\varphi\parallel_{C_{\gamma,H_0^1(\Omega),L^2(\Omega)}}^2+\frac{4k_1^2+8m_0^2}{(\alpha+2m_1)\alpha}\mid\Omega\mid e^{(2\gamma-\alpha)\theta}$$

$$+\frac{8}{(\alpha+2m_1)}e^{-\alpha t}e^{(2\gamma-\alpha)\theta}\int_{-\infty}^{t}e^{\alpha s}\parallel g(s)\parallel_{L^2(\Omega)}^2 ds$$

$$+6e^{-(\alpha-(\frac{4k_1^2}{(\alpha+2m_1)\lambda_1}+\frac{m_1}{\lambda_1}))(t-\tau)}e^{(2\gamma+(\frac{4k_1^2}{(\alpha+2m_1)\lambda_1}+\frac{m_1}{\lambda_1})-\alpha)\theta}\parallel\varphi\parallel_{C_{\gamma,H_0^1(\Omega),L^2(\Omega)}}^2$$

$$+\frac{2}{(\alpha+2m_1)}e^{-(\alpha-(\frac{4k_1^2}{(\alpha+2m_1)\lambda_1}+\frac{m_1}{\lambda_1}))t}e^{(2\gamma+(\frac{4k_1^2}{(\alpha+2m_1)\lambda_1}+\frac{m_1}{\lambda_1})-\alpha)\theta}$$

$$\times\int_{-\infty}^{t}e^{(\alpha-(\frac{4k_1^2}{(\alpha+2m_1)\lambda_1}+\frac{m_1}{\lambda_1}))s}\parallel g(s)\parallel_{L^2(\Omega)}^2 ds \tag{3. 4}$$

Then for every fixed $t\in\mathbb{R}$, and any $\varepsilon>0$, the condition (H3) implies that we can fixed T^* large enough such that for all $s\geq 0$ and $u_t\in U(t,t-s)D(t-s)$,

$$\sup_{\theta\in(-\infty,-T^*]}e^{2\gamma\theta}(\parallel u'(t+\theta)\parallel_{L^2(\Omega)}^2+\parallel \nabla u(t+\theta)\parallel_{L^2(\Omega)}^2)<\varepsilon. \tag{3. 5}$$

As follows, we divide the proof into two steps for any $\theta\in[-T^*,0]$.

Step 1. We first prove the process $\{U(t,\tau)\}$ on $C_{\gamma,H_0^1(\Omega),L^2(\Omega)}$ is $D_{C_{\gamma,H_0^1(\Omega),L^2(\Omega)}}$ pullback asymptotically compact. For any fixed $t\in\mathbb{R}$ and any $T>t-s$ with $s>0$, we consider that, $U(T,t-s)(\varphi)=\{u_\tau(\cdot;t-s,\varphi)\mid u(\cdot)$is a strong solution of (1. 1) with $\varphi\in Q(t-s)\}$, where $D_{C_{\gamma,H_0^1(\Omega),L^2(\Omega)}}$ is a $D_{C_{\gamma,H_0^1(\Omega),L^2(\Omega)}}$ pullback absorbing set in $C_{\gamma,H_0^1(\Omega),L^2(\Omega)}$. By condition (2. 3) and (57), we apply Gronwall lemma and conclude that

$$e^{2\gamma\theta}(\parallel u'(t+\theta)\parallel_{L^2(\Omega)}^2+\parallel \nabla u(t+\theta)\parallel_{L^2(\Omega)}^2)$$

$$\leq 6e^{-\alpha(t-\tau)}e^{(2\gamma-\alpha)\theta}\|\varphi\|^2_{C_{\gamma,H_0^1(\Omega),L^2(\Omega)}}+\frac{4k_1^2+8m_0^2}{(\alpha+2m_1)\alpha}|\Omega|e^{(2\gamma-\alpha)\theta}$$

$$+\frac{8}{(\alpha+2m_1)}e^{-\alpha t}e^{(2\gamma-\alpha)\theta}\int_{-\infty}^{t}e^{\alpha s}\|g(s)\|^2_{L^2(\Omega)}ds$$

$$+\frac{2}{(\alpha+2m_1)}e^{-(\alpha-(\frac{4k_1^2}{(\alpha+2m_1)\lambda_1}+\frac{m_1}{\lambda_1}))t}e^{(2\gamma+(\frac{4k_1^2}{(\alpha+2m_1)\lambda_1}+\frac{m_1}{\lambda_1})-\alpha)\theta}\times\int_{-\infty}^{t}e^{(\alpha-(\frac{4k_1^2}{(\alpha+2m_1)\lambda_1}+\frac{m_1}{\lambda_1}))s}\|g(s)\|^2_{L^2(\Omega)}ds.$$

$$+6e^{-(\alpha-(\frac{4k_1^2}{(\alpha+2m_1)\lambda_1}+\frac{m_1}{\lambda_1}))(t-\tau)}e^{(2\gamma+(\frac{4k_1^2}{(\alpha+2m_1)\lambda_1}+\frac{m_1}{\lambda_1})-\alpha)\theta}\|\varphi\|^2_{C_{\gamma,H_0^1(\Omega),L^2(\Omega)}} \tag{3.6}$$

Let $u=v+w$, we decompose equation (1.1) as follows:

$$\begin{cases}\frac{\partial^2 w}{\partial T^2}+\frac{\partial w}{\partial T}+\Delta w=f(T,u(T-\rho(T))) \\ +\int_{-\infty}^{0}G(T,r,u(T+r,x))dr,\text{in }[t-s,+\infty]\times\Omega, \\ w\mid\partial\Omega=0,\text{on }[t-s,+\infty]\times\partial\Omega, \\ w(T,x)=0,T\in(-\infty,t-s],x\in\Omega, \\ \frac{\partial w}{\partial T}=0,T\in(-\infty,t-s],x\in\Omega,\end{cases} \tag{3.7}$$

$$\begin{cases}\frac{\partial^2 v}{\partial T^2}+\frac{\partial v}{\partial T}+\Delta v=0\text{in }[t-s,+\infty]\times\Omega, \\ v\mid\partial\Omega=0,\text{on }[t-s,+\infty]\times\partial\Omega, \\ v(T,x)=\varphi(T-t+s),T\in(-\infty,t-s],x\in\Omega, \\ \frac{\partial v}{\partial T}=\frac{\partial\varphi(T-t+s)}{\partial T},T\in(-\infty,t-s],x\in\Omega.\end{cases} \tag{3.8}$$

Since $f=G=g=0$, by(3.1), we have

$$\|v_t\|^2_{C_{\gamma,H_0^1(\Omega),L^2(\Omega)}}\leq Ce^{-\alpha s}\|\varphi\|^2_{C_{\gamma,H_0^1(\Omega),L^2(\Omega)}}. \tag{3.9}$$

By Theorem 3.5, we only show (3.4) fulfills the conditions (2) in Theorem3.5. Consider two solutions u_T^i of system (1.1) corresponding to initial data $\varphi^i(i=1,2)$, and denote $z(T)=w^1(T)-w^2(T)$. *Then* $z(T)$ satisfies that

$$\begin{cases}\frac{\partial^2 z}{\partial T^2}+\frac{\partial z}{\partial T}+\Delta z=f(T,u^1(T-\rho(T)))-f(T,u^2(T-\rho(T))) \\ +\int_{-\infty}^{0}(G(T,r,u^1(T+r,x))-G(T,r,u^2(T+r,x)))dr,\text{in }[t-s,+\infty]\times\Omega, \\ z\mid\partial\Omega=0,\text{on }[t-s,+\infty]\times\partial\Omega, \\ z(T,x)=0,T\in(-\infty,t-s],x\in\Omega, \\ \frac{\partial z}{\partial T}=0,T\in(-\infty,t-s],x\in\Omega.\end{cases}$$

(3.10)

Taking the inner product in $L^2(\Omega)$ of (3.10) with $z'(T)$, we get

$$\frac{1}{2}\frac{d}{dT}(\|z'(T)\|^2_{L^2(\Omega)}+\|\nabla z(T)\|^2_{L^2(\Omega)})+\|z'(T)\|^2_{L^2(\Omega)}$$
$$=(f(T,u^1(T-\rho(T)))-f(T,u^2(T-\rho(T))),z'(T))$$
$$+\int_{-\infty}^{0}(G(T,r,u^1(T+r,x))-G(T,r,u^2(T+r,x)))dr,z'(T)).$$

Integrating from $t-s$ to t (where $\theta\in(-\infty,0]$) above T, we have

$$\|z'(t)\|^2_{L^2(\Omega)}+\|\nabla z(t)\|^2_{L^2(\Omega)}$$
$$\le 2\int_{t-s}^{t}(f(T,u^1(T-\rho(T)))-f(T,u^2(T-\rho(T))),z'(T))dT$$
$$+\int_{t-s}^{t}\int_{-\infty}^{0}(G(T,r,u^1(T+r,x))-G(T,s,u^2(T+r,x)))dr,z'(T))dT$$
$$\le 2\int_{t-s}^{t}\int_{\Omega}\left|f(T,u^1(T-\rho(T)))-f(T,u^2(T-\rho(T))\right|\left|z'(T)\right|dxdT$$
$$+2\int_{t-s}^{t}\int_{-\infty}^{0}\left|G(T,r,u^1(T+s,x))-G(T,r,u^2(T+r,x))\right|\left|z'(T)\right|drdxdT. \tag{3.11}$$

Replacing t by $t+\theta$, by the boundary condition of $z'(t)$ in (3.10), we deduce that

$$\|z'(t+\theta)\|^2_{L^2(\Omega)}+\|\nabla z(t+\theta)\|^2_{L^2(\Omega)}$$
$$\le 2\int_{t+\theta-s}^{t+\theta}\int_{\Omega}\left|f(T,u^1(T-\rho(T)))-f(T,u^2(T-\rho(T))\right|\left|z'(T)\right|dxdT$$
$$+2\int_{t+\theta-s}^{t+\theta}\int_{-\infty}^{0}\left|G(T,r,u^1(T+r,x))-G(T,r,u^2(T+r,x))\right|\left|z'(T)\right|drdxdT$$
$$\le 2\int_{t-s}^{t}\int_{\Omega}\left|f(T,u^1(T-\rho(T)))-f(T,u^2(T-\rho(T))\right|\left|z'(T)\right|dxdT$$
$$+2\int_{t-s}^{t}\int_{\Omega}\int_{-\infty}^{0}\left|G(T,r,u^1(T+r,x))-G(T,r,u^2(T+r,x))\right|\left|z'(T)\right|dsdxdT$$
$$\le 2\|f(T,u^1(T-\rho(T)))-f(T,u^2(T-\rho(T))\|_{L^2(\Omega\times[t-s,t])}\|z'(T)\|_{L^2(\Omega\times[t-s,t])}$$
$$+2\int_{t-s}^{t}\int_{\Omega}\int_{-\infty}^{0}\left|G(T,r,u^1(T+r,x))-G(T,r,u^2(T+r,x))\right|\left|z'(T)\right|drdxdT$$
$$\le 2\|f(T,u^1(T-\rho(T)))-f(T,u^2(T-\rho(T))\|_{L^2(\Omega\times[t-s,t])}\|z'(T)\|_{L^2(\Omega\times[t-s,t])}$$
$$+2\|\,\|G(T,r,u^1(T+r,x))-G(T,r,u^2(T+r,x))\|_{L^1(\mathbb{R}_-)}\|_{L^2(\Omega\times[t-s,t])}$$
$$\times\|z'(T)\|_{L^2(\Omega\times[t-s,t])}. \tag{3.12}$$

Let $u_{nT}\in U(T,t-s)\varphi_n$ with $\varphi_n\in Q(t-s)$.

Thanks to the step 4 of Theorem 2.3 and (3.1) and (3.5), we have

$u_n\to u$ weakly start in $L^\infty(t-s-T^*,t;H_0^1(\Omega))$, $u'_n\to u'$ weakly start in

$L^{\infty}(t-s-T^{*},t;L^{2}(\Omega))$.

Hence,

$$u_n \to u \quad \text{in} \quad L^2(t-s-T^*,t;L^2(\Omega)),$$

and

$$u_n(T,x) \to u(T,x) \quad \text{in} \quad L^2(\Omega) \text{a. e.} (T,x) \in [t-s-T^*,t] \times \Omega.$$

Since $f \in C(\mathbb{R} \times \mathbb{R};\mathbb{R})$, $G \in C(\mathbb{R} \times \mathbb{R} \times \mathbb{R};\mathbb{R})$, then

$$f(T,u_n(T,x)) \to f(T,u(T,x)) \text{in } L^2(\Omega) \text{a. e.} (T,x) \in [t-s-T^*,t] \times \Omega.$$

$G(T,r,u_n(T+r,x)) \to G(T,r,u(T+r,x))$ in $L^2(\Omega)$ a. e. $(T,r,x) \in [t-s,t] \times [-T^*,0] \times \Omega$. By Lebesgue dominated convergence theorem, we have

$$\lim_{n\to\infty}\lim_{m\to\infty} \| f(T,u_n(T-\rho(T))) - f(T,u_m(T-\rho(T))) \|_{L^2(\Omega\times[t-s,t])} = 0, \tag{3.13}$$

and

$$\lim_{n\to\infty}\lim_{m\to\infty} \| \| G(T,r,u_n(T+r,x)) - G(T,r,u_m(T+r,x)) \|_{L^1(\mathbb{R}_-)} \|_{L^2(\Omega\times[t-s,t])} = 0. \tag{3.14}$$

Thus, for ε above, and for every $s>0$, for all $t \in [t-s,\infty)$, one has

$$\begin{aligned} & \| z'(t+\theta) \|_{L^2(\Omega)}^2 + \| \nabla z(t+\theta) \|_{L^2(\Omega)}^2 \\ = & \max\{ \sup_{\theta\in(-\infty,-T^*]} e^{2\gamma\theta}(\| z'(t+\theta) \|_{L^2(\Omega)}^2 + \| \nabla z(t+\theta) \|_{L^2(\Omega)}^2), \\ & \sup_{\theta\in(-T^*,0]} \{ e^{2\gamma\theta} \| z'(t+\theta) \|_{L^2(\Omega)}^2 + \| \nabla z(t+\theta) \|_{L^2(\Omega)}^2 \} \} \\ \leq & \max\{ \varepsilon, \max_{\theta\in(-T^*,0]} \{ \| z'(\theta) \|_{L^2(\Omega)}^2 + \| \nabla z(\theta) \|_{L^2(\Omega)}^2 \} \}. \end{aligned} \tag{3.15}$$

Indeed, by(3.5), we have

$$\sup_{\theta\in(-\infty,-T^*]} e^{2\gamma\theta}(\| z'(t+\theta) \|_{L^2(\Omega)}^2 + \| \nabla z(t+\theta) \|_{L^2(\Omega)}^2) \leq \varepsilon.$$

Combining with(3.12)-(3.15) together, we conclude that

$$\begin{aligned} & e^{2\gamma\theta}(\| z_n'(t+\theta) - z_m'(t+\theta) \|_{L^2(\Omega)}^2 \\ & + \| \nabla z_n(t+\theta) - \nabla z_m(t+\theta) \|_{L^2(\Omega)}^2) e^{2\gamma\theta}(\| w_n'(t+\theta) - w_m'(t+\theta) \|_{L^2(\Omega)}^2 \\ & + \| \nabla w_n(t+\theta) - \nabla w_m(t+\theta) \|_{L^2(\Omega)}^2) \\ \leq & \max\{ \varepsilon, \max_{\theta\in(-T^*,0]} \{ (\| w_n'(\theta) - w_m'(\theta) \|_{L^2(\Omega)}^2 + \| \nabla w_n(\theta) - \nabla w_m(\theta) \|_{L^2(\Omega)}^2) \} \} \\ \leq & \varepsilon, \text{as } n \to \infty, m \to \infty, \end{aligned} \tag{3.16}$$

where

$$\begin{aligned} & \max_{\theta\in(-T^*,0]} \{ \| w_n'(\theta) - w_m'(\theta) \|_{L^2(\Omega)}^2 + \| \nabla w_n(\theta) - \nabla w_m(\theta) \|_{L^2(\Omega)}^2 \} \\ \leq & 2 \| f(T,u_n(T-\rho(T))) \\ & - f(T,u_m(T-\rho(T))) \|_{L^2(\Omega\times[t-s,t])} \| w_n'(T) \|_{L^2(\Omega\times[t-s,t])} \end{aligned}$$

$$+2\|\ \|G(T,s,u^1(T+s,x))-G(T,s,u^2(T+s,x))\|_{L^1(\mathbb{R})}\ \|_{L^2(\Omega\times[t-s,t])}$$
$$\times\|w_n'(T)\|_{L^2(\Omega\times[t-s,t])}\leq\varepsilon,\text{as } n\to\infty,m\to\infty.$$

Thus (2) in Theorem 3. 5 follows immediately.

Step 2. By lemma(3. 6) and the above statements, we can conclude that the process $\{U(t,\tau)\}$ is D —pullback asymptotically compact in $C_{\gamma,H_0^1(\Omega),L^2(\Omega)}$ and U has a closed bounded $D_{C_{\gamma,H^1(\Omega),L^2(\Omega)}}$ —pullback absorbing set $Q_{C_{\gamma,H^1(\Omega),L^2(\Omega)}}$

in$C_{\gamma,H_0^1(\Omega),L^2(\Omega)}$. Let

$$A_{D,C_{\gamma,H^1(\Omega),L^2(\Omega)}}(t)=\omega(Q_{C_{\gamma,H^1(\Omega),L^2(\Omega)}},t)=\bigcap_{T\in\mathbb{R}^-}\overline{\bigcup_{s\geq T}U(t,t-s)Q(t-s)},\forall t\in\mathbb{R}.$$

In order to show $\{A_{D,C_{\gamma,H^1(\Omega),L^2(\Omega)}}(t)\}_{t\in\mathbb{R}}$ is a D —pullback attractor, by Theorem (3. 4), it only remains to prove the negative invariance of $\{A_{D,C_{\gamma,H^1(\Omega),L^2(\Omega)}}(t)\}_{t\in\mathbb{R}}$.

Let$y\in A_{D,C_{\gamma,H^1(\Omega),L^2(\Omega)}}(t)$. Then there exist sequence$s_n\in\mathbb{R}^+$, $s_n\to+\infty$ as$n\to+\infty$, $x_n\in Q(t-s_n)$, and $y_n=U(t,t-s_n)x_n$ such that

$$y_n\to y\text{in}C_{\gamma,H_0^1(\Omega),L^2(\Omega)}\ \text{as}n\to+\infty. \tag{3. 17}$$

On the other hand, for$n$ sufficiently large,

$y_n=U(t,t-s_n)x_n=U(t,\tau)U(\tau,t-s_n)x_n$. Then by the D —pullback asymptotically compactness of the process of $\{U(t,\tau)\}$, there is a subsequence of $\tilde{x}_n=U(\tau,t-s_n)x_n=U(\tau,\tau-(\tau+s_n-t))x_n$ (relabelled the same), such that$y_n=U(t,t-s_n)\tilde{x}_n$ and

$$\tilde{x}_n\to x\text{in}C_{\gamma,H_0^1(\Omega),L^2(\Omega)}\ \text{as}n\to+\infty. \tag{3. 18}$$

Clearly, $x\in A_{D,C_{\gamma,H^1(\Omega),L^2(\Omega)}}(\tau)$.

We observe that

$$y_n(\theta)=\begin{cases}\tilde{x}_n(\theta+t-\tau),\text{if }\theta\in(-\infty,\tau-t),\\ u_n(\theta+t,\tau,\tilde{x}_n),\text{if }\theta\in[\tau-t,0],\end{cases} \tag{3. 19}$$

and y_n is bounded in $C_{\gamma,H_0^1(\Omega),L^2(\Omega)}$ for n sufficiently large.

Then $u_n(\theta+t,\tau,\tilde{x}_n)$ is bounded in $C_{[\tau-t,0];H_0^1(\Omega),L^2(\Omega)}$. By slightly modifying the proof of the existence of solutions of Theorem 2. 3, we can see that from (3. 18)

$$u_n(\theta+t,\tau,\tilde{x}_n)\rightharpoonup u(\theta+t,\tau,x_n),\text{in }L^2(\tau-t;H_0^1(\Omega))\cap W^{1,2}(\tau-t;L^2(\Omega)), \tag{3. 20}$$

where$u(\cdot)$ is a solution of (1. 1). This together with (3. 17)—(3. 20), we can deduce that$y=U(t,\tau)x\subset U(t,\tau)A_{D,C_{\gamma,H^1(\Omega),L^2(\Omega)}}(\tau)$. The proof is completed.

The connectedness follows from Propositions 15[15] and Propositions 2.5 and the fact the space $C_{\gamma,H_0^1(\Omega),L^2(\Omega)}$ is connected.

4 Existence of pullback attractors in $C_{\gamma,D(A),H_0^1(\Omega)}$

Theorem 4.1

Suppose that the hypotheses in Lemma(3.6) holds, $f\in C^1(\mathbb{R}\times\mathbb{R};\mathbb{R})$, $G\in C^1(\mathbb{R}\times\mathbb{R}\times\mathbb{R};\mathbb{R})$, satisfy (2.4)and (2.5),respectively,and that

$$\int_{-\infty}^{t}\left|\nabla g(r)\right|^2 e^{\delta r}<\infty,\forall t\in R,\delta>0. \tag{4.1}$$

Then the process $\{U(t,\tau)\}$ corresponding to equation (1.1) possesses a $D_{C_{\gamma,D(A),H_0^1(\Omega)}}$ −pullback absorbing sets in$D_{C_{\gamma,D(A),H_0^1(\Omega)}}$.

Proof

Our proof relies on an energy method,analogous to that employed in step 1 in the proof of Theorem 2.3.

Let$v=u'+\alpha u$. Taking the scalar product of (1.1) with$-\triangle v$, we have

$$\frac{1}{2}\frac{d}{dt}(\|\nabla v(t)\|_{L^2(\Omega)}^2+\|\triangle u(t)\|_{L^2(\Omega)}^2)+\alpha\|\triangle u(t)\|_{L^2(\Omega)}^2$$

$$+(1-\alpha)\|\nabla v(t)\|_{L^2(\Omega)}^2-\alpha(1-\alpha)(u(t),-\triangle v(t))$$

$$=(f(t,u(t-\rho(t)))+\int_{-\infty}^{0}G(t,r,u(t+r,x))dr+g(t,x),-\triangle v(t)).$$

By $\|\triangle u\|_{L^2(\Omega)}^2\geq\lambda_1\|\nabla u\|_{L^2(\Omega)}^2$, (2.4),(2.5) and Young's inequality,we can deduce that

$$\alpha\|\triangle u(t)\|_{L^2(\Omega)}^2+(1-\alpha)\|\nabla v(t)\|_{L^2(\Omega)}^2-\alpha(1-\alpha)(u(t),-\triangle v(t))$$

$$\geq\frac{\alpha}{2}\|\triangle u(t)\|_{L^2(\Omega)}^2+(1-\alpha-\frac{\alpha}{2\lambda_1})\|\nabla v(t)\|_{L^2(\Omega)}^2,$$

$$-\int_\Omega f(t,u(t-\rho(t)))\triangle v(t)dx=\int_\Omega f'_u(t,u(t-\rho(t)))\nabla u\nabla v(t)dx$$

$$\leq\widetilde{L}_f'\int_\Omega\left|u(t-\rho(t))\right||\nabla v(t)|dx\leq\frac{\widetilde{L}_f^{2\prime}}{2(1-\alpha-\frac{\alpha}{2\lambda_1})}$$

$$\|u_t\|_{C_{\gamma,L^2(\Omega)}}^2+\frac{1-\alpha-\frac{\alpha}{2\lambda_1}}{2}\|\nabla v(t)\|_{L^2(\Omega)}^2$$

$$\leq\frac{\widetilde{L}_f^{2\prime}}{(1-\alpha-\frac{\alpha}{2\lambda_1})}\|u_t\|_{C_{\gamma,L^2(\Omega)}}^2+\frac{1-\alpha-\frac{\alpha}{2\lambda_1}}{2}\|\nabla v(t)\|_{L^2(\Omega)}^2$$

$$-\int_{\Omega} g(t)\triangle v(t)dx\% = \int_{\Omega} \nabla g(t)\ \nabla v(t)dx$$

$$\leq \int_{\Omega} \left| \nabla g(t) \right| \left| \nabla v(t) \right| dx \leq \frac{2}{1-\alpha-\frac{\alpha}{2\lambda_1}} \| \nabla g(t) \|_{L^2(\Omega)}^2$$

$$+\frac{1-\alpha-\frac{\alpha}{2\lambda_1}}{8} \| \nabla v(t) \|_{L^2(\Omega)}^2,$$

$$-\int_{\Omega}\int_{-\infty}^{0} G(t,r,u(t+r,x))dr\triangle v(t)dx$$

$$=\int_{\Omega}\int_{-\infty}^{0} G'_u(t,r,u(t+r,x))\ \nabla u(t+r,x)dr\ \nabla v(t)dx$$

$$\leq (\int_{\Omega}(\int_{-\infty}^{0} G'_u(t,r,u(t+r,x))\ \nabla u(t+r,x)dr)^2 dx)^{\frac{1}{2}} \| \nabla v(t) \|_{L^2(\Omega)}$$

$$\leq \widetilde{L}_G' \| \nabla u(t+r,x) \|_{L^2(\Omega)} \| \nabla v(t) \|_{L^2(\Omega)} \leq \frac{2\widetilde{L}_G^{2\,\prime}}{1-\alpha-\frac{\alpha}{2\lambda_1}} \| \nabla u_t \|_{C_{\gamma,L^2(\Omega)}}^2$$

$$+\frac{1-\alpha-\frac{\alpha}{2\lambda_1}}{8} \| \nabla v(t) \|_{L^2(\Omega)}^2,$$

where

$$(\int_{\Omega}(\int_{-\infty}^{0} G'_u(t,r,u(t+r,x))\ \nabla u(t+r,x)dr)^2 dx)^{\frac{1}{2}}$$

$$\leq (\int_{\Omega}(\int_{-\infty}^{0} \left| G'_u(t,r,u(t+r,x)) \right| e^{-\frac{3\gamma r}{2}} e^{\frac{3\gamma r}{2}} \left| \nabla u(t+r,x) \right| dr)^2 dx)^{\frac{1}{2}}$$

$$\leq (\int_{\Omega}(\int_{-\infty}^{0} \left| G'_u(t,r,u(t+r,x)) \right|^2 e^{-3\gamma r}dr \int_{-\infty}^{0} e^{3\gamma r} \left| \nabla u(t+r,x) \right|^2 dr)dx)^{\frac{1}{2}}$$

$$\leq (\int_{-\infty}^{0} \left| G'_u(t,r,u(t+r,x)) \right|^2 e^{-3\gamma r}dr \int_{\Omega}(\int_{-\infty}^{0} e^{3\gamma r} \left| \nabla u(t+r,x) \right|^2 ds)dx)^{\frac{1}{2}}$$

$$\leq (\int_{-\infty}^{0} \left| G'_u(t,r,u(t+r,x)) \right|^2 e^{-3\gamma r}dr \int_{-\infty}^{0}\int_{\Omega}(e^{3\gamma r} \left| \nabla u(t+r,x) \right|^2 dxdr))^{\frac{1}{2}}$$

$$\leq (\int_{-\infty}^{0} \left| G'_u(t,r,u(t+r,x)) \right|^2 e^{-3\gamma r}dr \int_{-\infty}^{0}\int_{\Omega}(e^{3\gamma r} \left| \nabla u(t+r,x) \right|^2 dxdr))^{\frac{1}{2}}$$

$$\leq (\sup_{r\in(-\infty,0]} e^{2\gamma r} \| \nabla u(t+r,x) \|_{L^2(\Omega)}^2 \int_{-\infty}^{0} \left| G'_u(t,r,u(t+r,x)) \right|^2 e^{-3\gamma r}dr \int_{-\infty}^{0} e^{2\gamma r}dr)^{\frac{1}{2}}$$

$$\leq \widetilde{L}_G' \| \nabla u_t \|_{C_{\gamma,L^2(\Omega)}}.$$

Hence,

$$\frac{d}{dt}(\| \nabla v(t) \|_{L^2(\Omega)}^2 + \| \triangle u(t) \|_{L^2(\Omega)}^2)$$

$$+\alpha\|\triangle u(t)\|_{L^2(\Omega)}^2+(1-\alpha-\frac{\alpha}{2\lambda_1})\|\nabla v(t)\|_{L^2(\Omega)}^2$$

$$\leq\frac{4(\widetilde{L}_f^{2\prime}+\widetilde{L}_G^{2\prime})}{1-\alpha-\frac{\alpha}{2\lambda_1}}\|\nabla u_t\|_{C_{\gamma,L^2(\Omega)}}^2+\frac{4}{1-\alpha-\frac{\alpha}{2\lambda_1}}\|\nabla g\|_{L^2(\Omega)}^2.$$

Choosing $\alpha\leq\frac{2}{4+\frac{1}{\lambda_1}}$, we have $1-\frac{3\alpha}{2}-\frac{\alpha}{2\lambda_1}\geq\frac{\alpha}{2}$. Then

$$\frac{d}{dt}e^{\frac{\alpha}{2}t}(\|\nabla v(t)\|_{L^2(\Omega)}^2+\|\triangle u(t)\|_{L^2(\Omega)}^2)$$

$$=\frac{\alpha}{2}e^{\frac{\alpha}{2}t}(\|\nabla v(t)\|_{L^2(\Omega)}^2+\|\triangle u(t)\|_{L^2(\Omega)}^2)$$

$$+e^{\frac{\alpha}{2}t}\frac{d}{dt}(\|\nabla v(t)\|_{L^2(\Omega)}^2+\|\triangle u(t)\|_{L^2(\Omega)}^2)$$

$$\leq-\frac{\alpha}{2}e^{\frac{\alpha}{2}t}\|\triangle u(t)\|_{L^2(\Omega)}^2-(1-\frac{3}{2}\alpha-\frac{\alpha}{2\lambda_1})e^{\frac{\alpha}{2}t}\|\nabla v(t)\|_{L^2(\Omega)}^2$$

$$+\frac{4(\widetilde{L}_f^{2\prime}+\widetilde{L}_G^{2\prime})}{1-\alpha-\frac{\alpha}{2\lambda_1}}e^{\frac{\alpha}{2}t}\|\nabla u_t\|_{C_{\gamma,L^2(\Omega)}}^2+\frac{4}{1-\alpha-\frac{\alpha}{2\lambda_1}}\|\nabla g\|_{L^2(\Omega)}^2$$

$$\leq-\frac{\alpha}{2}e^{\frac{\alpha}{2}t}(\|\triangle u(t)\|_{L^2(\Omega)}^2+\|\nabla v(t)\|_{L^2(\Omega)}^2)$$

$$+Ce^{\frac{\alpha}{2}t}\|\nabla u_t\|_{C_{\gamma,L^2(\Omega)}}^2+Ce^{\frac{\alpha}{2}t}\|\nabla g\|_{L^2(\Omega)}^2.$$

By Gronwall Lemma and(3.1),it follows

$$\|\nabla v(t)\|_{L^2(\Omega)}^2+\|\triangle u(t)\|_{L^2(\Omega)}^2\leq e^{-\alpha t}e^{\alpha\tau}(\|\nabla v(\tau)\|_{L^2(\Omega)}^2+\|\triangle u(\tau)\|_{L^2(\Omega)}^2)$$

$$+Ce^{-\frac{\alpha}{2}t}\int_\tau^t e^{\frac{\alpha}{2}s}\|\nabla u_s\|_{C_{\gamma,L^2(\Omega)}}^2ds+Ce^{-\frac{\alpha}{2}t}\int_\tau^t e^{\frac{\alpha}{2}s}\|\nabla g\|_{L^2(\Omega)}^2ds$$

$$\leq e^{-\alpha t}e^{\alpha\tau}(\|\nabla v(\tau)\|_{L^2(\Omega)}^2+\|\triangle u(\tau)\|_{L^2(\Omega)}^2)+Ce^{-\alpha t}\|\varphi\|_{C_{\gamma,H(\Omega),L^2(\Omega)}}^2$$

$$+C+C\int_\tau^t e^{\alpha s}\|g(s)\|_{L^2(\Omega)}^2ds$$

$$+Ce^{-(\alpha-(\frac{4k_1^2}{(\alpha+2m_1)\lambda_1}+\frac{m_1}{\lambda_1}))t}\|\varphi\|_{C_{\gamma,H_0^1(\Omega),L^2(\Omega)}}^2$$

$$+Ce^{-(\alpha-(\frac{4k_1^2}{(\alpha+2m_1)\lambda_1}+\frac{m_1}{\lambda_1}))t}\int_\tau^t e^{(\alpha-(\frac{4k_1^2}{(\alpha+2m_1)\lambda_1}+\frac{m_1}{\lambda_1}))s}\|g(s)\|_{L^2(\Omega)}^2ds$$

$$+Ce^{-\frac{\alpha}{2}t}\int_\tau^t e^{\frac{\alpha}{2}s}\|\nabla g\|_{L^2(\Omega)}^2ds.$$

(2.2)Setting $t+\theta$ instead of t (where $\theta\leq0$),by and (2.3) it yields

$$e^{2\gamma\theta}\|\nabla u(t+\theta)\|_{L^2(\Omega)}^2+\|\triangle u(t+\theta)\|_{L^2(\Omega)}^2$$

$$\leq C^2(\alpha)e^{-\alpha t}e^{\alpha\tau}(\|\nabla u(\tau)\|_{L^2(\Omega)}^2+\|\triangle u(\tau)\|_{L^2(\Omega)}^2)+C(\alpha)Ce^{-\alpha t}\|\varphi\|_{C_{\gamma,H_0^1(\Omega),L^2(\Omega)}}^2$$

$$+CC(\alpha)+CC(\alpha)\int_{-\infty}^{t}e^{\alpha s}\|g(s)\|_{L^2(\Omega)}^2ds+CC(\alpha)e^{-(\alpha-(\frac{4k_1^2}{(\alpha+2m_1)\lambda_1}+\frac{m}{\lambda_1}))t}\|\varphi\|_{C_{\gamma,H^1(\Omega),L^2(\Omega)}}^2$$

$$+CC(\alpha)e^{-(\alpha-(\frac{4k_1^2}{(\alpha+2m_1)\lambda_1}+\frac{m}{\lambda_1}))t}\times\int_{-\infty}^{t}e^{(\alpha-(\frac{4k_1^2}{(\alpha+2m_1)\lambda_1}+\frac{m}{\lambda_1}))s}\|g(s)\|_{L^2(\Omega)}^2ds$$

$$+CC(\alpha)e^{-\frac{\alpha}{2}t}\int_{-\infty}^{t}e^{\frac{\alpha}{2}s}\|\nabla g\|_{L^2(\Omega)}^2ds.$$

Hence,

$$\|\nabla u_t\|_{C_{\gamma,L^2(\Omega)}}^2+\|\triangle u_t\|_{C_{\gamma,L^2(\Omega)}}^2\le(2+C^2(\alpha))e^{-\alpha t}e^{\alpha\tau}\|\varphi\|_{C_{\gamma,H^1(\Omega),L^2(\Omega)}}^2$$

$$+C(\alpha)Ce^{-\alpha t}\|\varphi\|_{C_{\gamma,H^1(\Omega),L^2(\Omega)}}^2+CC(\alpha)+CC(\alpha)\int_{-\infty}^{t}e^{\alpha s}\|g(s)\|_{L^2(\Omega)}^2ds$$

$$+CC(\alpha)e^{-(\alpha-(\frac{4k_1^2}{(\alpha+2m_1)\lambda_1}+\frac{m}{\lambda_1}))t}\|\varphi\|_{C_{\gamma,H^1(\Omega),L^2(\Omega)}}^2+CC(\alpha)e^{-(\alpha-(\frac{4k_1^2}{(\alpha+2m_1)\lambda_1}+\frac{m}{\lambda_1}))t}$$

$$\times\int_{-\infty}^{t}e^{(\alpha-(\frac{4k_1^2}{(\alpha+2m_1)\lambda_1}+\frac{m}{\lambda_1}))s}\|g(s)\|_{L^2(\Omega)}^2ds+CC(\alpha)e^{-\frac{\alpha}{2}t}\int_{-\infty}^{t}e^{\frac{\alpha}{2}s}\|\nabla g\|_{L^2(\Omega)}^2ds$$

$$\le(C+C^2(\alpha))e^{-\alpha t}e^{\alpha\tau}\|\varphi\|_{C_{\gamma,H^1(\Omega),L^2(\Omega)}}^2+CC(\alpha)+CC(\alpha)\int_{-\infty}^{t}e^{\alpha s}\|g(s)\|_{L^2(\Omega)}^2ds$$

$$+CC(\alpha)e^{-(\alpha-(\frac{4k_1^2}{(\alpha+2m_1)\lambda_1}+\frac{m}{\lambda_1}))t}\|\varphi\|_{C_{\gamma,H^1(\Omega),L^2(\Omega)}}^2+CC(\alpha)e^{-(\alpha-(\frac{4k_1^2}{(\alpha+2m_1)\lambda_1}+\frac{m}{\lambda_1}))t}$$

$$\times\int_{-\infty}^{t}e^{(\alpha-(\frac{4k_1^2}{(\alpha+2m_1)\lambda_1}+\frac{m}{\lambda_1}))s}\|g(s)\|_{L^2(\Omega)}^2ds+CC(\alpha)e^{-\frac{\alpha}{2}t}\int_{-\infty}^{t}e^{\frac{\alpha}{2}s}\|\nabla g\|_{L^2(\Omega)}^2ds.\ (4.2)$$

Denote the family $Q_{C_{\gamma,D(A),H_0^1(\Omega)}}=\{Q(t):t\in\mathbb{R}\}$, with$Q_t=\overline{Q_{C_{\gamma,D(A),H_0^1(\Omega)}}(0,\rho(t))}$, where

$$\rho^2(t)=C\|\varphi\|_{C_{\gamma,H^1(\Omega),L^2(\Omega)}}^2+C\int_{-\infty}^{t}e^{\alpha s}\|g(s)\|_{L^2(\Omega)}^2ds+C\int_{-\infty}^{t}e^{\frac{\alpha}{2}s}\|\nabla g\|_{L^2(\Omega)}^2ds$$

$$+Ce^{-(\alpha-(\frac{4k_1^2}{(\alpha+2m_1)\lambda_1}+\frac{m}{\lambda_1}))t}\int_{-\infty}^{t}e^{(\alpha-(\frac{4k_1^2}{(\alpha+2m_1)\lambda_1}+\frac{m}{\lambda_1}))s}\|g(s)\|_{L^2(\Omega)}^2ds,$$

is $D_{C_{\gamma,D(A),H_0^1(\Omega)}}$ −pullback absorbing for bounded sets for the process $\{U(t,\tau)\}$.

Theorem 4.2 Under the assumptions of Theorem4.1 and let Ω be a bounded domain in

$\mathbb{R}^n$with smooth boundary. The process $\{U(t,\tau)\}$ on $C_{\gamma,D(A),H_0^1(\Omega)}$ has a unique $D_{C_{\gamma,D(A),H_0^1(\Omega)}}$ −pullback attractor $\{A_{D,C_{\gamma,D(A),H_0^1(\Omega)}}(t)\}_{t\in\mathbb{R}}$.

Proof

By theorem4.1 and theorem 3.5, we need only show that the process $\{U(t,\tau)\}$ on $C_{\gamma,D(A),H_0^1(\Omega)}$ is $D_{C_{\gamma,D(A),H_0^1(\Omega)}}$ −pullback asymptotically compact, since the proof for the invariance of pullback attractors are similar to the arguments in Theorem 3.7.

First, the condition (H3) and(4.2) implies that we can fixed T^* large enough such that for all $s\ge0$ and$u_t\in U(t,t-s)D(t-s)$,

$$\sup_{\theta\in(-\infty,-T^*]} e^{2\gamma\theta}(\|\Delta u(t+\theta)\|^2_{L^2(\Omega)}+\|\nabla u'(t+\theta)\|^2_{L^2(\Omega)})<\varepsilon.$$

By(3.7) and (3.8) and (4.2), since $f=g=G=0$, it yields $\|v_t\|^2_{C_{\gamma,D(A),H_0^1(\Omega)}}\le Ce^{-\alpha s}\|\varphi\|^2_{C_{\gamma,D(A)}}$. This implies (1) in Theorem 3.5, now we only verify (2) in Theorem 3.5.

Taking the inner product in $L^2(\Omega)$ of(3.10) with $-\Delta z'(T)$, we get

$$\frac{1}{2}\frac{d}{dT}(\|\nabla z'(T)\|^2_{L^2(\Omega)}+\|\Delta z(T)\|^2_{L^2(\Omega)})+\|\nabla z'(T)\|^2_{L^2(\Omega)}$$
$$=(f(T,u^1(T-\rho(T)))-f(T,u^2(T-\rho(T)))$$
$$+\int_{-\infty}^0(G(T,s,u^1(T+s,x))-G(T,s,u^2(T+s,x)))ds,-\Delta z'(T)).$$

Integrating from $t-s$ to t (where $\theta\in(-\infty,0]$) above T, we have

$$\|\nabla z'(t)\|^2_{L^2(\Omega)}+\|\Delta z(t)\|^2_{L^2(\Omega)}$$
$$\le 2\int_{t-s}^t(f(T,u^1(T-\rho(T)))-f(T,u^2(T-\rho(T))),$$
$$-\Delta z'(T))dT+\int_{t-s}^t\int_{-\infty}^0(G(T,r,u^1(T+r,x))$$
$$-G(T,r,u^2(T+r,x)))dr,-\Delta z'(T))dT$$
$$\le 2\int_{t-s}^t\int_\Omega\left|f(T,u^1(T-\rho(T)))-f(T,u^2(T-\rho(T))\right|\,|-\Delta z'(T)|\,dxdT$$
$$+2\int_{t-s}^t\int_\Omega\int_{-\infty}^0|G(T,r,u^1(T+r,x))$$
$$-G(T,r,u^2(T+r,x))|\,|-\Delta z'(T)|\,drdxdT,$$

and

$$\int_{t-s}^t\int_\Omega f(T,u^1(T-\rho(T)))-f(T,u^2(T-\rho(T))(-\Delta z'(T))dxdT$$
$$=\int_{t-s}^t\int_\Omega(f'(T,u^1(T-\rho(T)))\nabla u^1(T-\rho(T))$$
$$-f'(T,u^2(T-\rho(T)))\nabla u^2(T-\rho(T)))$$
$$\times\nabla z'(T)dxdT=\int_{t-s}^t\int_\Omega f'(T,u^1(T-\rho(T)))(\nabla u^1(T-\rho(T))$$
$$-\nabla u^2(T-\rho(T)))\nabla z'(T)dxdT$$
$$+\int_{t-s}^t\int_\Omega(f'(T,u^1(T-\rho(T)))-f'(T,u^2(T-\rho(T))))$$
$$\nabla u^2(T-\rho(T))\nabla z'(T)dxdT$$
$$\le\widetilde{L}_f'\|\nabla u^1(T-\rho(T))-\nabla u^2(T-\rho(T))\|_{L^2(\Omega\times[t-s,t])}\|\nabla z'(T)\|_{L^2(\Omega\times[t-s,t])}$$
$$+\|f(T,u^1(T-\rho(T)))-f(T,u^2(T-\rho(T))\|_{L^3(\Omega\times[t-s,t])}$$

$$\times \| \nabla u^2(T-\rho(T)) \|_{L^6(\Omega\times[t-s,t])} \| \nabla z'(T) \|_{L^2(\Omega\times[t-s,t])},$$

and

$$\int_{t-s}^{t}\int_{\Omega}\int_{-\infty}^{0} G(T,r,u^1(T+r,x)) - G(T,r,u^2(T+r,x))(-\Delta z'(T))drdxdT$$

$$= \int_{t-s}^{t}\int_{\Omega}\int_{-\infty}^{0} (G'(T,r,u^1(T+r,x)) \nabla u^1(T+r,x) - G'(T,r,u^2(T+r,x))$$

$$\times \nabla u^1(T+r,x)) \nabla z'(T)drdxdT$$

$$= \int_{t-s}^{t}\int_{\Omega}\int_{-\infty}^{0} G'(T,r,u^1(T+r,x))(\nabla u^1(T+r,x)$$

$$- \nabla u^1(T+r,x)) \nabla z'(T)drdxdT$$

$$+ \int_{t-s}^{t}\int_{\Omega}\int_{-\infty}^{0} (G'(T,r,u^1(T+r,x)) - G'(T,r,u^2(T+r,x)))$$

$$\times \nabla u^2(T+r,x) \nabla z'(T)drdxdT$$

$$\le \widetilde{L}_G' \| \nabla u_T^1 - \nabla u_T^2 \|_{C_{\gamma,L^2(\Omega\times[t-s,t])}} \| \nabla z'(T) \|_{L^2(\Omega\times[t-s,t])}$$

$$+ \| \| G'(T,r,u^1(T+r,x)) - G'(T,r,u^2(T+r,x)) \|_{L^1([t-s,t])} \|_{L^3(\Omega\times[t-s,t])}$$

$$\times \| \nabla u_T^2 \|_{C_{\gamma,L^2(\Omega\times[t-s,t])}} \| \nabla z'(T) \|_{L^2(\Omega\times[t-s,t])}.$$

Replacing t by$t+\theta$, by the boundary condition of $z'(t)$ in(3. 10), we deduce that

$$\| \nabla z'(t+\theta) \|_{L^2(\Omega)}^2 + \| \Delta z(t+\theta) \|_{L^2(\Omega)}^2$$

$$\le -2\int_{t+\theta-s}^{t+\theta}\int_{\Omega} (f(T,u^1(T-\rho(T))) - f(T,u^2(T-\rho(T)))\Delta z'(T)dxdT$$

$$-2\int_{t+\theta-s}^{t+\theta}\int_{\Omega}\int_{-\infty}^{0} G(T,r,u^1(T+s,x)) - G(T,r,u^2(T+s,x))\Delta z'(T)dsdxdT$$

$$\le 2\widetilde{L}_f' \| \nabla u^1(T-\rho(T)) - \nabla u^2(T-\rho(T)) \|_{L^2(\Omega\times[t-s,t])} \| \nabla z'(T) \|_{L^2(\Omega\times[t-s,t])}$$

$$+2 \| f(T,u^1(T-\rho(T))) - f(T,u^2(T-\rho(T)) \|_{L^3(\Omega\times[t-s,t])}$$

$$\times \| \nabla u^2(T-\rho(T)) \|_{L^6(\Omega\times[t-s,t])} \| \nabla z'(T) \|_{L^2(\Omega\times[t-s,t])}$$

$$+ 2\widetilde{L}_G' \| \nabla u_T^1 - \nabla u_T^2 \|_{C_{\gamma,L^2(\Omega\times[t-s,t])}} \| \nabla z'(T) \|_{L^2(\Omega\times[t-s,t])}$$

$$+2 \| \| G'(T,r,u^1(T+s,x)) - G'(T,r,u^2(T+r,x)) \|_{L^1([t-s,t])} \|_{L^3(\Omega\times[t-s,t])}$$

$$\times \| \nabla u_T^2 \|_{C_{\gamma,L^2(\Omega\times[t-s,t])}} \| \nabla z'(T) \|_{L^2(\Omega\times[t-s,t])}.$$

Let $u_{nT} \in U(T,t-s)\varphi_n$ with $\varphi_n \in Q(t-s)$, where $Q_{C_{\gamma,D(A),H_0^1(\Omega)}} = \{Q(t)\}_{t\in\mathbb{R}}$ is a $D_{C_{\gamma,D(A),H_0^1(\Omega)}}$ $-$pullback absorbing set of the process $\{U(t,\tau)\}$ on $D_{C_{\gamma,D(A),H_0^1(\Omega)}}$.

Thanks to (2) of Theorem2. 3 and (4. 2), we have

$u_n \to u$ weakly start in $L^\infty(t-s-T^*,t;D(A))$, $u'_n \to u'$ weakly start in $L^\infty(t-s-T^*,t;H_0^1(\Omega))$.

Hence,

$$u_n \to u \quad \text{in} \quad L^2(t-s-T^*,t;L^2(\Omega)),$$

and

$$u_n(T,x) \to u(T,x) \text{in } L^2(\Omega) \text{a. e.} (T,x) \in [t-s-T^*,t] \times \Omega.$$

Since $f \in C(\mathbb{R} \times \mathbb{R};\mathbb{R})$, $G \in C(\mathbb{R} \times \mathbb{R} \times \mathbb{R};\mathbb{R})$, then

$$f(T,u_n(T,x)) \to f(T,u(T,x)) \text{ in } L^2(\Omega) \text{a. e.} (T,x) \in [t-s-T^*,t] \times \Omega.$$

$$G(T,r,u_n(T+r,x)) \to G(T,r,u(T+r,x))$$

$$\text{in} \quad C_{\gamma,L^2(\Omega)} \text{a. e.} (T,r,x) \in [t-s,t] \times [-T^*,0] \times \Omega.$$

By Lebesgue dominated convergence theorem, we have

$$\lim_{n\to\infty}\lim_{m\to\infty} \| f(T,u_n(T-\rho(T))) - f(T,u_m(T-\rho(T))) \|_{L^2(\Omega\times[t-s,t])} = 0, \tag{4.4}$$

and

$$\lim_{n\to\infty}\lim_{m\to\infty} \| \| G(T,r,u_n(T+r,x)) - G(T,r,u_m(T+r,x)) \|_{L^1(\mathbb{R})} \|_{L^2(\Omega\times[t-s,t])} = 0. \tag{4.5}$$

According to Aubin—Lions—Simon lemma, we know

$$\{L^\infty(0,T;H_0^1(\Omega)) \cap W^{1,\infty}(0,T;L^2(\Omega))\} L^m(0,T;L^2(\Omega))$$

is continuously embedding for any $1 < m < \infty$, $2 \le s \le \frac{2n}{n-2}$, since by (4.2), we know that

$$\| \nabla z'(T) \|^2_{L^2(\Omega\times[t-s,t])}, \quad \| \Delta u^2(T-\rho(T)) \|^2_{L^6(\Omega\times[t-s,t])} \text{are bounded.} \tag{4.6}$$

Thus, for $\varepsilon > 0$ above, and for every $s > 0$, for all $t \in [t-s,\infty)$, one has

$$\begin{aligned} & \| \nabla z'(t+\theta) \|^2_{L^2(\Omega)} + \| \Delta z(t+\theta) \|^2_{L^2(\Omega)} \\ = & \max\{ \sup_{\theta\in(-\infty,-T^*]} e^{2\gamma\theta}(\| \nabla z'(t+\theta) \|^2_{L^2(\Omega)} + \| \Delta z(t+\theta) \|^2_{L^2(\Omega)}), \\ & \sup_{\theta\in(-T^*,0]} \{e^{2\gamma\theta} \| \nabla z'(t+\theta) \|^2_{L^2(\Omega)} + \| \Delta z(t+\theta) \|^2_{L^2(\Omega)} \} \} \\ \le & \max\{\varepsilon, \max_{\theta\in(-T^*,0]} \{ \| \nabla z'(\theta) \|^2_{L^2(\Omega)} + \| \Delta z(\theta) \|^2_{L^2(\Omega)} \} \}. \end{aligned} \tag{4.7}$$

Indeed, by (4.3), we have

$$\sup_{\theta\in(-\infty,-T^*]} e^{2\gamma\theta}(\| \nabla z'(t+\theta) \|^2_{L^2(\Omega)} + \| \Delta z(t+\theta) \|^2_{L^2(\Omega)}) \le \varepsilon.$$

Combining with (4.4)—(4.7) together, we conclude that

$$\begin{aligned} & e^{2\gamma\theta}(\| \nabla w_n'(t+\theta) - \nabla w_m'(t+\theta) \|^2_{L^2(\Omega)} + \| \Delta w_n(t+\theta) - \Delta w_m(t+\theta) \|^2_{L^2(\Omega)}) \\ \le & \max\{\varepsilon, \max_{\theta\in(-T^*,0]} \{ \| \nabla w_n'(\theta) - \nabla w_m'(\theta) \|^2_{L^2(\Omega)} + \| \Delta w_n(\theta) - \Delta w_m(\theta) \|^2_{L^2(\Omega)} \} \} \\ \le & \varepsilon, \text{as} n \to \infty, m \to \infty, \end{aligned} \tag{4.8}$$

where

$$\max_{\theta\in(-T^*,0]}\{\|\nabla w_n'(\theta)-\nabla w_m'(\theta)\|^2_{L^2(\Omega)}+\|\Delta w_n(\theta)-\Delta w_m(\theta)\|^2_{L^2(\Omega)}\}$$

$$\le 2\tilde{L}_f'\|\nabla u^1(T-\rho(T))-\nabla u^2(T-\rho(T))\|_{L^2(\Omega\times[t-s,t])}\|\nabla z'(T)\|_{L^2(\Omega\times[t-s,t])}$$

$$+2\|f(T,u^1(T-\rho(T)))-f(T,u^2(T-\rho(T))\|_{L^3(\Omega\times[t-s,t])}$$

$$\times\|\nabla u^2(T-\rho(T))\|_{L^6(\Omega\times[t-s,t])}\|\nabla z'(T)\|_{L^2(\Omega\times[t-s,t])}$$

$$+2\tilde{L}_G'\|\nabla u_T^1-\nabla u_T^2\|_{C_{\gamma,L^2(\Omega\times[t-s,t])}}\|\nabla z'(T)\|_{L^2(\Omega\times[t-s,t])}$$

$$+2\|\|G'(T,r,u^1(T+s,x))-G'(T,r,u^2(T+r,x))\|_{L^1([t-s,t])}\|_{L^3(\Omega\times[t-s,t])}$$

$$\times\|\nabla u_T^2\|_{C_{\gamma,L^2(\Omega\times[t-s,t])}}\|\nabla z'(T)\|_{L^2(\Omega\times[t-s,t])}\le\varepsilon, \text{as} n\to\infty, m\to\infty.$$

The proof of Theorem(4. 2) is complete.

Acknowledgement

The authors are deeply grateful to the referee for his/her careful comments and valuable suggestions. The authors gratefully acknowledge the support of the National Natural Science Foundation of China (No. 11701244, 11772007, 11561060, 11661070, and 51468028), the Natural Science Foundation of Beijing (No. 1172002), the International Science and Technology Cooperation Program of China(No. 2014DFR61080), the Funding Project for Academic Human Resources Development in Institutions of Higher Learning under the Jurisdiction of Beijing Municipality(PHRIHLB). All authors wish to thank

professor Jing Li, Dr. Bin He and Yong Guan for many valuable suggestions leading to the improvement of this paper.

References

[1] Y. Kuang, Delay Differential Equations with Applications in Population Dynamics, Academic Press, Boston, 1993.

[2] J. D. Murray, Mathematical Biology, Springer—Verlag, Berlin, 1993.

[3] J. K. Hale, Asymptotic Behavior of Dissipative Systems, American Mathematical So—ciety, Providence, 1988.

[4] Y. Ran, Q. Shi, Klein—Gordon—Schröinger system: Dinucleon field, J. Math. Phys. 58(2017), 111509. 34

[5] A. Z. Manitius, Feedback controllers for a wind tunnel model involving a delay: an—alytical design and numerical simulation, IEEE Trans. Automat. Control 29 (1984): 1058—1068.

[6] Y. J. Wang, On the upper semicontinuity of pullback attractors for multi—valued pro

—cesses, Quart. Applied Math. ,71 (2013),369—399.

[7] T. Caraballo, P. Marn—Rubio, J. Valero, Attractors for differential equations with unbounded delays, J. Differential Equations 239 (2007) 311—342.

[8] G. Sell, Non—autonomous differential equations and dynamical systems, Amer. Math. Soc. 127(1967),241—283.

[9] G. Duvaut, J. L. Lions, Inequalities in Mechanics and Physics, Springer, 1976.

[10] V. Chepyzhov, M. Vishik, Attractors for equations of mathematical physics, American Mathematical Society Colloquium Publications, 49. American Mathematical Society, Providence, RI, (2002).

[11] T. Caraballo, J. Real, Attractors for 2D—Navier—Stokes models with delays, J. Differ—ential Equations 205 (2004) 271—297.

[12] T. Caraballo, P. E. Kloeden and J. Real, Pullback and forward attractors for a damped wave equation with delays, Stoch. Dyn. 4 (2004) 1—19.

[13] Y. J. Wang, Pullback attractors for a damped wave equation with delays, Stoch. Dyn. 4 (2015) 1—21.

[14] P. E. Kloeden and B. Schmalfuss, Asymptotical behaviour of nonautonomous differ—ence inclusions, Syst. Cont. Lett. 33 (1998) 275—280.

[15] P. Marn—Rubio, J. Real, J. Valero, Pullback attractors for a two—dimensional Navier—Stokes model in an infinite delay case, Nonlinear Anal. TMA 74 (2011) 2012—2030.

[16] D. N. Cheban, P. E. Kloeden and B. Schmalfuβ, The relationship between pullback, forwards and global attractors of non—autonomous dynamical systems, Nonlinear Dynamics Systems Theory, 2(2)(2002) 9—28.

[17] P. E. Kloeden and B. Schmalfuβ, Non—autonomous systems, cocycle attractors and variable time—step discretization, Numer. Algorithms, 14 (1997) 141—152.

[18] Y. Hino, S. Murakami, and T. Naito, Functional Differential Equations with Infinite Delay, Lecture Notes in Mathematics, Vol. 1473, Springer—Verlag, Berlin, 1991. 35

[19] J. M. Cushing, integro—differential Equations and Delay Models in Population Dynamics, Lecture Notes in Biomathematics 20, Heidelberg: Springer—Verlag, 1979.

[20] W. B. Ma, Z. Y. Lu, Delay Differential Equations—Introduction of Function Differential Equations, Chinese Science, 2013.

[21] J. C. Robinson, A. Rodrguez—Bernal, A. Vidal—Lpez, Pullback attractors and extremal complete trajectories for non—autonomous reaction—diffusion problems, J. Differential Equations 238 (2007) 289—337.

[22] O. Ladyzhenskaya, Attractors for Semigroups and Evolution Equations, vol. 25 of Lincei Lectures. Cambridge: Cambridge University Press, 1991.

[23] P. Marin—Rubio, J. Real, Pullback attractors for 2D—Navier—Stokes equations

with delay in continuous and sub－linear operators, Discrete Contin. Dyn. Syst. ,26 (2010), 989－1006.

[24] T. Caraballo, M. J. Garrido－Atienza, B. Schmalfuss and J. Valero, Global attractors for a non－autonomous integro－differential equation in materials with memory, Nonlinear Anal. ,73 (2010),183－201.

[25] C. Peng, Q. Shi, Stability of standing wave for the fractional nonlinear schrödinger equation, J. Math. Phys. ,59 (2018),011508.

[26] T. Caraballo, P. Marin－Rubio, J. Valero, Autonomous and non－autonomous attractors for differential equations with delays, J. Differential Equations, 208 (2005), 9－41.

[27] R. Temam, Navier－Stokes Equations, Theory and Numerical Analysis, 2nd ed. , North Holland, Amsterdam, 1979.

[28] J. K. Hale, S. M. Verduyn Lunel, Introduction to Functional Differential Equations, Springer－Verlag, 1993.

[29] T. Caraballo, G. Lukaszewicz, J. Real, Pullback attractors for asymptotically compact non－autonomous dynamical systems, Nonlinear Anal. 64 (2006) 484－498.

[30] P. Marn－Rubio, J. Real, On the relation between two different concepts of pullback attractors for non－autonomous dynamical systems, Nonlinear Anal. 71 (2009) 3956－3963.

[31] Y. J. Wang, P. E. Kloeden, Pullback attractors of a multi－valued process generated by parabolic differential equations with unbounded delays, Nonlinear Anal. 90 (2013)86－95. 36

[32] Y. J. Wang, P. E. Kloeden, The uniform attractor of a multi－valued process generated by reaction－diffusion delay equations on an unbounded domain, Discrete Contin. Dyn. Syst. 34 (2014)4343－4370.

(本文发表于 2017 年《Mathematical Methods in the Applied Sciences》第 41 卷 12 期)

Bifurcations of a new fractional—order system with a one—scroll chaotic attractor

Xiaojun Liu *

摘要:本文提出了一个具有单涡卷混沌吸引子结构的分数阶非线性系统。首先研究了系统的基本性质,分析了系统平衡点的稳定性,并基于此给出了该系统单涡卷吸引子的生成条件。从二维平面和三维空间的不同角度研究了系统的分岔问题。在等阶情况下,研究了在分数阶的阶数从 0.99 开始减小时,系统随参数变化的分岔问题。在不等阶的情况下,研究了其它阶数从 1 开始减小时,系统随一个分数阶的阶数变化的分岔问题。结果表明该系统具有丰富的动力学和分岔行为。

In this paper, a new fractional—order system which has a chaotic attractor of the one — scroll structure is presented. Firstly, the stability of the equilibrium points of the system is investigated. And based on the stability analysis, the generation conditions of the one—scroll structure for the attractor are determined. In a commensurate—order case, bifurcations with the variation of a system parameter are investigated as derivative orders decrease from 0.99. In an incommensurate—order case, bifurcations with the variation of a derivative order are analyzed as other orders decrease from 1.

1 Introduction

Fractional calculus is a field of applied mathematics that deals with derivatives and integrals of arbitrary orders. Although the seeds of fractional deriva-

* 作者简介:刘晓君(1980—),女,黑龙江富裕县人,天水师范学院数学与统计学院副教授、博士,主要从事分数阶非线性系统的动力学研究。

tives were planted over 300 years ago, the development of fractional calculus is very slow at an early stage for the absence of geometrical interpretation and applications. In the recent several decades, it has been applied to almost every field of science, engineering, economics, and secures communication and so on[1−5].

It is well known that fractional calculus is very suitable for the description of properties of various real materials. Meanwhile, fractional calculus provides an excellent instrument for the description of memory and hereditary properties of various materials and processes. This is the main advantage of fractional derivatives in comparison with classical integer−order models, in which such effects are in fact neglected. The advantages of fractional derivatives become apparentin modeling mechanical and electrical properties of real materials[6].

Many new fractional−order systems were presented in recent years. Meanwhile, rich and complex dynamics, such as periodic solutions, chaos windows, all kinds of bifurcations, boundary and interior crises were observed in these systems. For example, Chua system with a derivative order 2. 7 exists chaos motion [7]. Chaotic dynamics of a damped van der Pol equation with fractional order is investigated in[8]. Chen studied the nonlinear dynamics and chaos in a fractional −order financial system[9]. A periodically forced complex Duffing's oscillator was proposed, and chaos for the system was studied in details[10]. Bifurcation, chaos control and synchronization were investigated for a fractional − order Lorenz system with the complex variables[11]. In[12], boundary and interior crises were determined in a fractional−order Duffing system by a global numerical-computation method. A proposed standard for the publication of new chaotic systems was studied in[13]. The authors investigated a simple chaotic flow with a plane of equilibria in[14].

It is well known that bifurcation theory concerns the changes in qualitative or topologicalstructures of limiting motions such as equilibriums, periodic solutions, homoclinic orbits, heteroclinic orbits and invariant tori for nonlinear evolution equations as somerelevant parameters in the equations vary. Generally, the subject can be tracedback to the very earlier work of Poincarearound 1892[15]. Nowadays, it is afundamentaltool to analyze nonlinear problems which enables us to understand how and when asystem organizes new states and patterns near the original "trivial" one when a control parameter crosses a critical value. For fractional—order systems, a bifurcation implies a qualitative or topological change in dynamics with a variation of a system parameter or derivative order, and bifurcation analysis becomes harder due to the non—local property of the operator of fractional calculus. Many references have studied the bifurcations of fractional—order systems[16—20]. However, these investigations mainly focus on the bifurcation of a fractional—order system as a system parameter or a derivative order varies. To our knowledge, few works concerns bifurcations with the variation of both a system parameter and a derivative order or with the variation of both a derivative order and other orders.

Compared with integer—orderchaotic systems, fractional—order chaotic systems with more complex dynamic characteristics and more system parameters can provide higher security for secure communication[21,22]. In[23], the authors investigated the synchronization of a three—dimensional integer—order system. The differential equations of the system with simple structure were similar to those of the Lorenz system. It is well known that the Lorenz system has a chaotic attractor with double—scroll structure. The system in[23]has a chaotic attractor with only one—scroll and very abundant dynamic behaviors. Motivated by the above, in this paper, a corresponding fractional—order system is proposed and studied. Firstly, the stability of equilibrium points of the system is investigated. In a commensurate—order case, bifurcations with the variation of a system parameter are investigated as derivative orders decrease from 0. 99. In an incommensurate—order case, bifurcations with the variation of a derivative order are analyzed as the other orders decrease from 1. Period—doubling and saddle—node bifurcations can be observed from the bifurcation diagrams by numerical simulations.

The remainder of the paper is organized as follows. In section 2, thedefinitionsof the fractional calculus and related preliminariesare given. A new fractional—order system with one—scroll attractor is presented in section 3. In Section 4, bifurcations in the two cases of commensurate—order and incommensurate—order are analyzed respectively. Conclusions of the paper are drawn in section 5.

2 Fractional derivatives and preliminaries

2.1 Definitions

Fractional calculus can be considered as a generalization of integration and differentiation. The operator ${}_aD_t^q$ of fractional calculus can be defined by

$$ {}_aD_t^q = \begin{cases} \dfrac{d^q}{dt^q} & R(q) > 0 \\ 1 & R(q) = 0, \\ \displaystyle\int_a^t (d\tau)^q & R(q) < 0 \end{cases} \tag{1} $$

where q denotes the derivative order, $R(q)$ corresponds to the real part of q. The numbers a and t represent the limits of the operator.

In general, three definitions of fractional derivative are used frequently, namely, the Grunwald—Letnikov definition, the Riemann—Liouville and the Caputo definitions[6,24].

The Grunwald — Letnikov definition (GL) derivative with fractionalorder q can be described by

$$ {}_a^{GL}D_t^q = \lim_{h\to 0}\frac{1}{h^q}\sum_{j=0}^{\left[\frac{t-a}{h}\right]}(-1)^q\begin{pmatrix} q \\ j \end{pmatrix} f(t-jh), \tag{2} $$

where the symbol $[\,\cdot\,]$ represents the integer part.

TheRiemann—Liouville(RL)definition is

$$ {}_a^{RL}D_t^q f(t) = \frac{d^n}{dt^n}\frac{1}{\Gamma(n-q)}\int_a^t \frac{f(\tau)}{(t-\tau)^{q-n+1}}d\tau, \qquad n-1<q<n, \tag{3} $$

where $\Gamma(\,\cdot\,)$ denotes the gamma function.

The Caputo(C)fractional derivative is defined as follows:

$$ {}_a^{C}D_t^q f(t) = \frac{1}{\Gamma(n-q)}\int_a^t (t-\tau)^{n-q-1} f^{(n)}(\tau)d\tau, \qquad n-1<q<n. \tag{4} $$

For a fractional differential equation which is defined by Caputo derivatives, the initial condition takes on the same form as those for the integer—order ones, which can be measuredeasilyin applications. For this reason, the Caputo derivative will be adopted in the rest of the paper.

2.2 Numerical methods

Due to the non—local property of the operator of fractional calculus, it is not easy to obtain the numerical solutions for a fractional differential equation. Generally speaking, two approximation methods are frequently used, namely, an improved version of Adams－Bashforth－Moulton algorithm based on the predictor－correctors scheme and the frequency domain approximation[25－28]. For the accuracy[29], we will employ the improved predictor－corrector algorithm to solve a fractional differential equation in this paper.

In order to get the approximate solution of a fractional－order chaotic system by the improved predictor－corrector algorithm, the following equation is considered

$$\begin{cases} \dfrac{d^q x}{dt^q} = f(t,x), & 0 \leqslant t \leqslant T \\ x^k(0) = x_0^{(k)} & k = 0,1,\cdots,\lceil q \rceil - 1 \end{cases}, \tag{5}$$

which is equivalent to the Volterra integral equation

$$x(t) = \sum_{k=0}^{\lceil q \rceil - 1} x_0^{(k)} \frac{t^k}{k!} + \frac{1}{\Gamma(q)} \int_0^t (t-\tau)^{(q-1)} f(\tau, x(t)) d\tau. \tag{6}$$

Now, set $h = T/N$, $t_j = jh$, $(j = 0,1,\cdots,N)$. The corrector formula for Equation(6) can thus be discretized as follows:

$$x_h(t_{n+1} = \sum_{k=0}^{\lceil q \rceil - 1} x_0^{(k)} \frac{t_{n+1}^k}{k!} + \frac{h^q}{\Gamma(q+2)} f(t_{n+1}, x_h^p(t_{n+1})) + \frac{h^q}{\Gamma(q+2)} \sum_{j=0}^{n} \alpha_{j,n+1} f(t_j, x_h(t_j)), \tag{7}$$

where predicted values $x_h(t_{n+1})$ is determined by the following formula

$$x_h^p(t_{n+1}) = \sum_{k=0}^{\lceil q \rceil - 1} x_0^{(k)} \frac{t_{n+1}^k}{k!} + \frac{1}{\Gamma(q)} \sum_{j=0}^{n} \beta_{j,n+1} f(t_j, x_h(t_j)), \tag{8}$$

and

$$a_{j,n+1} = \begin{cases} n^{q+1} - (n-q)(n+1)^q, & j = 0 \\ (n-j+2)^{q+1} + (n-j)^{q+1} - 2(n-j+1)^{q+1}, & 1 \# j \quad n. \\ 1, & j = n+1 \end{cases}$$

$$b_{j,n+1} = \frac{h^q}{q}((n-j+1)^q - (n-j)^q), 1 \# j \quad n \tag{9}$$

The error estimate of this approach is $\max_{j=0,1,\cdots,N} |x(t_j) - x_h(t_j)| = O(h^p)$, where $p = \min(2, 1+q)$.

2.3 The stability of a fractional－order system

For fractional－order systems, the stability analysis of equilibrium points is complex and difficult due to the non－local property of fractional calculus. Here, the definitions of commensurate-order and incommensurate-order fractional-order systems will be given firstly.

Definition 1 For a fractional-order system, which can be described by $d^q x/dt^q = f(x)$, where $x = (x_1, x_2, \cdots, x_n)^T$ is the state vector, $q = (q_1, q_2, \cdots, q_n)^T$ is the fractional derivative orders vector, and $q_i > 0$. The fractional-order system is commensurate-order when all the derivative orders satisfy $q_1 = q_2 = \cdots = q_n$, otherwise it is an incommensurate-order system[30].

In order to investigate the stability of equilibrium points for fractional-order systems, the following lemma is used frequently.

Lemma 1 For a commensurate fractional - order system, the equilibrium points of the system are asymptotically stable if all the eigenvalues at the equilibrium E^*, satisfy the following condition:

$$|\arg(\mathrm{eig}(J))| = |\arg(\lambda_j)| > \frac{\pi}{2}q, j = 1, 2, \cdots, n \tag{10}$$

where J is theJacobian matrix of the system evaluated at the equilibria E^* [31].

3 A new fractional-order system

In this section, a new fractional-order system which consist of three differential equations is proposed, and can be denoted as follows,

$$\begin{cases} D^{q1} x = ax - y \\ D^{q2} y = x - z - by \\ D_z^{q3} = ax + 4z(y - 2) \end{cases}, \tag{11}$$

where x, y, z are state variables of the system, a, b the system parameters, and q_1, q_2, q_3 derivative orders.

When the derivative orders are selected as $q_1 = q_2 = q_3 = q = 0.99$, the system parameters $a = 0.3$, $b = 0.02$, and initial conditions $(x_0, y_0, z_0) = (-0.5, -1.1)$, the system(11)is chaotic. In Figs. 1(a)and(b), a chaotic attractor with a one-scroll on three-dimensional space and projected onto $x-y$ planeare depicted. In this case, the corresponding Lyapunov exponents are $\lambda_1 = 0.117$, $\lambda_2 = 0$, $\lambda_3 = -10.49$. When the derivative order $q = 0.985$ and $q = 0.965$, the corresponding attractors are displayed in Figs. 1(c)to(f).

The equilibrium points of the system(11)can be calculated by solving the equations $D^{q_1} x = 0$, $D^{q_2} y = 0$, $D^{q_3} z = 0$. The system contains two equilibriums, i. e.

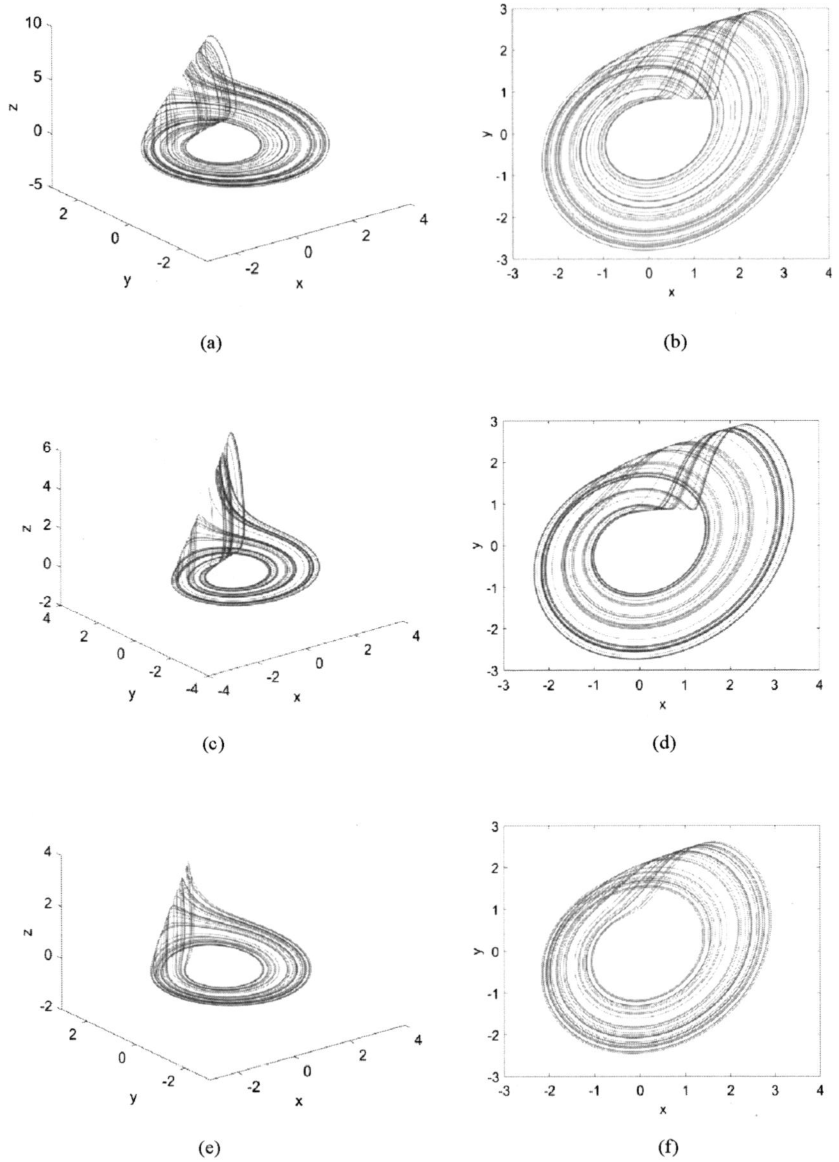

Fig. 1. The chaotic attractor of the system(11). (a)3D plot of the attractorwhen $q=0.99$; (b)the attractor projected onto $x-y$ plane when $q=0.99$; (c)3D plot of the attractor when $q=0.985$; (d)the attractor projected onto $x-y$ plane when $q=0.985$. (e)3D plot of the attractorwhen $q=0.965$; (d)the attractor projected onto $x-y$ plane when $q=0.965$.

$$\begin{cases} E_1(0,0,0) \\ E_2\left(\frac{2}{a}-\frac{1}{4(1-ab)},a\left(\frac{2}{a}-\frac{1}{4(1-ab)}\right),(1-ab)\left(\frac{2}{a}-\frac{1}{4(1-ab)}\right)\right) \end{cases}.$$

The equilibrium E_2 exists when the system parameters satisfy the condition $1-ab \neq 0$.

The Jacobin matrix for the system (11) evaluated at the equilibrium point (x^*, y^*, z^*) is given by

$$J(x^*, y^*, z^*) = \begin{bmatrix} a & -1 & 0 \\ 1 & -b & -1 \\ a & 4z^* & 4(y^*-2) \end{bmatrix}.$$

Based on the matrix, the characteristic equation at the equilibriums E_1 is

$$\lambda^3 + (8+b-a)\lambda^2 + [8(b-a)+1-ab]\lambda + 8(1-ab) - a = 0 \quad (12)$$

By the Routh—Hurwitz test, the equilibrium point E_1 has three roots with the negative real parts if and only if $8+b-a>0$ and $0<8(1-ab)-a<(8+b-a)[8(b-a)+1-ab]$. In our case, when the values of the system parameters are taken as $a=0.3, b=0.02$, the corresponding eigenvalues for the equilibrium point E_1 are $\lambda_1=-8.0047, \lambda_{2,3}=-0.1576\pm1.0089i$. According to the Lemma 1, the equilibrium point E_1 is locally stable.

The characteristic equation at equilibrium E_2 is

$$\lambda^3 + [(b-a)-4(y-2)]\lambda^2 - [4(y-2)(b-a)+ab-(4z+1)]\lambda \\ + 4(y-2)(ab-1) - a(1+4z) = 0. \quad (13)$$

By the Routh—Hurwitz test, the equilibrium point E_2 has three roots with the negative real parts if and only if $(b-a)-4(y-2)>0$ and

$$0 < 4(y-2)(ab-1) - a(1+4z) \\ < [(b-a)-4(y-2)][4(y-2)(b-a)+ab-(4z+1)].$$

The corresponding eigenvalues for the equilibrium point E_2 are $\lambda_1=-7.4038, \lambda_{2,3}=0.1029\pm0.9653i$ when the system parameters take the same values as before. Then based on the Lemma 1, we can get the fixed point E_2 is unstable.

In a nonlinear dynamical system, a saddle point is an equilibrium point on which the equivalent linearized model has at least one eigenvalue in the stable region and one eigenvalue in the unstable region. In the same system, a saddle point is called saddle point of index 1 if one of the eigenvalues is unstable and other eigenvalues are stable. Also, a saddle point of index 2 is a saddle point with one stable eigenvalue and two unstable eigenvalues. In chaotic systems, it is proved that scrolls are generated only around the saddle points of index 2. Moreover, saddle points of index 1 are responsible only for connecting scrolls [32, 33].

From the above analysis we can see that the equilibrium point E_2 is a saddle point of index 2. In chaotic systems, it is proved that the scrolls of a chaotic attractor are generated only around the saddle points of index 2. Moreover, the saddle points of index 1 are responsible for connecting the scrolls. In the fractional—order system (11), the equilibriumE_1 is not a saddle point of index 1. The necessary condition for the existence of double—scroll attractor in system (11) cannot be satisfied. Therefore, the chaotic attractor of the system (11) is one—scroll.

Compared with an integer—order system, the derivative order is an important parameter for a fractional—order system. For the system (11), the system parametersand the initial conditions are fixed. The phase diagrams with different values of the derivative order $q$are employed to demonstrate the behavior of the system (11), as shown in Fig. 2. From which it can be seen that the system (11) converges to a fixed point as $q = 0.90$, and it is period—1 for $q = 0.92$, period —2 for $q = 0.95$, and period—4 for $q = 0.955$.

4 Bifurcations of the new system

Bifurcations play a vital role in dynamics research for fractional—order systems. Therefore, in this section, bifurcation analysis is conducted to study the rich dynamical behavior of the fractional—order system (11) in the two cases of commensurate—order and incommensurate—order, respectively.

4.1 The commensurate—order case

Firstly, with the system parameters fixed and the derivative order varying on the closed interval$q \in [0.9, 1]$, bifurcation diagram for the system (11) is depicted in Fig. 3. Clearly, the evolution of the period—doubling scenario and saddle—node bifurcation can be observedfrom this figure. When the order satisfies $q \leq 0.958$, the route of leading to chaos for the system (11)is period—doubling bifurcation. When $q = 0.978$, the chaotic solutions disappear suddenly, and two new period—1 solutions appear, which means the saddle—node bifurcation occurs as

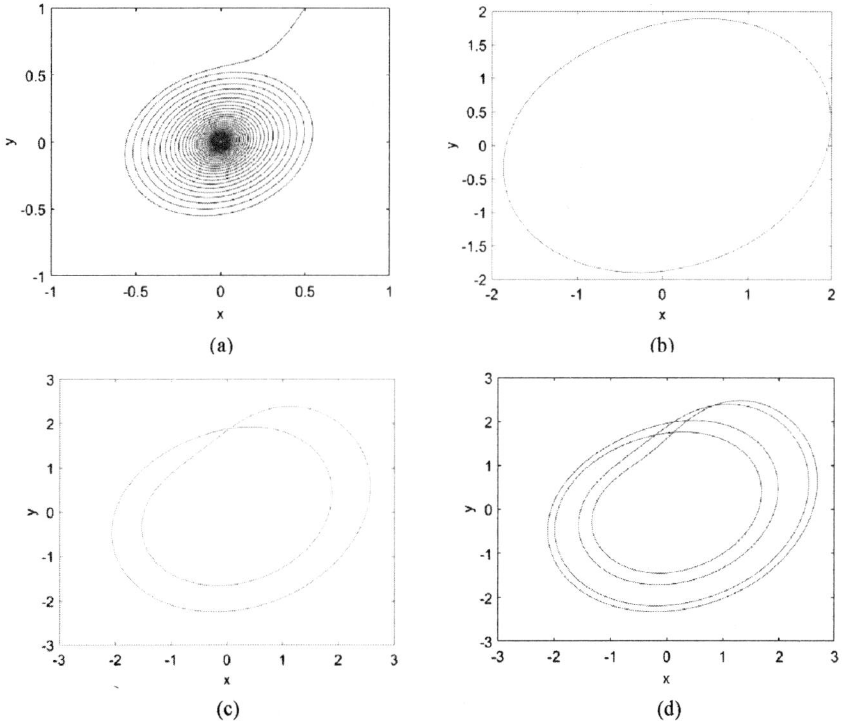

Fig. 2. The attractors for different values of the derivative order$q$. (a) $q=0.90$; (b) $q=0.92$; (c) $q=0.95$; (d) $q=0.955$.

the order $q$varies.

Secondly, bifurcations with the variation of the system parameter$a$are studied for $q = 0.99$ and$b = 0.02$. In Fig. 4, it can be seen that a series of period—doubling bifurcations occurs as the system parameter $a$decreases from 0. 3 to 0. 1. Meanwhile, the period—3 and chaoswindows can also be observed from the bifurcation diagram.

In order to get abetter understanding of the dynamics of the system (11), bifurcations with the variation of the parameter a for the different values of derivative order q are given. With the parameter a varying, the corresponding bifurcation diagrams with several specified values of the derivative order like$q = 0.98$, $q = 0.97$, $q = 0.96$, $q = 0.95$, $q = 0.94$, and $q = 0.93$ are plotted in Fig. 5 for the system (11). Compared these figures, it can be seen that the bifurcation structure of the system (11) changes qualitatively with the variation of the pa-

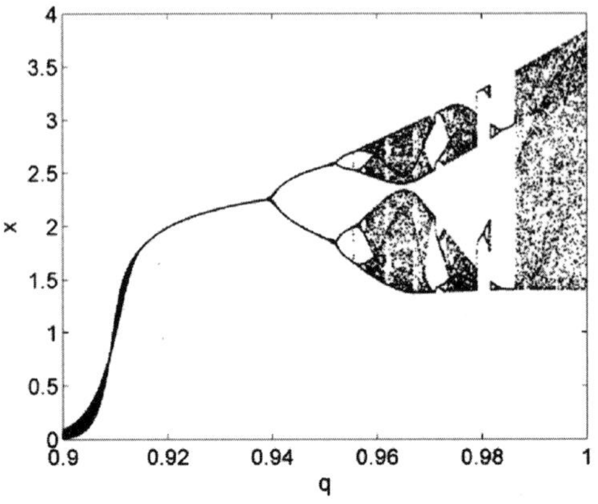

Fig. 3. The bifurcation diagram of the system (11) with the variation of the derivative order *q*.

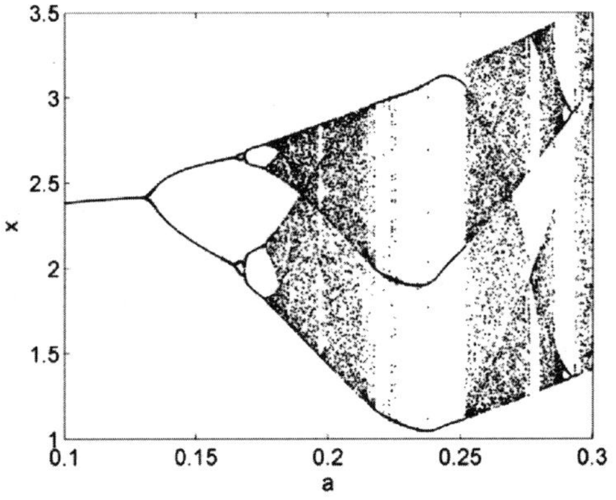

Fig. 4. The bifurcation diagram of the system (11) with the variation of the system parameter *a*.

rameter a and the order q. The area of the chaotic motion decreases with the augment of the periodic motion as the derivative order decreases. From Fig. 5(a) to Fig. 5(c), it is clear that the route leading to chaos is period—doubling bifurcation. Meanwhile, a typical period—doubling bifurcation, which numerical solutions change from period—1 to period—2, can be seen clearly from Figs. 5(d) and 5(e). The system becomes totally periodic as the order $q$decreases to 0.93, see Fig. 5(f). In order to further discuss the bifurcations with the variation of both the system parameter a and derivativeorder q, a bifurcation diagram in the three—dimensional space is depicted in Fig. 6. It can be seen that dynamics of the system (11) becomes simple as the derivative order decreases from 0.98 to 0.93.

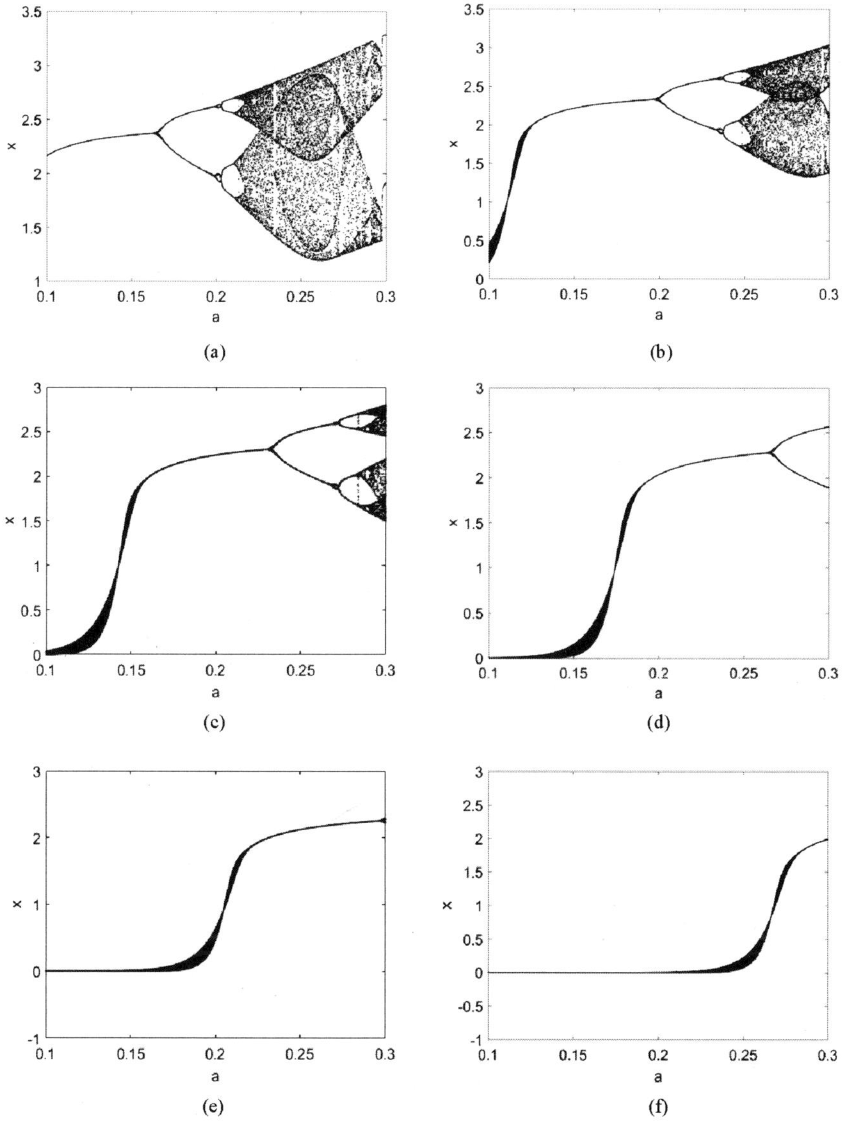

Fig. 5. The bifurcation diagrams of the system (11) when the parameter *a*varied with the different values of the order *q* (a)*q*=0.98; (b)*q*=0.97; (c)*q*=0.96; (d)*q*=0.95; (e)*q*=0.94; (f)*q*=0.93.

Thirdly, the bifurcation of the system (11) with the variation of the parame-

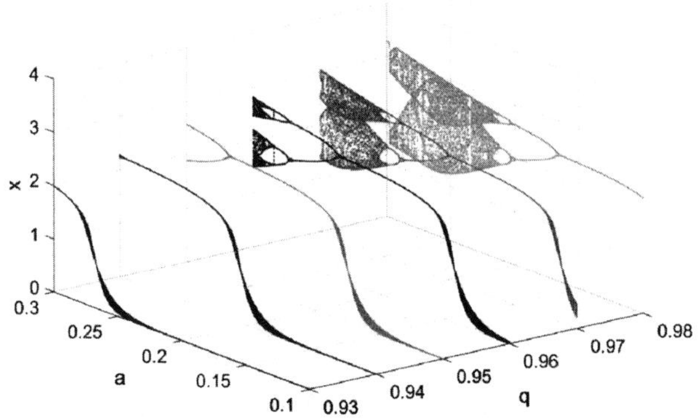

Fig. 6. The bifurcation diagram of the system (11) in three-dimensional space with the variation of both the parameter *a* and the order *q*.

ter b is studied for $q = 0.99$ and $a = 0.3$. In Fig. 7, there is a long time of the chaotic window for $b \leq -0.008$. Meanwhile, the route leading to chaos is the period-doublingbifurcation for the system (11).

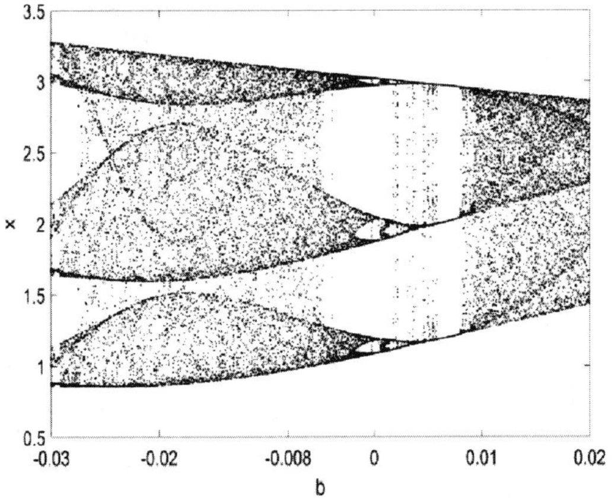

Fig. 7. The bifurcation diagram of the system (11) with the variation of the parameter *b*.

Bifurcations with the variation of the parameter b are studied for different values of the derivative order q. When the derivative order q decreases from 0.98 to 0.93, the corresponding bifurcation diagrams are plotted for the fractional-order system (11) when the parameter $b \in [-0.03, 0.02]$, see Fig. 8. From which, it is clear that structure of dynamics of the system (11) evolves as the order $q$varies. The interval of the chaotic motion increases and that of the periodic motion decreases. Meanwhile, the system (11) is periodic completely when the order $q = 0.94$, and period—4 and period—2motions can be obtained from Fig. 8 (e). When the order $q = 0.93$, only period—2 and period—1motions exist, see

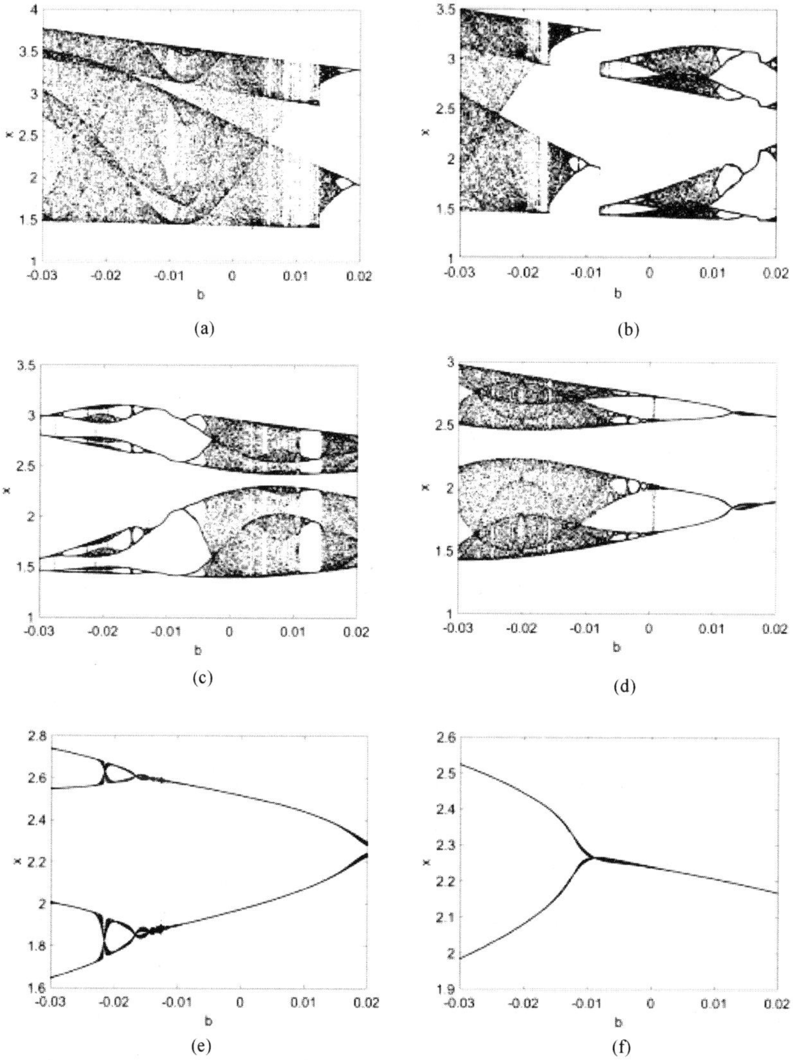

Fig. 8. The bifurcation diagrams of the system (11) when the parameter *b* varied with the different values of the order *q* (a)$q=0.98$; (b)$q=0.97$; (c)$q=0.96$; (d)$q=0.95$; (e)$q=0.94$; (f)$q=0.93$.

Fig. 8(f). In order to further discuss the bifurcations with the variation of both the system parameter b and derivativeorder q , a bifurcation diagram in the three—dimensional space is plotted in Fig. 9. It can be seen that dynamics of the system (11) becomes complex as the derivative order increases from 0.93 to 0.98.

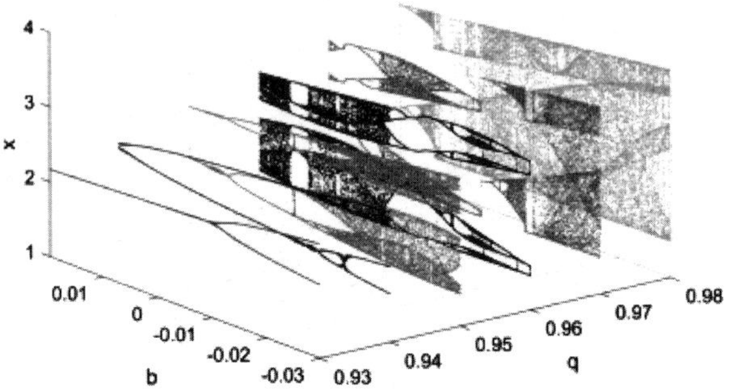

Fig. 9. The bifurcation diagram of the system (11) in three—dimensional space with the variation of both the parameter *b* and the order *q*.

4.2 The incommensurate—order case

In order to learn more about the characteristics of the new fractional—order-system (11), the dynamics of the system (11) with the variation of the different derivative orders will be investigated in this subsection. In the following work, the system parameters are taken as $a = 0.3$ and $b = 0.02$.

Firstly, thebifurcation diagram with the derivative order $q_1 \in [0.9, 1]$ as the others two derivative orders q_2, q_3 are both fixed and $q_2 = q_3 = q = 1$ is plotted. A series of period—doubling bifurcations can be seen in Fig. 10. Based on this, bifurcation diagrams versus the order q_1 when the others two derivative orders decrease from 0.98 to 0.93 are given in Fig. 11. For Fig. 11(a)to Fig. 11(c), it can be seen that the two branches of the bifurcation gradually decouple, and the area of the chaos gradually decreases with that of the period increases. For-Fig. 11(d) to Fig. 11(f), it is clearthat the fractional—order system (11)is periodic completely. A bifurcation diagram with the variation of both the order q_1 and derivative order q in the three—dimensional space is plotted in Fig. 12. It is clear that dynamics of the system (11) becomes simple as the derivative order decreases from 0.98 to 0.93.

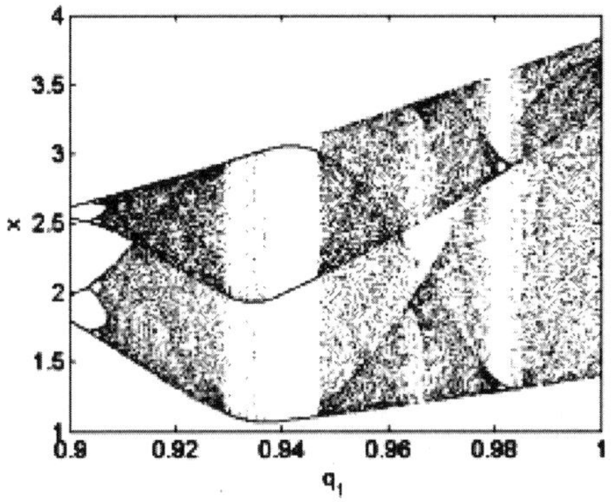

Fig. 10. The bifurcation diagram of the system (11) with the variation of the order q_1.

Secondly, thebifurcation diagram versus the order $q_2 \in [0.9, 1]$ when the others two derivative orders q_1, q_3 are both fixed and $q_1 = q_3 = q = 1$ is plotted in Fig. 13. Meanwhile, a series of bifurcation diagrams versus the order q_2 when the others two derivative orders decrease from 0.98 to 0.85 are given in Fig. 14. By comparing Fig. 14(a) and Fig. 14(b), it is clear that the dynamics of the system (11) becomes simple as the order q decreases. For Fig. 14(b) and Fig. 14(c), it can be observed that the area of chaos increases and that of period decreases. From Fig. 14(c) and Fig. 14(d), it can be seen that the two branches of the bifurcation gradually couple, and the area of the chaos gradually decreases with that of the period increases. For the rest of the figures in the Fig. 11, it can be seen the system (11) is totally periodic when the order $q = 0.90$. Meanwhile, the typical period—doubling bifurcation can be seen in the Fig. 14(g). When the order $q = 0.85$, only period—1 motion for the system (11) exists. A bifurcation diagram with the variation of both the order q_2 and derivative order$q$ in the three—dimensional space is plotted in Fig. 15. It is clear that dynamics of the system (11) becomes simple as the derivative order decreases from 0.98 to 0.85.

The bifurcation diagrams with the variation of the orderq_3 when the other two orders q_1, q_2 decrease from 0.98 to 0.93 will not be given in here for the similarity.

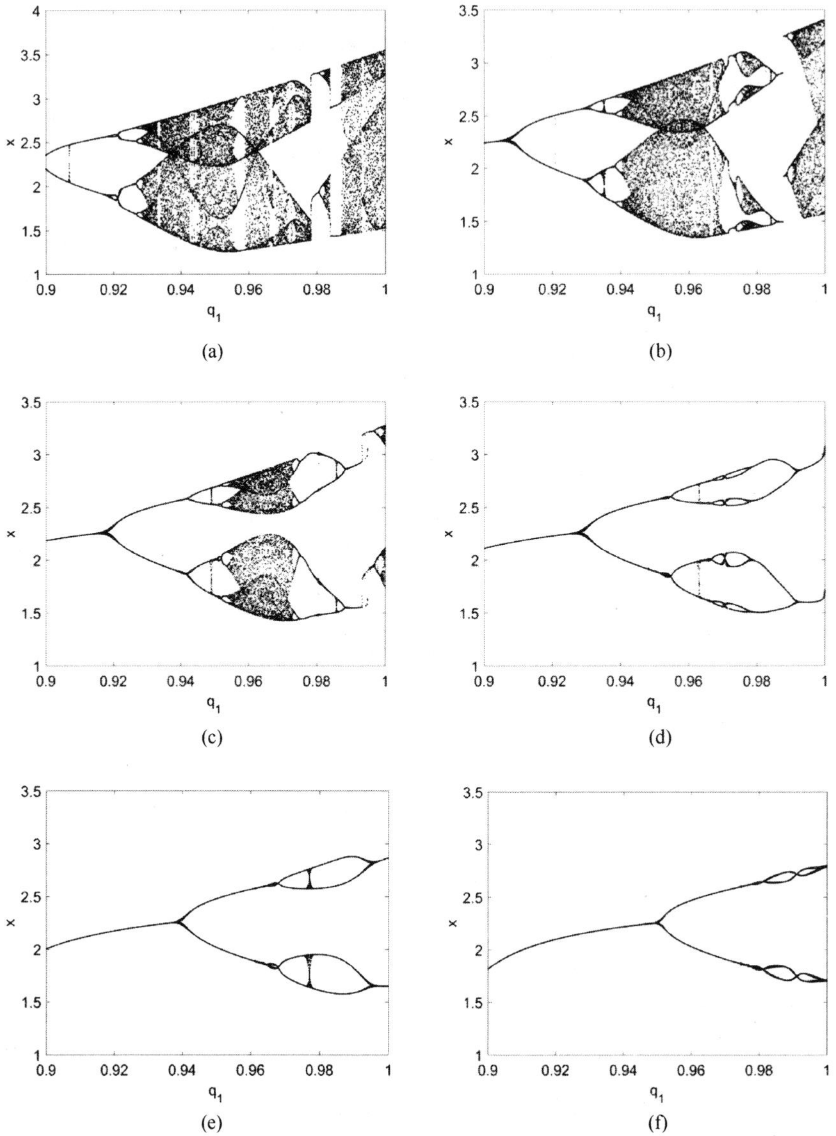

Fig. 11. The bifurcation diagram of the system (11) when the order q_1 varied with the different values of the order q (a)$q=0.98$; (b)$q=0.97$; (c)$q=0.96$; (d)$q=0.95$; (e)$q=0.94$; (f)$q=0.93$.

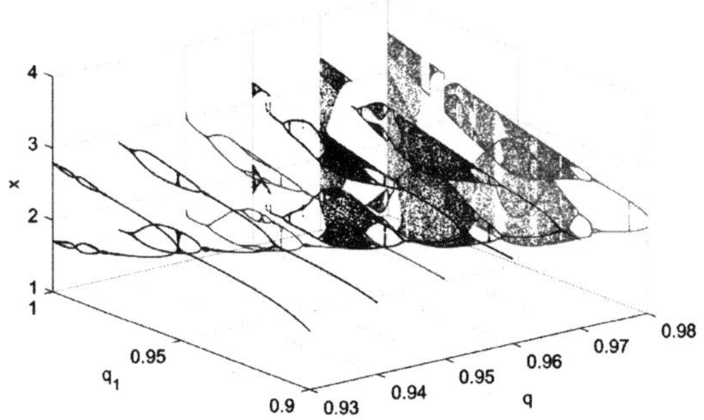

Fig. 12. The bifurcation diagram of the system (11) in three−dimensional space with the variation of both the orders q_1 and q.

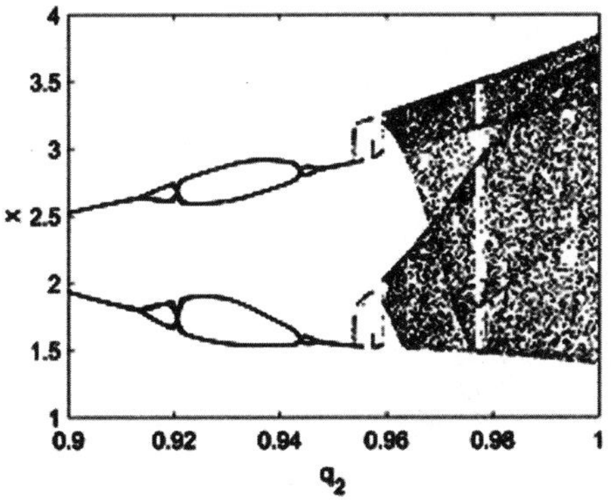

Fig. 13. The bifurcation diagram of the system (11) with the variation of the derivative order q_2.

5 Conclusions

In this paper, a new fractional−order system is presented. Firstly, the stability of the equilibriums isanalyzed. Based on the stability analysis, the reason of the generation for the one scroll of the attractor is analyzed. Phase diagrams for the different values of the derivative order are obtained to show the rich dynamics of the new fractional−order system.

Bifurcations of the new fractional−order system in commensurate−order and incommensurate−order cases are studied in details. In the commensurate−order case, when the derivative order decreases from 0.99, bifurcations with the variation of a system parameter are investigated. In the incommensurate−order

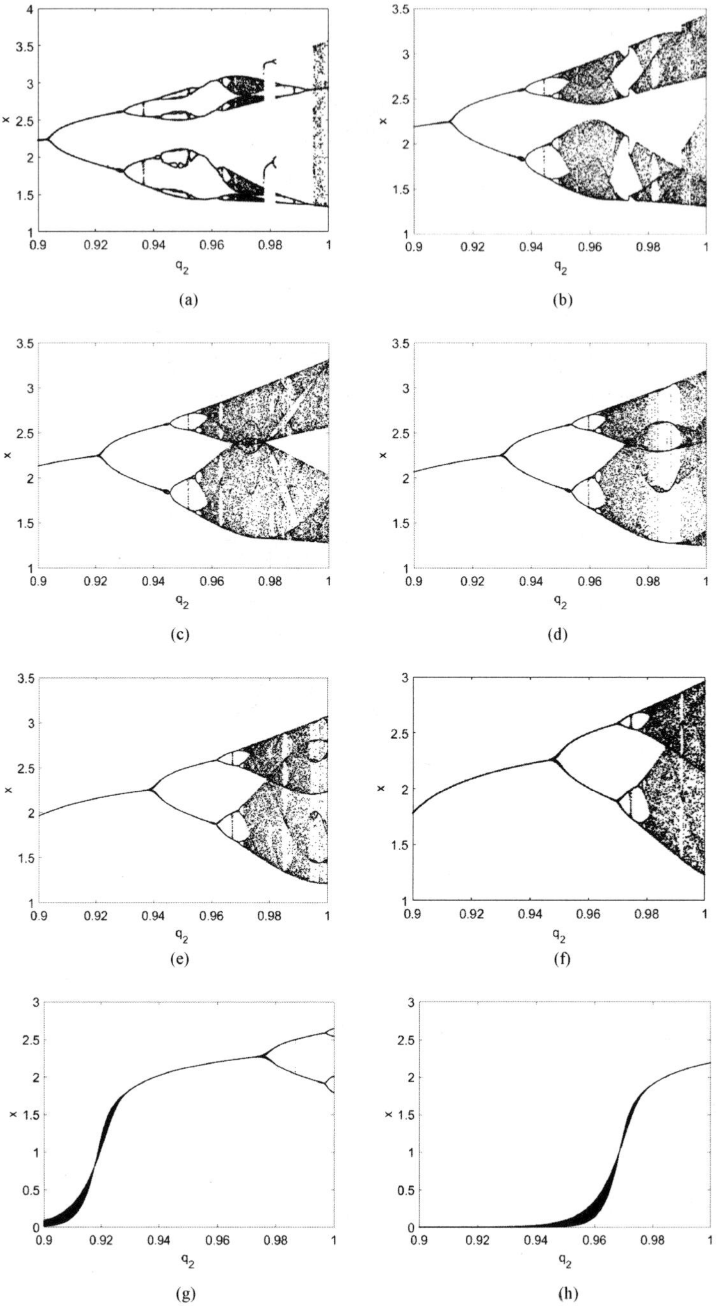

Fig. 14. The bifurcation diagram of the system (11) when the order q_2 varied with the different values of the order q (a)$q=0.98$; (b)$q=0.97$; (c)$q=0.96$; (d)$q=0.95$; (e)$q=0.94$; (*f*)q=0.93; (*e*)q=0.90; (*f*)q=0.85.

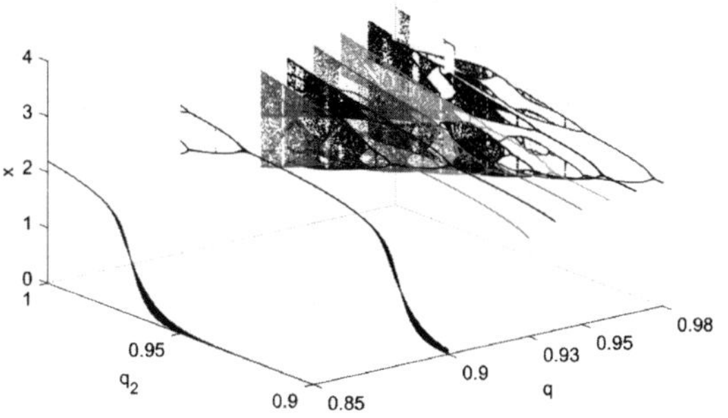

Fig. 15. The bifurcation diagram of the system (11) in three-dimensional space with the variation of both the orders q_2 and q.

case, bifurcations with the variation of a derivative order when the other orders decreases from 1 are analyzed. Period-doubling and saddle-node bifurcations can be observed.

What's more, it can be concluded that the dynamics of the new fractional-order system becomes periodic or simple when the derivative order approaches to 0, and chaotic or complex when the order approaches to 1 from a global perspective. These results obtained in this paper can be referenced for the bifurcation control of fractional-order systems. The generalization of the conclusion for other fractional-order systems will be our future work.

Acknowledgments

This work is supported by the National Natural Science Foundation of China (NSFC) underthe grant Nos. 11702194 and 11702195.

References

[1]O Lyubomudrov, M Edelman, and G M Zaslavsky *Int. J. Modern Physics B* 17 4149 (2003)

[2]J T Machado and A M Galhano *Commun. Nonlinear Sci. Numer. Simulat* 55 174 (2018)

[3]J T Machado, M Mata and A Lopes *Entropy* 17 5402(2015)

[4]LJSheu, WCChen, YCChen and WTWeng *World Academy of Science, Engineering and Technology* 411057(2010)

[5]AKiani-B, KFallahi, NPariz and HLeung *Commun. Nonlinear Sci. Numer. Simulat*

223709(2008)

[6]I Podlubny *Fractional differential equations*(San Diego:Academic Press)(1999)

[7]TTHartley,CFLorenzo and HKQammer *IEEETrans Circuits Syst*－1:*Fund ThAppl* 42485(1995)

[8]J H Chen and W Ch Chen *Chaos SolitonFract* 25 188(2008)

[9]W Ch Chen *Chaos SolitonFract* 36 1305(2008)

[10]X Gao and J Yu *Chaos SolitonFract* 24 1097(2005)

[11]C Luo and X Y Wang *Nonlinear Dyn* 71241(2013)

[12]X J Liu L Hong and J Jiang *Chaos* 26 084304(2016)

[13]JCSprott *Int. J. Bifurcation chaos* 212391(2011)

[14]SJafari,JCSprott and MMolaie *Int. J. Bifurcation chaos* 26 1650098(2016)

[15]H Poincaré *Les MéthodesNouvelles de la MécaniqueCéleste* (Pairs:Gauthier－Villars)1892

[16]X J Liu and L Hong *Int. J. Bifurcation chaos* 231350175(2013)

[17]D Cafagna and G Grass *Int. J. Bifurcation chaos* 18 1845(2008)

[18]H A EI－Saka,E Ahmed,Mi. Shehata and A M A EI－Sayed *Nonlinear Dyn* 56 121 (2008)

[19]A Y T Leung,H X Yang and P Zhu *Commun. Nonlinear Sci. Numer. Simulat* 19 1142(2014)

[20]Y D Ji,L Lai,S C Zhong and L. Zhang *Commun. Nonlinear Sci. Numer. Simulat* 57 352(2018)

[21]J Palanivel,K Suresh,S Sabarathinam and K. Thamilmaran *Chaos SolitonFract* 95 33(2017)

[22]LJ Sheu,WC ChenYCChen and W TWeng *World Academy of Science,Engineering and Technology* 651057(2010)

[23]A Kiani－B,K Fallahi,N ParizandHLeung *Commun. Nonlinear Sci. Numer. Simulat* 14 863(2009)

[24]L Cheng *Synchronization and circuit implementation for chaotic and hyperchaotic systems*, (Chuangchun:Northeast Normal University)(2002)(in Chinese)

[25]A AKilbas,H M Srivastava and J J Trujillo, *Theory and Applications of Fractional Differential Equations* (Singapore:Elsevier)(2006)

[26]EHairer,SPNørsett and GWanner *Solving Ordinary Differential Equations I*: *NonstiffProblems* (Berlin:Springer－Verlag)(1993)

[27]EHairer and GWanner *Solving Ordinary Differential Equations II*: *Stiff and Differential－Algebraic Problems* (Springer－Verlag)(1991)

[28]KDiethelm,NJFord and ADFreed *Nonlinear Dyn* 293(2002)

[29]ACharef,HHSun,YYTsao and BOnaral *IEEE TransAutom Control* 371465(1992)

[30]MSTavazoei and MHaeri *Nonlinear Anal:Theory MethodsAppl* 691299(2008)

[31]DMatignon *IMACS,IEEE-SMC Proceedings,Lille,France* 2963(1996)

[32]MSTavazoei and MHaeri *Physics Letters A* 367102(2007)

[33]A Nasrollahi and N Bigdeli *Int. J. General Systems* 43 880(2014)

(本文 2019 年发表于《Discrete Dynamics in Society and Nature》2019 卷)

Influence of medium correction of nucleon－nucleoncross－section on the fragmentationand nucleon emission

Yong－Zhong Xing　Jian－Ye Liu　Wen－Jun Guo＊

摘要：本文利用同位旋相关的量子分子动力学模型（IQMD），研究了同位旋相关的核子－核子碰撞截面的介质关联对于中能重离子碰撞的多重碎裂和核子发射的影响。研究发现：介质关联增强了中等质量碎块的多重性（N_{imf}）和核子的发射（N_n）对于核子－核子碰撞截面的同位旋效应的依赖性。同时发现动量相关作用（MDI）对于增强两体碰撞中介质关联和核子发射的同位旋依赖性起着至关重要的作用。这一结论为我们进一步认识极端条件下核子－核子相互作用有着重要意义。

The influence of medium correction from an isospin dependent nucleon－nucleon cross－sectionon the fragmentation and nucleon emission in the intermediate energy heavy ion collisions wasstudied by using an isospin dependent quantum molecular dynamical model（IQMD）. We found that the medium correction enhances the dependence of multiplicity of intermediate mass fragment N_{imf} and the number of nucleon emission N_n on the isospin effect of the nucleon－nucleon cross－section，while the momentum dependent interaction（MDI）produces also an important role for enhancing the influence of the medium correction on the isospin dependence of two－body collision in the fragmentation and nucleon emission processes. After considering the medium correction and the role of momentum dependent interaction the increasefor the dependence of N_{imf} and N_n on the

＊　作者简介：邢永忠（1963—），男，甘肃天水人，理学博士，天水师范学院电子信息与电气工程学院教授，主要从事量子混沌与重离子核物理的理论研究。

isospin effect of two－body collision is favorable to learn the information about the isospin dependent nucleon－nucleon cross－section.

2003 Published by Elsevier Science B. V.

1. Introduction

The isospin physics in heavy－ion collision(HIC)at intermediate energies has been an important topic in recent years[1－3,18]. These studies is not only important for understanding the collision mechanism and nuclear structure but also for getting the knowledge about the isospin asymmetric nuclear matter equation of state(EOS)and isospin dependent nucleon－nucleon cross－section. Recently,some observables were found to be the good probes for extracting the information of an isospin asymmetric nuclear matter EOS and the in－medium nucleon－nucleon cross－section at intermediate energies[4－29]because they are sensitive only to one of the isospin dependent mean field and the isospin dependent in－medium nucleon－nucleon cross－section. Bao－An Li[22]has pointed out,for instance,that the proton－neutron differential collective flow and proton ellipse flow can be used to probe the isospin asymmetric nuclear matter equation of state. He also found recently that the isospin asymmetry of the high－density nuclear matter formed in high－energy heavy－ion collision is uniquely determined by the high－density behavior of the nuclear symmetry energy[23]. Our studies in last few years indicated that the nuclear stopping,the number of nucleon emission and the multiplicity of intermediate mass fragments in HIC at intermediate energies can be used to probe the isospin－dependent in－medium nucleon－nucleon cross－section[24,25]. In this work we investigate further the influence of the medium correction of the isospin dependent nucleon－nucleon cross－section on the fragmentation and the nucleon emission in the heavy ion collisions at intermediate energies by using an isospin dependent quantum molecular dynamical model. We found that the multiplicity of intermediate mass fragments Nimf and the number of nucleon emission Nn for the medium effect $\alpha=-0.2$(see in Eq. (7))are always less than those for $\alpha=0.0$(free nucleon－nucleon collision) in the intermediate energy heavy ion collisions. In particular,the differences between Nimf s or Nns from an isospin dependent nucleon－nucleon cross－section and an isospin independent one with $\alpha=-0.2$ are always larger than those with

$\alpha=0.0$, i. e. , the medium correction of two－body collision increases the dependence of Nimf and Nn on the isospin effect of nucleon－nucleon cross－section, while MDI enhances also the influence of the medium correction on the isospin effect of two－body collision in the fragmentation and nucleon emission processes.

2. IQMD model

The quantum molecular dynamics(QMD)[30]contains two ingredients: density dependent mean field and in－medium nucleon－nucleon cross－section. To describe isospin effects appropriately, QMD should be modified properly: the density dependent mean field should contain correct isospin terms including symmetry potential and Coulomb potential, the in－medium nucleon－nucleon cross－sections should be different for neutron－neutron(proton－proton)and neutron－proton collisions, in which the Pauli blocking should be counted by distinguishing neutrons and protons. In addition, the initial condition of the ground state of two colliding nuclei should also contain isospin information.

Considering the above ingredients, we have made important modifications in QMD to obtain an isospin－dependent quantum molecular dynamics(IQMD)[1, 3]. The initial

density distributions of the colliding nuclei in IQMD are obtained from the calculations of the Skyrme－Hatree－Fock with parameter set SKM[39]. The initial code of IQMD was used to determine the ground state properties of the colliding nuclei, such as the binding energies and RMS radii, which agree with the experimental data for obtaining the parameters of interaction potential as an input data for the collision dynamics calculations by using the code of IQMD.

The interaction potential is

$$U(\rho)=U^{Sky}+U^{C}+U^{sym}+U^{Yuk}+U^{MDI}+U^{Pauli} \tag{1}$$

where U C is Coulomb potential.

The density dependent Skyrme potential U Sky, the Yukawa potential U Yuk, the momentum dependent interaction U MDI and the Pauli potential U Pauli[30, 37]are given by the following equations, respectively

$$U^{Sky}=\alpha\left(\frac{\rho}{\rho_0}\right)+\beta\left(\frac{\rho}{\rho_0}\right)^{\gamma}$$

$$U^{Yuk} = t_3 \exp\left(\frac{|\bar{r}_1 - \bar{r}_2|}{m}\right) \Big/ \frac{|\bar{r}_1 - \bar{r}_2|}{m}$$

$$U^{MDI} = t_r \ln^2[t_5(\bar{p}_1 - \bar{p}_2)^2 + 1] \frac{\rho}{\rho_0}$$

And

$$U^{Pauli} = V_P \left\{ \left(\frac{\hbar}{p_0 q_0}\right)^3 \exp\left(-\frac{(\vec{r}_1 - \vec{r}_2)^2}{2{q_0}^2} - \frac{(\vec{p}_i - \vec{p}_j)^2}{2{p_0}^2}\right) \right\} \delta_{p_i p_j}, \tag{5}$$

With

$$\delta_{pipj} = \begin{cases} 1, for \quad neutron - neugron\ or\ proton - proyou \\ 0, for \quad neutron - proton \end{cases}$$

In this work we used following different symmetry potentials[1,3]:

$$U_1^{sym} = \pm 2e_a u\delta, U_2^{sym} = \pm 2e_a u^2\delta + e_a u^2 \delta^2, U_0^{sym} = 0.0. \tag{6}$$

Here ea is the strength of symmetry potential taking the value of 16 MeV and $U_0^{sym} = 0.0$ indicates the case without any symmetry potential. $u = \frac{\rho}{\rho_0}$, δ is the relative neutron excess $\delta = \frac{\rho_n - \rho_p}{\rho_n + \rho_p} = \frac{\rho_n - \rho_p}{\rho}$. Here ρ, ρ_0, ρ_n, *and* ρ_p are total, normal, neutron, and proton densities, respectively. In the first terms on the right-hand side of Eq. (6) the upper + means repulsive for neutrons and the lower − is attractive for protons. The parameters of interaction potentials are in Table 1.

The NOMDI in Table 1 means without MDI. The influence of medium correction on the nucleon−nucleon cross−section is an important topic in HIC at intermediate energies[31,32]. D. Klakow et al. proposed that the in−medium nucleon−nucleon cross−section should be a function of the nucleon distribution density as follows[33]

$$\sigma_{NN} = \left(1 + \alpha \frac{\rho}{\rho_0}\right) \sigma_{NN}^{free} \tag{7}$$

Table 1 The parameters of the interaction potential

	α (MeV)	β (MeV)	γ (MeV)	t_3 (MeV)	m (fm)	t_4 (MeV)	t_5 (MeV^{-2})	V_p (MeV)	p_0 (MeV/c)	q_0 (fm)
MDI	−390.1	320.3	1.14	7.5	0.8	1.57	5×10^{-4}	30	400	5.64
NOMDI	−356	303	1.1667	7.5	0.8	0.0	0×10^{-4}	30	400	5.64

The parameter $\alpha = -0.2$ has been found to reproduce the flow data[34,35]. The free neutron−proton cross−section is about a factor of 3 times larger than the free proton−proton or the free neutron−neutron one below 400 MeV, which contributes the main isospin effect from nucleon—nucleon collisions at intermediate heavy ion collisions. In fact, the ratio of the neutron—proton cross—section to proton—proton (or neutron—neutron) cross—section in the medium, σnp / σpp, depends sensitively on the evolution of the nuclear density distribution and beam energy. We used Eq. (7) to take into account the medium effects, in which the neutron—proton cross—section is always larger than the neutron—neutron or proton—proton cross—section in the medium at the beam energies in this paper. Here σNNfree is the experimental nucleon—nucleon cross—section[36].

$$\sigma_{np}^{free}=\begin{cases}381.0(mb), & E\leqslant 25(MeV),\\ \frac{5067.4}{E^2}+\frac{9069.2}{E}+6.9466(mb), & 25<E\leqslant 40(MeV),\\ \frac{239380.0}{E^2}+\frac{1802.0}{E}27.14(mb), & 40<E\leqslant 310(MeV),\\ 34.5(mb), & 310<E\leqslant 800(MeV)\end{cases} \quad (8)$$

$$\sigma_{nn}^{free}=\sigma_{pp}^{free}=\begin{cases}80.6(mb), & \\ -\frac{1174.8}{E^2}+\frac{3088.5}{E}+5.3107(mb), & 25<E\leqslant 40(MeV),\\ \frac{93074.0}{E^2}+\frac{11.148}{E}+22.429(mb), & 40<E\leqslant 310(MeV),\\ \frac{887.37}{E^2}+0.05331E+3.5475(mb), & 310<E\leqslant 800(MeV),\end{cases}$$

(9)

We constructed the clusters by means of a modified coalescence model[37], in which particle relative momentum is smaller than $p_0=300$ MeV/c and relative distance is smaller than $R_0=3.5$ fm. The restructured aggregation model[38] has been applied to avoid the nonphysical clusters after constructing the clusters, until there were not any nonphysical clusters to be produced.

3. Results and discussions

The isospin effect of the in—medium nucleon—nucleon cross—section on the observables is defined by the difference between the observables for an isospin dependent nucleon—nucleon cross—section σ iso and for an isospin independent one σ noiso in the medium. Here σ^{iso} is defined as $\sigma_{np}\geqslant\sigma_{nn}=\sigma_{pp}$ and σ^{noiso} means $\sigma_{np}=\sigma_{nn}=\sigma_{pp}$, where, σ_{np}, σ_{nn}, and σ_{pp} are the neutron—proton, neutron—

neutron, and proton—proton cross—sections in medium, respectively.

3. 1. Influence of medium correction of two—body(collision) on the Nn and Nimf

In order to study the influence from the medium correction of the isospin dependent nucleon—nucleon cross—section on the isospin effects of the fragmentation and the nucleon emission, we investigated the number of nucleon emission Nn as a function of the beam energy at impact parameter b=4. 0 fm for the mass symmetry system 76Kr+76Kr(top panels) and mass asymmetry system 112Sn+40Ca(bottom panels). Two colliding systems are the same system mass At+Ap =152, where At and Ap are projectile mass and target mass, respectively. The different symmetry potentials U1sym, U2sym, and U0sym as well as different kinds of nucleon—nucleon cross—sections, i. e. , the isospin dependent in—medium nucleon—nucleon cross—section σ iso and the isospin independent one σ noiso are used here. Namely, there are six cases: U0sym + σ iso, U1sym + σ iso, U2sym+σ iso, U1sym+σ noiso, andU2sym+σ noiso with α=—0. 2(left panels) and α=0. 0(right panels) in Fig. 1. The detail explanation about line symbols in Fig. 1 is given in figure. It is clear to see that all of lines with filled symbols are larger than those with open symbols, i. e. , all of Nns with σ iso are larger than those with σ noiso because the collision number is larger for σ iso than that for σ noiso. We also found that the gaps between lines with the filled symbols and with the open symbols are larger but the variations among lines in each group are smaller. Because the large gaps come from the isospin effect of two—body collision and small variations are produced from the symmetry potential, i. e. , Nn depends sensitively on the isospin effect of in—medium nucleon—nucleon cross—section and weakly on the symmetry potential. In particular, the gaps between two group lines with α=—0. 2 are larger than those with, i. e. , the medium correction of two—body collision enhances the dependence of Nn on isospin effect of two—body collision.

In order to investigate quantitatively the influence of the medium correction of two—body collision on the nucleon emission, it is to define

$$\Delta N_N(\Delta\sigma,\alpha) = N_n(\sigma^{iso},\alpha) - N_n(\sigma^{noiso},\alpha) \qquad (10)$$

Fig. 2shows the time evolution of(solid line) and(dashed line) at impact parameter b=4. 0 fm for 76Kr+76Kr at E=100 MeV/nucle—on(top panels) and

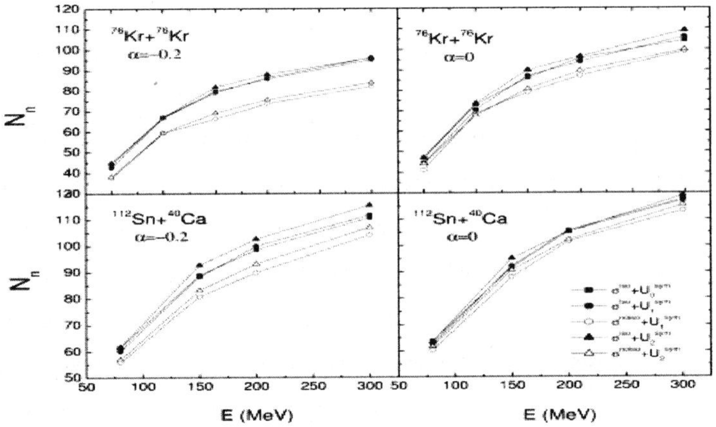

Fig. 1. The nucleon emission number Nn as a function of the beam energy for systems 76Kr+76 Kr and 112 Sn+40 Ca for six cases(see text).

112Sn+40Ca at E=200 MeV/nucleon(bottompanels). In this case, there are about the same center of mass energy per nucleon for two colliding systems. The symmetry potentials are U1sym(left panels)and U2sym(right panels). It is clear to see that all of solid lines are higher than dashed lines, i. e. , the medium correction of nucleon—nucleon cross—section increases sensitively the dependence of Nn on the isospin effect of two—body collision as above mentioned.

To investigate the evolution of above dependence with increasing beam energy for two different colliding systems, the variation $\Delta N_n(\Delta\sigma,\Delta\alpha)$ as a function of beam energy at impact parameter of 4. 0 fm for symmetry potential U1sym is given in Fig. 3. Where $\Delta N_n(\Delta\sigma,\Delta\alpha)$ is defined as

$$\Delta N_n(\Delta\sigma,\Delta\alpha) = \Delta N_n(\Delta\sigma,\alpha = -0.2) - \Delta N_n(\Delta\sigma,\alpha = 0.0) \tag{11}$$

In which $\Delta N_n(\Delta\sigma,\alpha)$ is taken from Eq. (10). From Fig. 3 we can see that all of $\Delta N_n(\Delta\sigma,\Delta\alpha)$ are larger than zero, i. e. , the medium correction of the nucleon—nucleon cross—section increases the dependence of Nn on the isospin effect of two—body collision in the beam energy region from 50 MeV/nucleon to 300 MeV/nucleon. we also find that $\Delta N_n(\Delta\sigma,\Delta\alpha)$s for mass symmetry system are always larger than those for mass asymmetry system due to more collision number for the mass symmetry system with the same system mass.

In order to investigate the contributions from all of impact parameters to Nn, Fig. 4 shows the impact parameter average values of Nn b(from at equilibri-

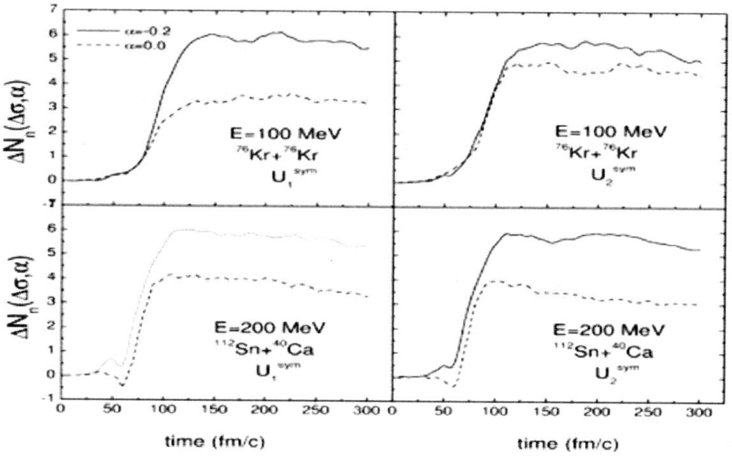

Fig. 2. The time evolution of $\Delta N_n(\Delta\sigma,\alpha)$ for the systems as the same as Fig. 1 at E=100 MeV/nucleon(top panels) and 200 MeV/nucleon(bottom panels) for two symmetry potentials(see text)

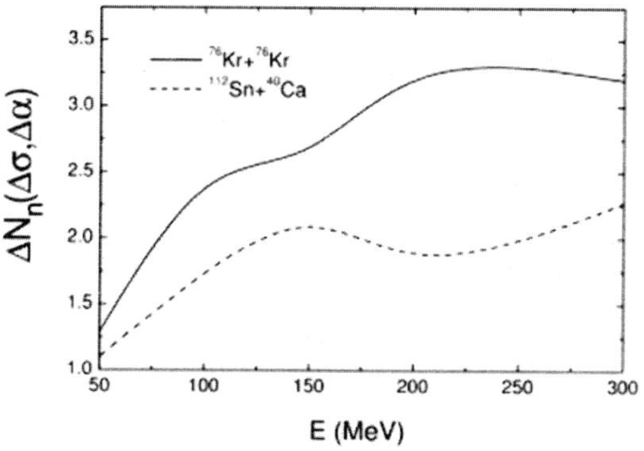

Fig. 3. $\Delta N_n(\Delta\sigma,\Delta\alpha)$ as a function of the beam energy for the systems as the same as Fig. 1 in two cases(see text).

um time 200 fm/c)as a function of the beam energy for above two colliding systems with the same incident channel conditions and line symbols as in Fig. 1. The same conclusion as mentioned in Fig. 1 is also obtained here, i. e. , the medium correction of nucleon—nucleon cross—section enhances the dependence of Nn b on the isospin effect of two—body collision.

Fig. 5 shows the impact parameter average values of the multiplicity of intermediate mass fragment, Nimf b(at equilibrium time 200 fm/c), as a function of the beam energy for the same incident channel conditions and line symbols as in Fig. 4. Where the charge number of intermediate mass fragment is taken from 3 to 13. From Fig. 5 we got the same conclusions as Nn b, namely, Nimf b depends sensitively on the isospin effect of in—medium nucleon—nucleon cross—section and weakly on the symmetry potential. In particular, the gaps between two group lines with $\alpha=-0.2$ are larger than those with $\alpha=0.0$ which indicates that the medium correction enhances also the dependence of $\langle N_{imf}\rangle_b$ on the isospin effect of the two—body collision.

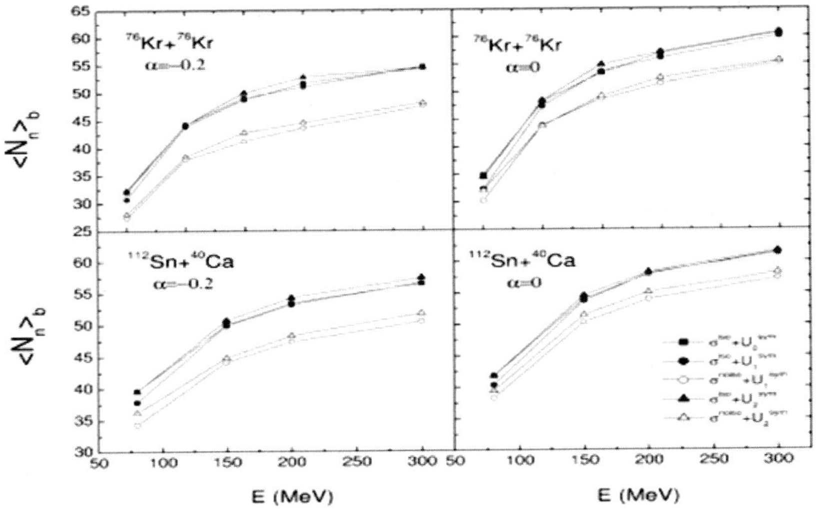

Fig. 4. The impact parameter average value of the number of nucleon emission $\langle N_n\rangle_b$ as a function of beam energy for the same incident channel conditions and line symbols as Fig. 1(see text).

3.2. Important role of MDI on Nimf and Nn in the medium corrections of two—body collision

We also found an important role of the MDI on Nimf and Nn in the medium correction of two—body collision. Fig. 6 shows the time evolution of Nimf for the reaction 76Kr+76Kr with symmetry potentials U1sym at beam energy of 100 MeV/nucleon and impact parameter of 4.0 fm. They are four cases:

(1) $\alpha=-0.2+\sigma^{iso}$ (solid line),

(2) $\alpha=0.0+\sigma^{iso}$ (dashed line),

(3) $\alpha=-0.2+\sigma^{noiso}$ (dot line), and

(4) $\alpha=0.0+\sigma^{noiso}$ (dot—dashed line) with MDI in the left window and NOMDI in the right window.

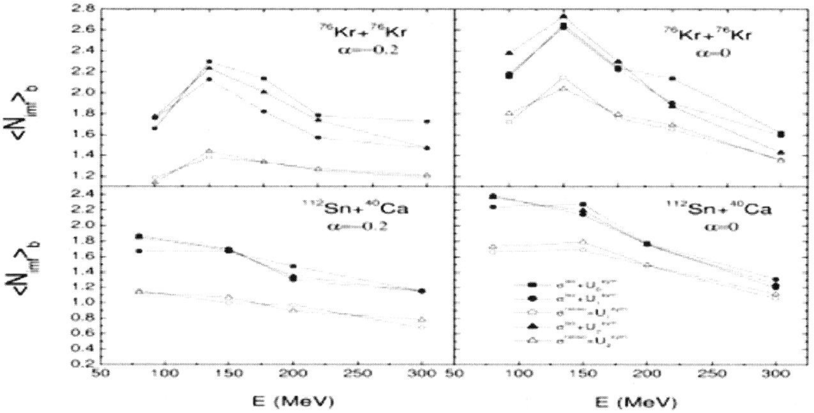

Fig. 5. The impact parameter average values of the multiplicity of intermediate mass fragment, as a function of the beam energy for the same incident channel conditions and line symbols as Fig. 4.

It is clear to see that the gap between the lines with MDI is larger than corresponding gap between lines with NOMDI in the medium ($\alpha=-0.2$), i. e., MDI increases the isospin effect of two—body collision on the Nimf in the medium because above gaps are produced from the isospin effect of nucleon—nucleon cross—section in the medium.

For the $\langle N_n \rangle_b$ we have gotten the same conclusion as $\langle N_{imf} \rangle_b$ in Fig. 6.

3.3. Explanations for the medium correction of two—body collision and the role of MDI on the Nimf and Nn

Why does the medium correction of nucleon—nucleon cross—section and MDI enhance the dependences of multiplicity of intermediate mass fragments Nimf and the number of nucleon emission Nn on the isospin effect of two—body collision? Physically there are three mechanisms at work here.

The average momentum of a particle in medium is higher in a heavy ion collision than in cold nuclear matter at the same density.

(2) MDI induces the transporting momentum more effectively from one part of the system to another, in which particles also move with a higher velocity in the medium than in free space for a given momentum.

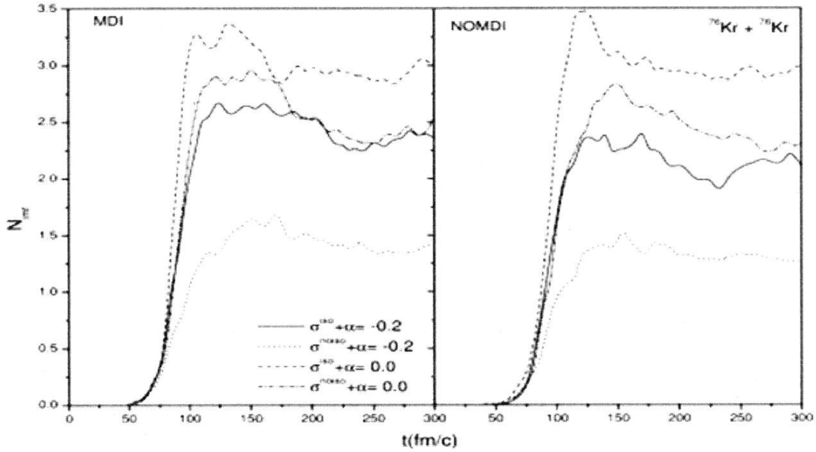

Fig. 6. The time evolution of the Nimf with MDI(solid line)and NOMDI(dot line)for systems 76 Kr+76Kr at E=100 MeV/nucleon and b=4. 0 fm(see text)

(3)As well know that the isospin dependent in—medium nucleon—nucleon cross—section is a sensitive function of the nuclear density distribution and beam energy as shown in Eq. (7).

Fig. 7 shows the time evolution of the ratio of nuclear density to normal one, $\frac{\rho}{\rho_0}$,for four cases:they are ρ(σ iso,α=—0. 2)(solid line),ρ(σ noiso,α=—0. 2)(dotted line),ρ(σ iso,α=0. 0)(dashed line),and ρ(σ noiso,α=0. 0)(dot—dashed line)for the reaction 76Kr+76Kr with symmetry potential U_1^{Sym} at E=150 MeV/nucleon and b=4. 0 fm.. From the values of peak for $\frac{\rho}{\rho_0}$ in the insert in Fig. 7 it is clear to see that ρ(σ iso,α=—0. 2)(solid line)is larger than ρ(σ noiso,α=—0. 2)(dot line)and ρ(σ iso,α=0. 0)(dashed line)is larger than ρ(σ noiso,α=0. 0)(dot—dashed line)because the larger collision number from σ iso increases the nuclear stopping and dissipation,which enhances the nuclear density,compared to the case with σ noiso. From Fig. 7 we can also see that $\frac{\rho}{\rho_0}$ decreases quickly with increasing the time after the peak of $\frac{\rho}{\rho_0}$. During above process the larger compression produces quick expanding process of the colliding system and the small compression induces slow expanding process,at the same time,the $\frac{\rho}{\rho_0}$ decreasequickiy with expanding process of system. But the decreasing velocity of $\frac{\rho}{\rho_0}$ is larger for the quick expansion system than that for the slow expansion system, up to about after 70 fm/c, on the contrary, ρ(σnoiso, α=−0.2)(dot line) is larger than ρ(αiso, α=−0.2)(solid line) and ρ(σnoiso, α=0.0)(dot−dashed line) is larger than ρ(σ iso, α=0.0)(dashed line). In particular, the gap between two lines for α =−0.2 is larger than that for α=0.0 after about 70 fm/c. This property is very similar to the Nifm and Nn, which means that the medium correction of an isospin dependent nucleon−nucleon cross−section enhances also the dependence of ρ on the isospin effect of two−body collision, which induces the same effects on Nifm and Nn through the nucleon−nucleon cross−section as a function of the nuclear density as shown in Eq. (7).

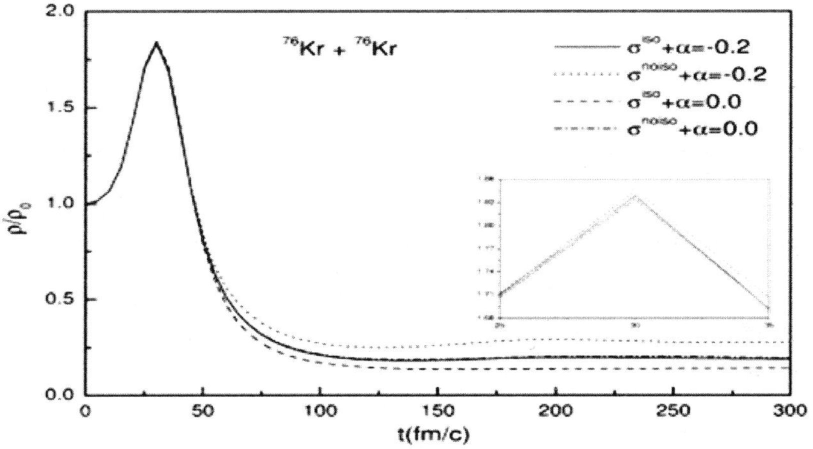

Fig. 7. The time evolution of the ratio of nuclear density to normal density $\frac{\rho(\sigma^{iso},\alpha)}{\rho_0}$ for four cases (see text)

4. Summary and conclusion

We studied the influences of the medium correction of the isospin dependent nucleon−nucleon cross−section and MDI on the fragmentation and nucleon emission in the heavy ion collisions at intermediate beam energies by using the IQMD. From the calculation results we can get the following conclusions:

(1) $\langle N_n \rangle_b$ and $\langle N_{imf} \rangle_b$ depend sensitively on the isospin effect of nucleon−nucleon cross−section and weakly on the symmetry potential.

(2)In particular, the medium correction of nucleon—nucleon cross—section enhances the dependence of $\langle N_n \rangle_b$ and $\langle N_{imf} \rangle_b$ on the isospin effect of nucleon—nucleon cross—section in the intermediate beam energy region.

(3)MDI produces an important role for enhancing the isospin effect of two—body collision on the $\langle N_n \rangle_b$ and $(N_{imf} \rangle_b$ due to the medium correction.

Acknowledgements

We thank Prof. Bao—An Li for helpful discussions. This work is supported by the Major State Basic Research Development Programme of China under Grant No. G2000077400, the National NatureScience Foundation of China under Grant Nos. 10175080, 10004012, 19847002, and 10175082; 100 person project of Chinese Academy of Sciences and Knowledge Innovation Project of Chinese A-cademy of Sciences under Grant No. KJCX2—SW—N02.

References

[1]B. A. Li, W. Udo Schroder, Isospin Physics in Heavy—Ion Collision at Intermediate Energies, Nova Science, New York, 2001.

[2]M. S. Hussein, R. A. Rego, C. A. Bertulani, Phys. Rep. 201(1993)279.

[3]B. —A. Li, C. —M. Ko, W. Bauer, Int. J. Mod. Phys. E 7(1998)147.

[4]R. Wada, et al., Phys. Rev. Lett. 58(1987)1829.

[5]S. J. Tennello, et al., Phys. Lett. B 321(1994)14; S. J. Tennello, et al., Nucl. Phys. A 681(2001)317c.

[6]R. Pak, et al., Phys. Rev. Lett. 78(1997)1022; R. Pak, et al., Phys. Rev. Lett. 78 (1997)1026.

[7]G. D. Westfall, Nucl. Phys. A 630(1998)27c; G. D. Westfall, Nucl. Phys. A 681(2001) 343c.

[8]G. J. Kunde, et al., Phys. Rev. Lett. 77(1996)2897.

[9]M. L. Miller, et al., Phys. Rev. Lett. 82(1999)1399.

[10]H. Xu, et al., Phys. Rev. Lett. 85(2000)716; M. B. Tsang, et al., Phys. Rev. Lett. 86 (2001)5023.

[11]W. Udo Schroder, et al., Nucl. Phys. A 681(2001)418c.

[12]L. G. Sobotka, et al., Phys. Rev. C 55(1994)R1272.

[13]F. Rami, et al., Phys. Rev. Lett. 84(2000)1120.

[14]M. Farine, T. Sami, B. Remaud, et al., Z. Phys. 339(1991)363.

[15]H. Muller, B. D. Serot, Phys. Rev. C 52(1995)2072.

[16]B. A. Li, et al., Phys. Rev. Lett. 76(1996)4492; B. A. Li, et al., Phys. Rev. Lett. 78(1997)1644.

[17]G. Kortmeyer, W. Bauer, G. J. Kunde, Phys. Rev. C 55(1997)2730.

[18]M. Colonna, et al., Phys. Lett. B 428(1998)1; V. Baran, et al., Nucl. Phys. A 632(1998)287; V. Baran, M. Colonna, M. DiToro, et al., Nucl. Phys. A 703(2002)603

[19]J. Pan, S. Das Gupta, Phys. Rev. C 57(1998)1839.

[20]P. Chomaz, F. Gulminelli, Phys. Lett. B 447(1999)221.

[21]J. Y. Liu, et al., Phys. Rev. C 63(2001)054612; J. Y. Liu, et al., Nucl. Phys. A687(2001)475.

[22]B. A. Li, Phys. Rev. Lett. 85(2000)4221; B. A. Li, Phys. Rev. C 64(2001)054604.

[23]B. -A. Li, Phys. Rev. Lett. 88(2002)192701.

[24]J. Y. Liu, et al., Phys. Rev. Lett. 86(2001)975;

J. Y. Liu, Y. Z. Xing, W. J. Guo, et al., Chin. Phys. Lett. 19(8)(2002)1078.

[25]J. Y. Liu, W. J. Guo, Y. Z. Xing, Phys. Lett. B 540(2002)213.

[26]J. Y. Liu, W. J. Guo, Y. Z. Xing, et al., Phys. Rev. C 67(2003)024608.

[27]D. R. Bowman, C. M. Mader, et al., Phys. Rev. C 46(1992)1834.

[28]M. L. Miller, O. Bjarki, et al., Phys. Rev. Lett. 82(1999)1399.

[29]R. Pak, B. -A. Li, W. Benenson, et al., Phys. Rev. Lett. 78(1997)1026.

[30]J. Aichelin, G. Peilert, A. Bohnet, et al., Phys. Rev. C 37(1988)2451.

[31]G. Q. Li, R. Machleidt, Phys. Rev. C 49(1994)566, and references therein.

[32]Q. F. Li, Z. X. Li, G. J. Mao, Phys. Rev. C 62(2000)014606; Q. F. Li, Z. X. Li, Chin. Phys. Lett. 19(3)(2002)321.

[33]D. Klakow, G. Welke, W. Bauer, Phys. Rev. C 48(1993)1982.

[34]M. J. Huang, et al., Phys. Rev. Lett. 77(1996)3739.

[35]G. D. Westfall, et al., Phys. Rev. Lett. 71(1993)1986.

[36]K. Chen, Z. Fraenkel, G. Friedlander, et al., Phys. Rev. 166(1968)949.

[37]G. F. Bertsch, S. D. Gupta, Phys. Rep. 160(1988)1991.

[38]C. Ngo, H. Ngo, S. Leray, et al., Phys. Rep. A 499(1989)148.

[39]P. G. Reinhard, in: Computational Nuclear Physics, Vol. 1, Springer, Berlin, 1991, pp. 28-50.

(本文发表于 2003 年《Nuclear Physics A》第 723 卷)

The 2p photoionization of ground—state sodium in the vicinity of Cooper minima

Xiaobin Liu　Yinglong Shi　Chenzhong Dong*

摘要:利用多组态 Dirac—Fock 方法研究了基态 Na 的 2p 光电离过程.本文的计算值与其他的实验和理论结果进行了分析比较,显示出良好的一致性,充分说明了多组态 Dirac—Fock 方法的可靠性.在近电离阈的能量范围,由于 Cooper 极小的存在光电离截面随入射光子能量的变化显示出明显的非单调性,Cooper 极小源于主矩阵元的符号变化且通常对电子关联非常敏感.随入射光子能量的连续增加,出射光电子的径向波函数将朝靠近核的方向移动.结果表明:光电离截面及 Cooper 极小的位置主要依赖于初束缚电子 $2p_{1/2,3/2}$ 与末态出射连续光电子径向波函数的相对位置.

The photoionization processes of ground—state sodium have been investigated with the multiconfiguration Dirac—Fock method. The results are in good or at least reasonable agreement with available experimental and theoretical data. In the energy region near the threshold, the cross sections show non—monotonic changes because of Cooper minima, which due to the sign changes of dominant dipole matrix elements and are very sensitive to electron correlations. As the energy increases continuously, the radial wave functions of the photoelectrons will move towards the nucleus. The values of the cross sections, and hence the Cooper minima, mainly depend on the relative positions of the one—electron radial wave functions of the initial bound electrons $2p_{1/2,3/2}$ and the continuum photoelectrons.

---

* 作者简介:刘晓斌(1973—),男,甘肃天水人,理学博士,天水师范学院电子信息与电气工程学院教授,主要从事原子结构与光电离方面的研究。

1 INTRODUCTION

The photoionization of atoms is one of the fundamental processes of atomic physics, and is very important in various studies of the stellar atmosphere, controlled thermonuclear plasma, highly correlated materials, radiation protection and laser technology. Usually, the understanding and modeling of matter in the plasma state, whether natural or artificial, requires the knowledge of photoionization, and the atom's photoionization cross sections are needed for the determination of plasma opacities and plasma diagnostics [1]. For ground- and excited-state sodium there have been both experimental [2-11] and theoretical [12-22] reports over the last two decades, but comparatively little is known about 2p photoionization in the energy region close to the ionization thresholds. Therefore, it would be worthwhile to close this gap, and there are sophisticated many-electron atomic codes available today that could greatly facilitate the task.

One important feature of photoionization processes, which is sensitive to relativistic effects and electron correlations, is the Cooper minimum [23-27]. The cross sections near the ionization thresholds become interesting due to these minima, which are not due to any sort of resonance effects, but rather are due to the sign changes of dominant dipole matrix elements as they pass through zeroes. The Cooper minima, which occur extensively in the photoionization of outer and near-outer subshells of atoms throughout the periodic table [28], have significant effects on the shape of the photoionization cross sections, the angular distributions, and the ratios [29,30]. After Cooper's pioneering work [23,24], many experimental and theoretical studies were performed to investigate the Cooper minimum; in one recent experiment conducted at the University of Wisconsin Synchrotron Radiation Center [31], the 2p photoionization cross sections from the ground state $1s^2 2s^2 2p^6 3s^1 (^2S_{1/2})$ were calculated using a complete relativistic multiconfiguration Dirac-Fock (MCDF) method. In addition, the photon energy dependences of the cross section ratios $^3P_0/^3P_2$ and $^3P_1/^3P_2$ for different final ionic states $1s^2 2s^2 2p^5 3s^1 (^3P_{0,1,2})$ were studied and compared with the experimental intensity ratios [9].

The aim of our work is to provide detailed results for the 2p photoionization processes near the ionization thresholds. Calculations are performed in a relativistic framework and theoretical analysis of the results is presented. Finally, a brief summary and conclusions are given. Atomic units ($e = m = \hbar = 1$) are used throughout this paper, unless otherwise indicated.

2 THEORY

The present work is based on the MCDF packages GRASP92 [32,33] and RATIP [34], with the recently developed Relphoto08 component, which has already been used to study the photoionization processes in our previous papers [35—37]. Within this approach, the atomic state functions(ASFs)are formed as linear combinations from the jj—coupled configuration state functions(CSFs) with the same parity P, same total angular momentum J, and one of its projections M

$$| \alpha(PJM) \rangle | = \sum_{r=1}^{n_c} c_r(\alpha) \mid \gamma_r(PJM) \rangle \quad (1)$$

where n_c is the number of CSFs. The mixing coefficients$cr(a)$, which can be obtained by diagonalizing the electron—electron interaction matrix, represent the electron correlations, and γ_r represents the occupation of the shells as well as all further quantum numbers from the coupling of these shells that are required for a unique specification of the N—electron basis. In most standard computations, the CSFs are constructed as antisymmetrized products of a common set of orthonormal orbitals and are optimized on the basis of the Dirac—Coulomb Hamiltonianin [33]. Because of the good LSJ—coupling conditions in Na, the inherently jj—coupled ASFs are interpreted in the LSJ basis. The unitary transform between the two bases is performed by applying the LSJ program [38]. Further relativistic contributions to the representation $\{cr(a)\}$ could be added for medium and heavy elements, which often helps to improve the level structure and transition matrix elements, but they are found to play a negligible role for Na in our present calculations.

The photoionization cross section from an initial state i to a final state f is calculated as follows[39]

$$\sigma_{if}^{PI} = 4\pi^2 a_0^2 \alpha \frac{df_{if}}{d\varepsilon} \quad (2)$$

where α is the fine structure constant, a_0 is the Bohr radius and ε is the kinetic energy of the outgoing photoelectron. The oscillator strength density $df_{if}/d\varepsilon$ can be calculated by

$$\frac{df_{ij}}{d\varepsilon}=\frac{\pi c}{(2L+1)\omega^2}\left|\langle\alpha_f^{N-1}(P'_fJ'_fM'_f),\varepsilon\kappa;\alpha_f^N(P_fJ_fM_f)\parallel\hat{O}^{(L)}\parallel\alpha_i^N(P_iJ_iM_i)\rangle\right|^2 \tag{3}$$

where ω is the photon energy, $|\alpha_i^N(P_iJ_iM_i|$ is the ASF of the initial state, and $|\alpha_f^{N-1}(P'_fJ'_fM'_f),\varepsilon\kappa;\alpha_f^N(P_fJ_fM_f)\rangle$ is the ASF of the final state constructed by antisymmetrical products of the wave functions of the final ionic state $|\alpha_f^{N-1}(P'_fJ'_fM'_f)\rangle$ and photoelectron continuum state $|\varepsilon\kappa\rangle$, while $\hat{O}^{(L)}$ is the radiative field multiple operator with order L.

Relphoto08 calculates partial photoionization cross sections by decomposing the atom—photon interaction matrix into one—electron matrix elements for each pair of CSFs within the given basis [40]. The photoionization cross section is then obtained by summation of partial cross sections over all allowed final states $|\alpha_f^{N-1}(P'_fJ'_fM'_f),\varepsilon\kappa;\alpha_f^N(P_fJ_fM_f)\rangle$. Apart from the photoemission of just an inner—shell electron, rearrangement of the residual bound—state electron density is likely to occur, which will affect the photoionization processes. In the following, we shall employ the initial and final(bound) states separately in order to include the main parts of the relaxation effects. Such relaxations are found to be important for the inner—shell photoionization, and may modify the cross sections by as much as 30% or even more [41]. The Babushkin gauge, which reduces to the length form in the nonrelativistic limit, was chosen for the present calculations.

3 RESULTS AND DISCUSSIONS

In the 2p photoionization processes, the initial atomic state and final ionic states were generated in three different computational models: (A) by using all CSF from the single reference configurations $1s^2 2s^2 2p^6 3s$ and $1s^2 2s^2 2p^5 3s$, (B) from $1s^2 2s^2 2p^6 (3s+4s)$ and $1s^2 2s^2 2p^5 (3s+3d+4s)$, and (C) from $1s^2 2s^2 2p^6 (3s+4s)$ and $1s^2 2s^2 2p^5 (3s+3d+4s+4d)$. Usually, the main limitations of the MCDF approach occur for open—shell structures from missing parts of electron correlations due to the restricted size of the wave function expansion, but further attempts to enlarge the CSF basis fail because of the convergence problems. In the following, model (C) was used especially for the photoionization calculations in order to include more electron correlations.

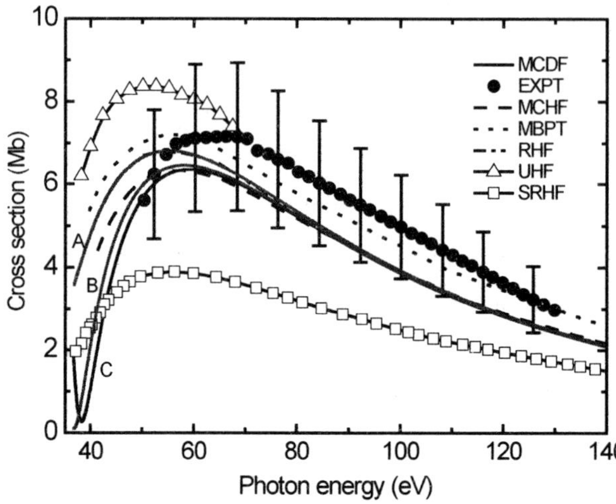

Figure1. Total photoionization cross sections of the final configuration $1s^2 2s^2 2p^5 3s^1$. Solid lines are the results which come from our MCDF calculations in the Babushkin gauge with the wave functions from computational models (A)—(C). The other theoretical results in the length gauge are the MCHF [13] (dashed line), MBPT [42] (dotted line), relaxed Hartree—Fock [43] (dash—double—dotted line), unrelaxed Hartree—Fock [44] (open triangles), and single channel relaxed Hartree—Fock calculations [19] (open squares). The experimental results (closed circles) are from [7], and their error bars are shown at ten photon energies.

Electron correlations are essential for an understanding of photoionization processes, and thereby to obtain reliable theoretical values for the associated cross sections and rates that can compete with the corresponding experimental measurements and other theories. Figure 1 shows the 2p total photoionization cross sections in the photon energy range 35—140 eV. Our theoretical results, based on the wave functions from models(A)—(C), are presented as the sum over four cross sections of thefinal configuration$1s^2 2s^2 2p^5 3s^1$ ($^3P_{0,1,2}$, $^1P_1$) since the photoelectron experiments did not resolve the fine structure of the final ionic states[7]. For comparison, the other five theoretical results, calculated in the length gauge, are also shown in figure 1. It is seen that the total cross sections from the relaxed Hartree—Fock(RHF) calculations [43] coincide with ours for model(A) near the peak. The MCDF calculations of models(B) and (C) are closer to the experimental data than the multiconfiguration Hartree-Fock (MCHF) [13] and RHF results in the energy range 55-140 eV. Although the results from the many-body perturbation theory (MBPT) [42] at low energies are somewhat higher than the corresponding experimental values, the agreement gets better as function of increasing energies as seen in the figure 1. Generally, all theoretical results, except for those of the single-channel relaxed Hartree-Fock (SRHF) [19] and unrelaxed Hartree-Fock (UHF) models [44], lie within the typical 25%-30% experimental uncertainty. The theoretical results which do not include any orbital relaxation effects, as in the UHF calculations, are not expected to agree with the experiments. The total cross section of our calculations are generally smaller than the experimental values, but a large portion can still be considered to lie within the experimental error bars. A better result may be obtained if some higher configurations are included in the models, but this is beyond the scope of our present work.

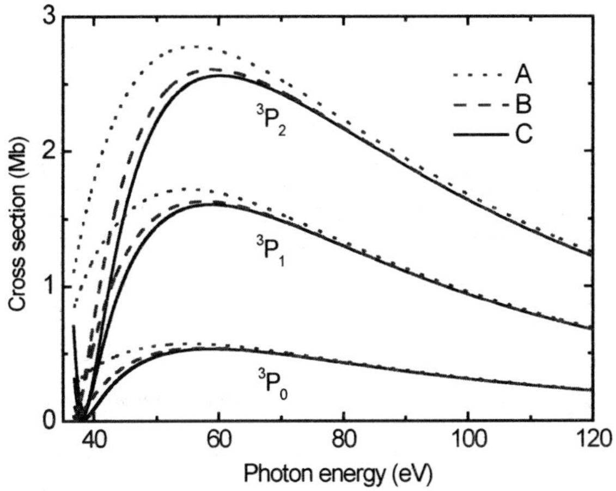

Figure 2. Photoionization cross sections of the final ionic states $1s^2 2s^2 2p^5 3s^1$ ($^3P_{0,1,2}$). Dotted lines are the cross sections with the wave functions from computational model (A), dashed lines are those from model (B), and solid lines from model (C).

The cross sections of the final ionic states ($^3P_{0,1,2}$) are shown in figure 2, but those of ($^1P_1$) are not included as they nearly coincide with those of ($^3P_1$). The outstanding characteristics of these theoretic results are the large deviations from monotonic changes in the low energy region, and the minima can be seen from models(B) and (C). For each final ionic state, the cross sections from different models are in good agreement with each other when the photon energy is large, but this is not the same case near the thresholds. The cross sections calculated from models(B) and (C) have their minima at different photon energies. These minima are identifiable as Cooper minima(Cooper 1962, Kim *et al* 1981, Johnson *et a*l 1982, Manson *et al* 1983) around which the dominant partial cross sections pass through zeroes due to the sign changes of the corresponding matrix elements. For different models, the minima positions are different from each other. Therefore, the study of Cooper minima provides a particularly sensitive test of how well the calculations describe the electron correlations in photoionization, and some care has to be taken to include the main parts of the correlations by using appropriate reference configurations.

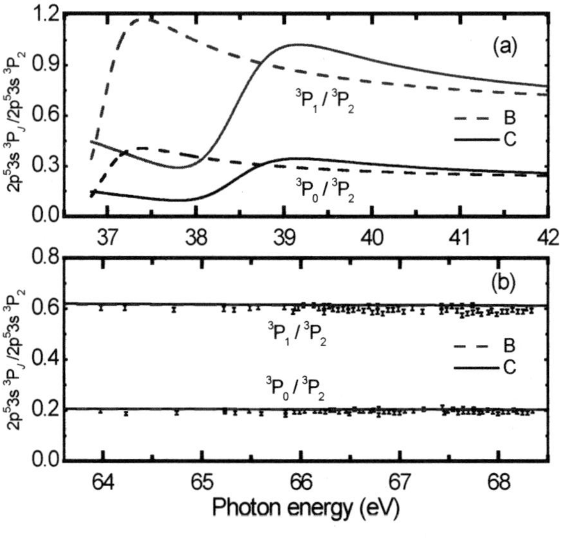

Figure3. The branching ratios of the cross sections of final ionic states $1s^2 2s^2 2p^5 3s^1$ ($^3P_{0,1,2}$) in different energy regions. Dashed lines show the ratios with the wave functions from computational model(B), solid lines are those from model (C), and the points with error bars are experimental data from [9].

Since the experimental values are relative [9], we cannot compare theoretical absolute cross sections with the experimental results. However, we can scrutinize the branching ratios, the cross sections ratios for the alternate final ionic states $1s^2 2s^2 2p^5 3s^1$ ($^3P_{0,1,2}$), for which experiment and theory can be compared on an absolute basis. The branching ratios from models (B) and (C) as a functions of photon energy are presented in figure 3, and the experimental data of the intensity ratios are also provided for comparison [9]. Figure 3(a) shows our calculated ratios in the energy range of 36.5—42 eV where Cooper minima occurred; we can see that the photoionization characteristics of various final ionic states differ from each other significantly. Our calculated ratios are certainly not statistical, i. e., they are not proportional to $2J'_f+I$ of the final ionic states, which confirms the fact that the photoionization cross sections pass through Cooper minima in the low energy region. Even for the same final ionic states, the differences between models (B) and (C) are large. These indicate that the electron correlations are very important in the energy region. In figure 3(b), it can be seen that the overall agreement between the experimental and theoretical branching ratios $^3P_0/^3P_2$ and $^3P_1/^3P_2$ is good. In sharp contrast to the results indicated in figure 3(a), not only do the ratios from models (B) and (C) nearly coincide with each other, but they are flat and just the statistical values of 0.2 and 0.6, respectively. This suggests the physical interpretation: if the differences in the $2J'_f+I$ quantum states can be neglected as well as the further many—electron effects, the photoionization cross sections should be proportional to $2J'_f+I$ of the final ionic state $^3P_{J'_f}$. We can therefore be confident that the pure fine—structure and correlation effects are not so important for the photionization processes if the photon energy is far above the threshold. Even so, the comparison of high—resolution measurements with advanced MCDF calculations shows some discrepancies makes it clear that our present understanding of the electron correlations is still far from being complete, even for a relatively simple system such as sodium. Figure 4 shows the cross sections of the final ionic states $1s^2 2s^2 2p^5 3s^1$ ($^3P_{0,1,2}$, $^1P_1$) with the wave functions from models (B) and (C) in the energy region near the thresholds. For a certain final state, the differences between the photoionization cross sections from different models are very obvious, especially in the energy region very close to the corresponding threshold despite the fact that the cross sections of $^3P_1$ nearly coincide with those of $^1P_1$. The positions of the Cooper minima are very sensitive to the accuracy of the wave functions by which the cross sections are calculated. Indeed, the cross sections from models (B) and (C) do not go through the Cooper minima at the same photon energy, though the same trends can be seen from the curves.

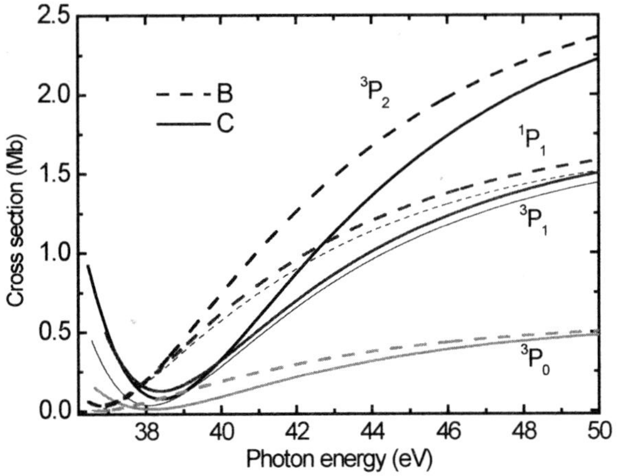

Figure4. The photoionization cross sections of $1s^2 2s^2 2p^5 3s^1$ ($^3P_{0,1,2}$, $^1P_1$) in the photon energy region near the thresholds. Dashed lines show the cross sections with the wave functions from model (B), and solid lines are from model (C).

For the $1s^2 2s^2 2p^6 3s^1$ ($^2S_{1/2}$)→$1s^2 2s^2 2p^5 3s^1$ ($^3P_2$) transition, the total angular momentum quantum number of the initial state is 1/2, and that of the final ionic state is 2. Within the electric dipole (E1) approximation, the angular momentum quantum numbers of the photoelectrons are $l=0$ ($\varepsilon s_{1/2}$) and $l=2$ ($\varepsilon d_{3/2}$, $\varepsilon d_{5/2}$), then the total quantum numbers for the final states are formed by coupling the total angular quantum numbers 1/2, 3/2 and 5/2 of the photoelectrons to the 2 of the final ionic state. To comply with the selection rule $\Delta j=0, \pm 1$ for electric dipole transitions, the total angular quantum number of the final state is 3/2, which is obtained from the coupling of the photoelectron $\varepsilon s_{1/2}$ with the final ionic state($^3P_2$), so we have 1/2 and 3/2 from the coupling of photoelectron $\varepsilon d3/2$ with the final ionic state($^3P_2$), and the other 1/2 and 3/2 from photoelectron $\varepsilon d_{5/2}$. Thus, the allowed photoelectrons are $\varepsilon s_{1/2}$, $\varepsilon d_{3/2}$, $\varepsilon d_{3/2}$, $\varepsilon_{5/2}$ and $\varepsilon d_{5/2}$. It should be noted that while the angular momentum and total angular quantum numbers of some photoelectrons are the same, the corresponding radial wave functions are trivially different because the photoelectrons come from different relativistic dipole coupling channels. The radial probability density of the continuum photoelectrons should be further studied to understand the appearance of Cooper minima. In the relativistic framework [32], the one-electron wave functions of bound electrons are spinors of rank 4:

$$\psi_{\varepsilon\kappa m}(\hat{r}) = \frac{1}{r}\begin{pmatrix} P_{\varepsilon\kappa}(r) & \chi_{\kappa m}(\theta,\phi) \\ iQ_{\varepsilon\kappa}(r) & \chi_{-\kappa m}(\theta,\phi) \end{pmatrix} \tag{4}$$

where the magnetic quantum numbers $m = -j \ldots j$, $P_{n\kappa}(r)$ and $Q_{n\kappa}(r)$ are the large and small component radial wave functions. The functions $\chi_{\pm\kappa m}(\theta,\phi)$ are spinors of rank 2 made up of spherical harmonics and Clebsch-Gordan coefficients. For a continuum photoelectron, the basic form is the same except that the kinetic energy is used instead of the principal quantum number n.

For photoionization of the final ionic state $1s^2 2s^2 2p^5 3s^1$ $(^3P_2)$, the radial probability density $|P_nk(r)|^2 + |Qnk(r)|^2$ of the initial bound electrons $2p_{1/2,3/2}$ from model (C), and that of continuum photoelectrons $\varepsilon s_{1/2}$, $\varepsilon d_{3/2}$, and $\varepsilon d_{5/2}$ for different photon energies are shown in figure 5. At 36.56 eV photon energy, clearly their overlap is small, especially for photoelectron $\varepsilon s_{1/2}$. This leads to small dipole matrix elements and, hence, a small cross section for this photon energy. As the energy increases to 38.38 eV, which is the calculated position of the Cooper minimum from model (C), the overlap is nearly zero for $\varepsilon d_{3/2}$ and $\varepsilon d_{5/2}$, but is relatively large for $\varepsilon s_{1/2}$. However, there are two nodes at $r \approx 0.2$ a.u. and $r \approx 1.0$ a.u., which implies there must be considerable cancellation of the positive and negative portions of the dipole matrix element since the left- and right-hand sides of the node have opposite signs, hence the photoionization cross section is very small. A similar explanation can also be used for 54.46 eV, where the overlap is large despite the fact that the radial probability density of photoelectron $\varepsilon s_{1/2}$ has three nodes.

Intuitively, the radial wave functions will *move towards* the nucleus as the photon energy (kinetic energy of the photoelectron) increases. The partial photoionization cross section of a specific relativistic dipole coupling channel, as a function of photon energy, thus reaches a maximum value and then starts to decrease due to the cancellation. At a particular energy point, the cancellation is complete when the positive and negative portions of the dipole matrix element are equal,

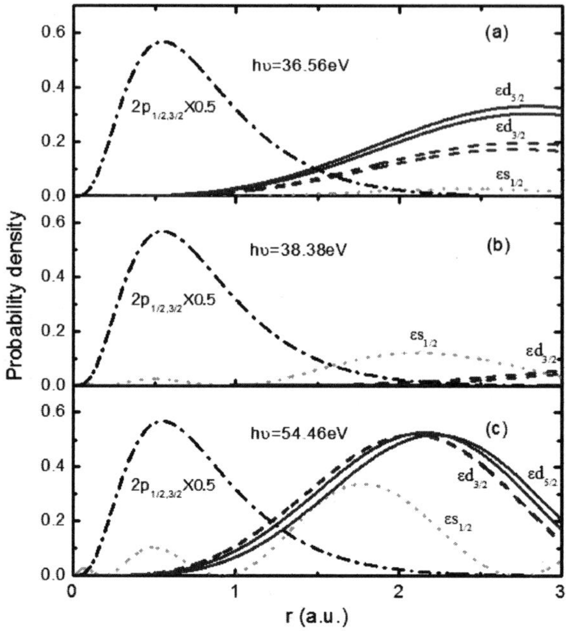

Figure5. For the transition $1s^2 2s^2 2p^6 3s^1\ (^2S_{1/2}) \rightarrow 1s^2 2s^2 2p^5 3s^1\ (^3P_2)$, the radial probability density of the initial bound electrons $2p_{1/2,3/2}$ from model (C), and that of continuum photoelectrons $\varepsilon s_{1/2}$, $\varepsilon d_{3/2}$, $\varepsilon d_{5/2}$ permitted by the electric dipole selection rule for different photon energies $h\nu = 36.56, 38.38$ and 54.46 eV. Solid lines show the radial probability density of photoelectrons $\varepsilon d_{5/2}$, dashed lines are those for $\varepsilon d_{3/2}$, dotted lines for $\varepsilon s_{1/2}$ and dash—dotted lines for the initial bound electrons $2p_{1/2,3/2}$. For comparison, the radial probability density of the initial bound electrons is multiplied by 0.5.

then the partial cross section of the channel will become zero and a Cooper minimum can be seen. Generally, when the photon energy increases continuously, the radial wave functions penetrate deeper and deeper into the core, and their oscillations become more and more rapid, as indicated in figure 6.

We can give the following qualitative interpretation for such behavior: the radial wave functions of continuum photoelectrons are essentially plane waves, i.e., they are sine or cosine functions at an infinite distance from the nucleus, and the influence from the final ion can be neglected. However, it must be emphasized that at very large but finite distances from the nucleus, they are not perfect plane waves. As the distance becomes closer, they lose their plane wave characteristics

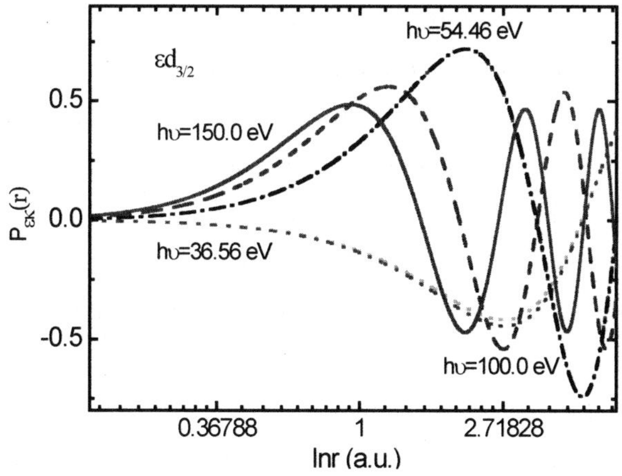

Figure6. For photoionization of $1s^2 2s^2 2p^5 3s^1$ ($^3P_2$), the large component radial wave functions of photoelectrons $\varepsilon d_{3/2}$ from model (C). Solid lines show the radial wave functions for photon energy $h\nu = 150$ eV, dashed lines are those for 100 eV, dotted lines for 36. 56 eV and dash—dotted lines for 54. 46 eV.

gradually and become relativistic distorted—waves due to the increasing importance of the electron correlations and effects of the nucleus. In general, the latter become significant in the short range if the kinetic energy of the photoelectrons is small. On the other hand, for large kinetic energies, the oscillations of the radial wave functions are very rapid, so the influences on the photoelectrons is not so important and the plane wave characteristics are still prominent even at near—nucleus distances, despite the fact that the radial wave functions are essentially insensitive to energy change. As a result, the continuum radial wave functions will become more compact and their nodes *move in* towards the nucleus as the kinetic energy of the photoelectrons (photon energy) increases. The rapid oscillations of the continuum radial wave functions intersect with those of the initial bound electrons $2p_{1/2,3/2}$ where the dipole matrix element has alternate signs. Thus, although the radial wave functions of the initial bound electrons $2p_{1/2,3/2}$ are invariant, as deduced from figure 5, the average value of the dipole matrix element for each relativistic dipole coupling channel will be close to zero as the energy increases continuously. It can therefore be interpreted that the Cooper minimum, located near the ionization threshold, originates from the cancellation of the positive and negative contributions to the dipole matrix element.

As expected from the discussions above, there arefive relativistic dipole coupling channels for the $1s^2 2s^2 2p^6 3s^1 (^2S_{1/2}) \rightarrow 1s^2 2s^2 2p^5 3s^1 (^3P_2)$ transition, so the photoionization cross section is obtained as a simple sum of the five partial cross sections as presented in figure 7. The curves confirm the fact that the partial cross sections of different relativistic channels do not go through zero at the same photon energy, which is, of course, due to the radial wave functions of photoelectrons since the initial state is exactly the same for different photon energies. Thus the cross section does not really vanish, i. e. , the Cooper minimum for the transition is not really zero, as can be seen from figures 2 and 4. We further note from figure 7 that the photoionization cross section is dominated by the relativistic channels of the $\varepsilon d_{5/2}$ photoelectrons near the threshold, and their Cooper minima play a crucial role in the overall shape of the cross section in the low energy region.

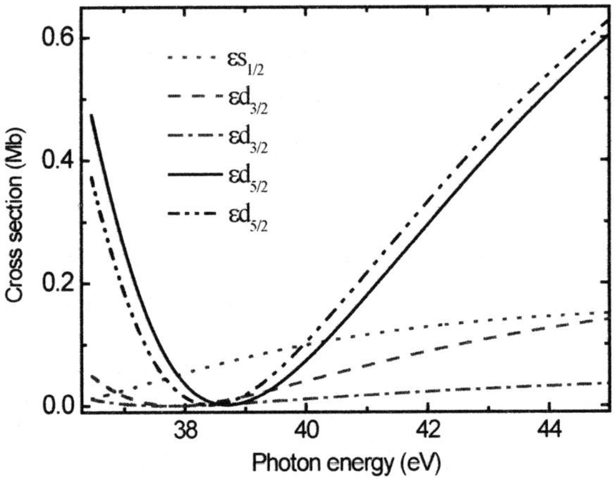

Figure7. The partial cross sections of $1s^2 2s^2 2p^5 3s^1 (^3P_2)$. Dotted line indicates the partial cross section for photoelectron $\varepsilon s_{1/2}$, dashed and dash—dotted lines are those for $\varepsilon d_{3/2}$, solid and dash—double—dotted lines are for $\varepsilon d_{5/2}$.

4 CONCLUSIONS

We have calculated the 2p photoionization cross sections from the ground state of neutral sodium using the MCDF method. The results are compared with previously reported theoretical calculations and experimental measurements, and are in good or reasonable agreement in most cases if models include more electron correlations. In the low energy region, the non—monotonic changes of the photoionization cross sections are very obvious, and the cross section ratios $^3P_0/^3P_2$ and $^3P_1/^3P_2$ for different final ionic states $1s^2 2s^2 2p^5 3s^1$ ($^3P_{0,1,2}$) are not statistical. However, in higher energy regions, the cross sections are proportional to the number of final quantum states. The radial wave functions of continuum photoelectrons will *move towards* the nucleus as the kinetic energy increases continuously. Thus, the existence of Cooper minima mainly depends on the relative positions of the bound and continuum one—electron radial wave functions. The positions of the minima, which distort the energy dependence of the photoionization cross sections, are very sensitive to the electron correlations.

Acknowledgments

This work has been supported by the National Natural Science Foundation of China (Grants Nos. 11264033, 11464040 and 10774122), the Science Research Foundation of Tianshui Normal University (Grant No. TSA1108) and the Scientific Research Foundation of Higher Education of Gansu Province (Grant No. 2014A—104). We thank Dr XB Ding of Northwest Normal University for his continuous support and fruitful discussions in the course of our studies.

References

[1]A. Müller, et al. J. Phys. B: At. Mol. Opt. Phys. 47, 135201(2014).

[2]D. Cubaynes, J. M. Bizau, F. J. Wuilleumier, B. Carré, F. Gounand. Phys. Rev. Lett. 63, 2460(1989).

[3]S. Cavalieri, F. S. Pavone, M. Matera. Phys. Rev. Lett. 67, 3673(1991).

[4]M. A. Baig, S. A. Bhatti. Phys. Rev. A 50, 2750(1994).

[5]B. Rouvellou, et al. Phys. Rev. Lett. 75, 33(1995).

[6]R. Eramo, S. Cavalieri, L. Fini, M. Matera, L. F. DiMauro. J. Phys. B: At. Mol. Opt. Phys. 30, 3789(1997).

[7]D. Cubaynes, et al. Phys. Rev. A 57, 4432(1998).

[8]D. Cubaynes, et al. Phys. Rev. Lett. 92, 233002(2004).

[9]D. Cubaynes, et al. J. Phys. B: At. Mol. Opt. Phys. 40, F121(2007).

[10]D. Cubaynes, et al. Phys. Rev. A 80, 023410(2009).

[11]M. A. Baigts, M. S. Mahmoodf, K. Sommefl, J. Honnes. J. Phys. B: At. Mol. Opt. Phys. 27,389(1994).

[12]A. K. Jain,K. C. Mathur. J. Phys. B:At. Mol. Opt. Phys. 26,433(1993).

[13]H. P. Saha. Phys. Rev. A 50,3157(1994).

[14]S. S. Tayal,A. Z. Msezane,S. T. Manson. Phys. Rev. A 49,956(1994).

[15]S. Baier,et al. J. Phys. B:At. Mol. Opt. Phys. 27,1341(1994).

[16]J. C. Liu,Z. W. Liu. J. Phys. B:At. Mol. Opt. Phys. 27,4531(1994).

[17]A. Kupliauskienė. Phys. Scr. 53,149(1996).

[18]A. Kupliauskienė. J. Phys. B:At. Mol. Opt. Phys. 30,1865(1997).

[19]N. Rakštikas,A. Kupliauskienė. Phys. Scr. 58,587(1998).

[20]X. D. Yang,J. C. Liu,Y. . S Cheng. J. Phys. B:At. Mol. Opt. Phys. 31,465(1998).

[21]A. Kupliauskienė. J. Phys. B:At. Mol. Opt. Phys. 31,2885(1998).

[22]K. Miculis,W. Meyer. J. Phys. B:At. Mol. Opt. Phys. 38,2097(2005).

[23]J. W. Cooper. Phys. Rev. 128,681(1962).

[24]J. W. Cooper. Phys. Rev. Lett. 13,762(1964).

[25]Y. S. Kim,R. H. Pratt,A. Ron. Phys. Rev. A 24,1626(1981).

[26]X. Y. Han,X. Gao,J. M. Li,L. Voky,N. Feautrier. Phys. Rev. A 74,062710(2006).

[27]G. B. Pradhan,et al. J. Phys. B:At. Mol. Opt. Phys. 44,201001(2011).

[28]Y. S. Kim,A. Ron,R. H. Pratt,B. R. Tambe,S. T. Manson. Phys. Rev. Lett. 46,1326 (1981).

[29]W. R. Johnson, V. Radojevi ? , P. Deshmukh, K. T. Cheng. Phys. Rev. A 25, 337 (1982).

[30]N. Shanthi. J. Phys. B:At. Mol. Opt. Phys. 21,L427(1988).

[31]S. B. Whitfield,J. Kane,R Wehlitz. J. Phys. B:At. Mol. Opt. Phys. 40,3647(2007).

[32]K. G. Dyall,I. P. Grant,C. T. Johnson,F. A. Parpia,E. P. Plummer. Comput. Phys. Commun. 55,425(1989).

[33]F. A. Parpia,C. F. Fische,I. P. Grant. Comput. Phys. Commun. 94,249(1996).

[34]S. Fritzsche,J. Electron. Spectrosc. Relat. Phenom. 114－116,1155(2001).

[35]X. B. Liu,et al. J. Phys. B:At. Mol. Opt. Phys. 44,115001(2011).

[36]C. C. Sang,X. B. Ding,C. Z. Dong. Chin. Phys. Lett. 25,3624(2008).

[37]J. J. Wan,et al. Phys. Rev. A 79,022707(2009).

[38]G. Gaigalas,T. Zalandauskas,S. Fritzsche. Comput. Phys. Commun. 157,239(2004).

[39]R. D. Cowan. *The Theory of Atomic Structure and Spectra*(Berkeley,CA:University of California Press,1981)p 525.

[40]K. Jänkälä,et al. J. Phys. B:At. Mol. Opt. Phys. 40,3435(2007).

[41]M. Kutzner,V. Radojevi ? . Phys. Rev. A 49,2574(1994).

[42]E. M. Isenberg, S. L. Carter, H. P. Kelly, S. Salomonson. Phys. Rev. A 32, 1472 (1985).

[43]B. I. Craig, F. P. Larkins. J. Phys. B: At. Mol. Opt. Phys. 18, 3569(1985).

[44]C. E. Theodosiou, W. Fielder. J. Phys. B: At. Mol. Opt. Phys. 15, 4113(1982).

（本文发表于 2016 年《Journal of physics B: Atomic, Molecular and optical physics》49 卷 13 期）

Location－dependent Raman Transition in Gravity－gradient Measurements Using Dual Atom Interferometers

Yuping Wang　Jiaqi Zhong　Hongwei Song　Lei Zhu　Yimin Li
Xi Chen　Runbing Li　Jin Wang　Mingsheng Zhan *

摘要：本文调查了用于重力梯度测量的双原子干涉仪中的拉曼跃迁的位置依赖性。拉曼激光光束中的由电光调制产生的拉曼光束对之间的串扰使得拉曼跃迁依赖于原子的位置。因此，干涉条纹的对比度也依赖于拉曼光脉冲和原子作用的位置，所以调节拉曼光脉冲和原子作用的位置对优化重力梯度测量的精度至关重要。为了进一步降低剩余的串扰影响，用于^{85}Rb原子的拉曼光的失谐使用^{87}Rb的饱和吸收谱锁频来实现。理论分析和调制实验用来确定优化的拉曼跃迁的位置，并用于重力梯度的测量。通过阿伦方差评估，重力差的分辨率达到4.9×10^{-9} m/s^2@15000s，对应的重力梯度测量的分辨率是7.4E @15 000s。

The location－dependent Raman transition was investigated based on dual atom interferometers which were designed for gravity－gradient measurements. The cross talk between the Raman beam pairs generated by an electro－optical modulator make the Raman transition location dependent. Therefore, the fringe contrast also depends on the location where the Raman pulses interact with the atoms, so it is important to adjust the interaction location to optimize the precision of the gravity－gradient measurements. To further reduce the residual cross talk, the detuning of the Raman beams for ^{85}Rb atoms was controlled using the saturated absorption spectra of the ^{87}Rb atoms. The optimal location for the Ra-

* 作者简介：王玉平（1976—），男，甘肃文县人，天水师范学院电子信息与电气工程学院副教授、博士，主要从事冷原子应用和激光技术的研究。

man transition was determined by theoretical analysis and modulation experiments, and the atomic trajectory was optimized and applied to the gravity-gradient measurements. The resolution of the differential gravity measurement evaluated by the Allan deviation was 4.9×10^{-9} m/s^2@15 000s, and the corresponding resolution of the gravity gradient measurement was 7.4 E @ 15 000 s.

1. INTRODUCTION

Atom interferometers (AIs) have been applied in precision measurements, such as the testing of the weak equivalence principle[1], measurement of rotation [2,3], and measurement of gravity[4,5]. However, the vibrational noise present in atom interferometry limits its sensitivity in these measurements. Fortunately, the atom gravity gradiometers (AGGs) based on dual AIs[6] and single AIs[7,8] have been demonstrated to eliminate the influence of the common-mode vibrational noise, and thesensitivities have been improved by several orders of magnitude. Further, AGGs have been applied as powerful tools in the determination of the gravitational constant[9-11] and measurement of the Earth's gravity gradient[12]. A compact movable AGG for navigation, geodesy, underground structure detection, and mineral exploration was demonstrated in 2009[13]. Also, an AGG for mapping the global gravity field from space is still ongoing[14]. In gravity-gradient measurements with dual AIs, the Raman beam pairs interact simultaneously with atom clouds at different locations. The Raman transition is very important for the performance of an AGG, and a Raman beam pair for Raman transition is usually produced by employing either optical phase-locking loops[15, 16], acousto-optic modulators (AOMs), or electro-optic modulators (EOMs). The EOM method is popularly used to produce low-noise, high-intensity, stable Raman beams[17-19] for a compact AI. However, multiple Raman beam pairs consisting of the carrier and sidebands[20-22] are present in the EOM method, and each pair of Raman beams drives a Raman transition. Consequently, the cross talk of the multiple Raman beam pairs cause the effective Rabi frequency to be location dependent in atom interferometry[20,21]; thus the fringe contrast also depends on the location where the Raman pulses interact with the atoms. To optimize the precision of the gravity-gradient measurement, it is important to investigate the location-dependent Raman transition that exists in dual

AIs. In this paper, we study the location-dependent Raman transition in AIs, and investigate its effect on the gravity-gradient measurement. Unlike the gravity gradiometer that adopts two AIs with separated vacuum chambers[12,14], our compact gravity gradiometer is composed of two identical ^{85}Rb AIs that share the same vacuum chamber. This configuration is useful for eliminating the wave-front distortion and the steering noise[23] of the Raman beams caused by windows and air flux. To reduce the residual cross talk, we control the detuning of the Raman beams for ^{85}Rb atoms using the saturated absorption spectra of ^{87}Rb atoms, which also improved the stability and reliability of the Raman beams. Based on theoretical analysis and modulation experiments, the flight trajectory of the atoms was carefully designed, the Raman transition was optimized, measurement of the gravity gradient was demonstrated, and the results were evaluated.

2. LOCATION-DEPENDENT RAMAN TRANSITION IN THE GRAVITY-GRADIENT MEASUREMENT

First we consider the dual ^{85}Rb AI used for measuring the vertical component of the gravity-gradient tensor. The vertical separation between the two AIs was L. The Raman beam pairs were generated by a phase-modulation method wherein two ^{85}Rb atom clouds were prepared in the individual AIs and were launched vertically. Next, a sequence of stimulated Raman transitions were applied to the atom clouds to split, redirect, and recombine the atomic wave packets, and thereby to form two interference loops. The gravity-gradient information was extracted by measuring the phase difference existing between the interference fringes of the two interference loops. The Raman transition plays an important role in the gravity gradient measurement. Typically, the Raman transition in atom interferometry is independent of location where Raman beams interact with atoms if the Raman beams are ideal. For Raman beams generated by a phase modulation, the performance of the Raman transition depends on the location, and this location-dependent Raman transition influences the accuracy of the gravity-gradient measurement. To analyze the dependence of the Raman transition on location, we consider an atom with three states: a lower ground state $|g\rangle$ (for ^{85}Rb, $|F=2, m_F=0\rangle$), an upper ground state $|e\rangle$ (for ^{85}Rb, $|F=3, m_F=0\rangle$) and an excited state $|i\rangle$. The frequency separation between the ground states $|e\rangle$

and |g> is ω_{eg}. The Raman beams are generated by an EOM and then retroreflected by a mirror, where the frequency of the EOMs driving the microwave is ωm(where $\omega_m \approx \omega_{eg}$), and the distance between the atoms and the mirror is l. In the weak phase-modulation limit, suppose two pairs of Raman beams exist that consist of the carrier with the frequency ω_c and the ± 1 order sidebands with the frequency $\omega_{\pm 1}$. One Raman pair consists of a carrier(ω_c, propagating upward) and the +1-order sideband($\omega_{+1} = \omega_c + \omega_m$, propagating downward), where the single-photon detuning of the Raman transition is Δ_{+1} (according to the excited state |i>). The other Raman pair consists of the carrier(ω_c, propagating downward) and the -1-order sideband($\omega_c - \omega_m$, propagating upward), where the single-photon detuning is Δ_{-1} ($\Delta_{-1} = \Delta_{+1} + \omega_m$). The electric field that drives the resonant two-photon transition between the ground states |g> and |e> is written as

$$E = E_{+1}\cos(\omega_{+1}t - k_{+1}z + \phi_0) + E_c\cos(\omega_c t - k_c z + \phi_c) + E_c\cos(\omega_c t - k_c z + \phi_c) + E_{-1}\cos(\omega_{-1}t - k_{-1}z + \phi_{-1}), \quad (1)$$

where $k_j = \omega_j / c$ is the wave vector, ϕ_0 is the initial phase of the incident Raman beams, $\phi_j = -2k_j l + \phi_0$ is the phase of the retroreflected Raman beams, and $-2k_j l$ is the delayed phase introduced by an additional distance $2l$.

The atom interacts with the two Raman beam pairs, and each Raman beam pair drives a two-photon Raman transition between the two ground states |g> and |e> via the intermediate state |i>. Consequently, the cross talk between different Raman beam pairs causes the Raman transition to be location dependent. Both Raman pairs are resonant with the hyperfine transition of atoms, and the effective Rabi frequency is the combination of the individual Rabi frequencies. Considering the phase of each Rabi frequency[24], the effective Rabi frequency can be written as

$$\Omega(l)_{eff} = \frac{\Omega_{+1}^{*}\Omega_c}{2\Delta_{+1}} e^{i2k_c l} + \frac{\Omega_c^{*}\Omega_{-1}}{2\Delta_{-1}} e^{i2k_{-1} l}, \quad (2)$$

where the first term on the right-hand side $2k_j l = \phi_0 - \phi_j$ is the Rabi frequency of the Raman beam pair ω_c and ω_{+1}, and the second term is the Rabi frequency of the Raman beam pair ω_c and ω_{-1}. Taking into account the symmetry of the sideband, $\Omega_{+1} * \Omega c = \Omega c * \Omega_{-1}$, the square of the effective Rabi frequency can be written as

$$\Omega(l)_{eff}^2 = \Omega_0^2 \left[1 + \left(\frac{\Delta_{+1}}{\Delta_{-1}} \right)^2 + 2\frac{\Delta_{+1}}{\Delta_{-1}} \cos\left(\frac{2\pi l}{\lambda_m / 2} \right) \right] \quad (3)$$

where $\Omega_0 = \frac{\Omega_{+1}^{*}\Omega_c}{2\Delta_{+1}}$ is the Rabi frequency of the Raman beam pair(ω_c and$\omega_c + \omega_m$)and$\lambda_m = 2\pi/(k_c - k_{-1})$is the wavelength of the microwave. Obviously,$\Omega(l)$eff is a periodic function of$\lambda_m/2$, and its amplitude depends on Δ_{+1}. When a Raman pulse with a duration of$\tau$is applied to atoms in the initial state |g>, the probability of atoms being in the state |e> is given as

$$P_e(l) = \frac{\Omega(l)_{eff}^2}{\Omega(l)_{eff}^2 + \delta^2}\sin^2\left(\sqrt{\Omega(l)_{eff}^2 + \delta^2}\,\frac{\tau}{2}\right), \tag{4}$$

where δ is the difference between the two-photon detuning and the relative ac Stark shift of the two levels. Compared to $\Omega(l)$eff, $\delta$is negligible. Therefore, ignoring δ in Eq. (4), the probability of atoms being in state |e> is simplified as

$$P_e(l) = \frac{1}{2}[1 - \cos(|\Omega(l)_{eff}|\tau)], \tag{5}$$

Obviously, when the duration$\tau$is a constant, also has a spatial periodicity of$\lambda_m/2$. for ^{85}Rb atoms, $\lambda m/2$ is 5 cm.

In dual AIs using two Raman beam pairs, the gravity gradient is measured using two atom clouds separated by L. When two atom clouds interact with Raman beam pairs, the distances from atom cloud 2(AC2)and atom cloud 1(AC1)to the mirror are l and $l+L$, respectively. Then, the effective Rabi frequency of AC2 is the same as that given in Eq. (3), and the probability of AC2 is the same as that given in Eq. (5). Therefore, substituting $l+L$ for l and taking $L = n\lambda_m/2 + \Delta l$ into Eqs. (3)and(5), the effective Rabi frequency of AC1 can be written as

$$|\Omega(l+L)_{eff}| = |\Omega(l+\Delta l)_{eff}| \tag{6}$$

and the probability of AC1 can be written as

$$P_e(l+L) = P_e(l+\Delta l) \tag{7}$$

where n is an integer and $0 \leqslant \Delta l < \lambda_m/2$. According to Eqs. (6)and(7), if the separation between two atom clouds, L, is an integer multiple of the spatial wavelength $\lambda_m/2(\Delta l = 0)$, the effective Rabi frequency of AC1 is obtained by shifting that of AC2 by an integer multiple of the spatial wavelength $\lambda_m/2$; that is, $|\Omega(l+L)_{eff}| = |\Omega(l)_{eff}|$. Similarly, the probability of AC1 is $P_e(l+L) = P_e(l)$. This means a Raman pulse synchronously interacts with two atom clouds at different locations, and aπpulse(or$\pi/2$ pulse) for one atom cloud is also aπpulse(or$\pi/2$ pulse)for the other atom cloud. If L is not an integer multiple of $\lambda_m/2(0 < \Delta l < \lambda_m/2)$, however, the effective Rabi frequency and the probability of AC1 have a delayed phase, $2k_m\Delta l$, with respect to those of AC2 for an arbitrary location l, where $k_m = 2\pi/\lambda_m$ is the microwave wave vector. This means that a Raman pulse cannot synchronously operate on two atom clouds at different locations, and aπ-pulse(or$\pi/2$ pulse)for one atom cloud will definitely not be aπpulse(or$\pi/2$ pulse) for the other atom cloud. When a Ramanπpulse interacts with two atom clouds, there is a big difference(even up to 30%)between the probabilities of AC2 and AC1 in $|e\rangle$. Considering the discussion above, the location—dependent Raman transition affects the fringe contrast of both AIs, and the separation between the two AIs determines if their fringe contrasts are simultaneously optimized. To improve the accuracy of the gravity—gradient measurement, the separation between two atom clouds, L, should be designed as an integer multiple of $\lambda_m/2$. Unfortunately, in an actual AGG, L is not an exact integer multiple of $\lambda_m/2$. According to Eqs. (3)—(7), one can optimize the location—dependent Raman transition in the gravity—gradient measurement using dual differential AIs.

3. EXPERIMENTAL SETUP

The atom gravity gradiometer consisted of a dual AIwith vertical configuration; its schematic diagram is shown in Fig. 1. Two AIs shared an ultrahigh—vacuum chamber(pressure$<10^{-8}$ Pa). The Raman beams were delivered by polarization—maintaining optical fibers that traveled from the laser module to the top window of the vacuum chamber, and were then retroreflected by a mirror mounted under the bottom window. The Raman beams were expanded into collimated beams with a $1/e^2$ diameter of 2 cm and a power of 100 mW. To maintain the orthogonal polarization of incident and retroreflected Raman beams, a quarter—wave plate was inserted between the mirror and the bottom window. Because the Raman beams and the mirror were shared in this configuration, it was immune to common phase noise. Each AI comprised a two—dimensional magneto—optical trap(2D—MOT)[25], a three—dimensional magneto—optical trap(3D—MOT), an interference zone, and a detecting zone. The 2D—MOT provided a radially cooled ^{85}Rb beam for the 3D—MOT, and a push beam was applied to enhance the atomic flux so that the 3D—MOT could trap more than 1×10^8 atoms within 1 s while maintaining an ultrahigh vacuum. Precooled ^{85}Rb atoms inthe 2D—MOT were pushed to and trapped in the center of the 3D—MOT within 1 s, and then were further cooled to 5μK and launched vertically by amoving optical molasses. Atom clouds trapped in both MOTs were launched to interference zones simultaneously, and a Raman pulse sequence was applied to achieve atom interferometry. A bias magnetic field of 200 mG was applied to define the quantization axis, where the stray magnetic fields around the interference zones were shielded by a factor of 50 usingμ—metal films. The detecting zones were set at 11 cm above the 3D—MOTs, and the separation between the two 3D—MOTs was 66. 35 cm.

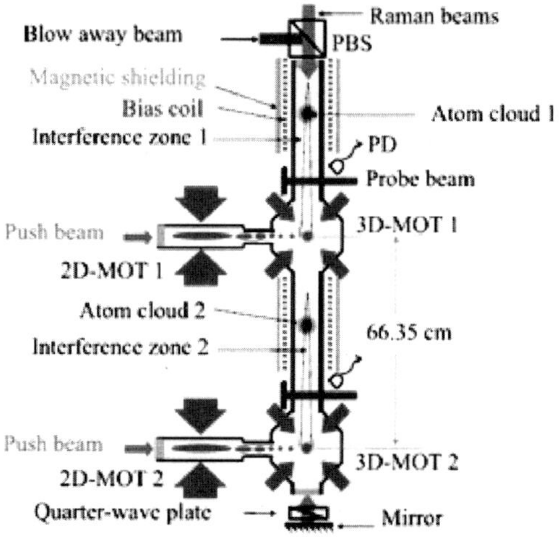

FIG. 1. The schematic diagram of the atom gravity gradiometer. PBS: polarizing beam splitter; PD: photo detector; 2D-MOT: two-dimensional magneto-optical trap; 3D-MOT: three-dimensional magneto-optical trap.

A schematic diagram of the Raman beams is shown in Fig. 2. The Raman beams were taken from the output of a master laser(Toptica, DL Pro), whose linewidth was 200 kHz and output power was 60 mW. To obtain a large detuning (1. 3 GHz)of the Raman beams for ^{85}Rb, we locked the master laser to the cross peak($5S_{1/2}$, $F=2$ to $5P_{3/2}$, $F=1,3$ transitions)of the ^{87}Rb atoms via saturated absorption spectroscopy, and generated Raman beams with a frequency difference of 3. 04 GHz by a fiber EOM(Eospace, PM—0K5—10—PFU—PFA—780)[21]. The Raman beams were amplified by a homemade tapered amplifier(TA), and their frequencies were slightly adjusted by an 80—MHz AOM with a double—pass configuration. The 3. 04 GHz driving signal for thefiber EOM was generated by mixing a 2. 96—GHz microwave signal and an 80—MHz radio signal, where the microwave signal was generated by a signal generator (Rigol, DSG3060) and the radio signal was supplied by an arbitrary waveform generator (Agilent, 33250A). A microwave amplifier (Mini−Circuits ZVE−8GX+) was used to amplify the 3.04 GHz signal. Finally, the radio source was chirped to compensate for the gravity−induced Doppler shift.

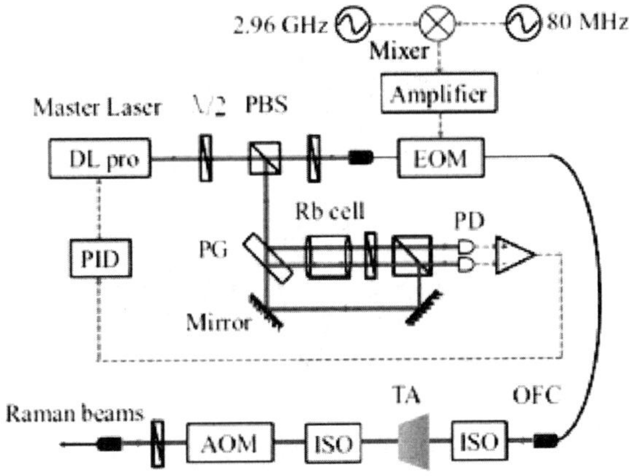

FIG. 2. The schematic diagram of Raman beams. PD: photodetector; PBS: polarized beam splitter; PG: plate glass; OFC: optical fiber coupler.

4. OPTIMIZATION OF LOCATION−DEPENDENT RAMANTRANSITION IN A DUAL ATOM INTERFEROMETER

We experimentally investigate the location−dependent Raman transition by monitoring the population of $F=3, m_F=0$ versus the interaction position. First, the atom clouds are launched to the top of the interference zones, whereupon a Raman pulse is applied to interact with the free−fall atom clouds. The differential frequency of the Raman pulse is scanned with a step of 10 kHz, and the resonance signal of the transition from $F=2, m_F=0$ to$F=3, m_F=0$ is obtained. Figure 3 plots the populations of the$F=3, m_F=0$ state after applying the Raman pulse with a duration of 20μs to interference zone 1(IZ1)(blue squares in Fig. 3) and interference zone 2(IZ2)(green dots in Fig. 3), and also plotted are the fits based on Eq. (4)(black lines in Fig. 3; $\omega_m=2\pi\times$ 3.035 GHz, $\Delta_{+1}=2\pi\times$ 1.312 GHz, $\Delta_{-1}=2\pi\times$ 4.347 GHz, $\tau=20\mu$s). The fitted parameters$\delta$and Ω_0 are 0.04 Hz and $2\pi\times$ 22.6 kHz, respectively. According to the population in Fig. 3, there are three minima spaced by 5 cm, as expected, while the theoretical fits indicate that the population varies simultaneously with a spatial period of$\lambda_m/2$. The baseline of the gravity gradiometer used is 66. 35 cm, which is not an integer multiple of 5 cm. This causes the population of IZ1 to be shifted by 1. 23 cm according to that of IZ2. This roughly agrees with the theoretical value of 1. 35 cm.

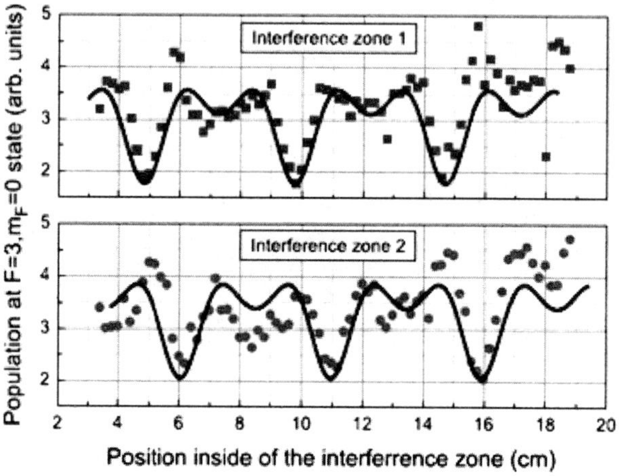

FIG. 3. Population of the $F=3$, $m_F=0$ state after applying the Raman pulse with duration of 20 μs along interference zone 1 (blue squares) and interference zone 2 (green dots). The position is referenced to the bottom of the individual interference zone.

The Rabi oscillation also varies with the interaction location. We select two typical positions inside the interference zones, $z1=7.78$ cm and $z2=14.78$ cm, where according to Fig. 3, $z1$($z2$) corresponds to the location where the population is near its maximum(minimum) for IZ1. The experimental results are shown in Fig. 4. For IZ1, the population of the $F=3$ state at $z1$(blue squares in Fig. 4) is a damped oscillation with a frequency of 20 kHz, while that at $z2$(blue boxes in Fig. 4) slowly increases with the duration of the Raman pulse. These results imply that the efficiency of the Raman transition at $z1$ is high, while it is at its lowest at $z2$; i. e. , a Ramanπpulse can coherently transfer atoms between two hyperfine ground states at $z1$ but not at $z2$. For IZ2, the population of the$F=3$ state at $z1$(green dots in Fig. 4) is a damped oscillation with a frequency of 22 kHz, while that at $z2$(green circles in Fig. 4) is a critically damped oscillation. Therefore, the coherence of the atoms at $z1$ is better than that at $z2$, and a Ramanπpulse can coherently transfer atoms between the two hyperfine ground states at both $z1$ and

$z2$. Consequently, the positions where each Raman pulse interacts with atoms can be used to optimize the Raman transition. Substituting Δ_{+1}, Δ_{-1}, Ω_0, and ω_m into Eq. (3), we find that $\Omega_{eff}(l)$ reaches a minimum (16 kHz) at the positions where the population reaches a minimum, and it reaches a maximum (29 kHz) at positions where the population is at a secondary maximum, according to Fig. 3. The experimental and theoretical values are roughly consistent except that the damped oscillation is not observed at $z2$.

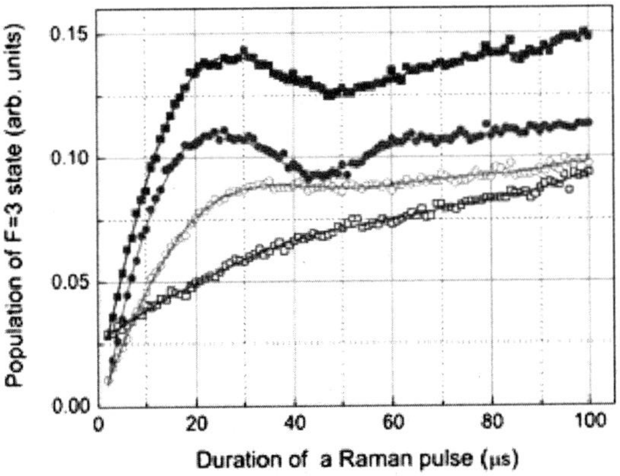

FIG. 4. Comparison of Rabi oscillations of atoms at two locations. The blue squares and blue boxes are Rabi oscillations at $z_1 = 7.78$ cm and $z_2 = 14.78$ cm in interference zone 1, respectively. The green dots and green circles are Rabi oscillations at z_1 and z_2 in interference zone 2, respectively. The solid curves are the fits to the individual data.

We thereby design an optimal flight parabola for atoms under two conditions: (1) each of three Raman pulses interacts with atoms with higher Raman transition efficiency; (2) the time interval T of the Raman pulses is as long as possible. The Raman beam that propagates downward is red－detuned by 2.08 MHz, while the Raman beam that propagates upward is blue－detuned by the same amount. Consequently, the atoms are launched at a velocity of 2.3 m/s, and the apex of their parabolic trajectory is at a position 12.7 cm inside the interference zones. The Raman π pulse is applied 10ms before the apex, and T is 120 ms, corresponding to the positions where the Raman pulse interacts with the atoms successively at 4.2, 12.6, and 6.7 cm inside the interference zones.

5. GRAVITY-GRADIENT MEASUREMENTS

We test the designed flight trajectory for atoms by observing the interference fringes. First, we choose a chirp rate of 25 MHz/s for the Raman laser to compensate for gravity, and the duration of the Raman π pulse is set as 20μs. Once the atoms are launched to the interference zones, they are transferred from the $F=3, m_F=0$ to the $F=2, m_F=0$ state by a Raman π pulse, and then the atoms in the $F=3$ state are blown away by a 1-ms pulse that is resonant with the $F=3$ to $F'=4$ transition. Subsequently, the Raman $\pi/2-\pi-\pi/2$ pulse sequence is applied to form atom interference loops, where the population in the $F=3$ state is detected by a 5-ms pulse that is resonant with the $F=3$ to $F'=4$ transition, the phase of the second $\pi/2$ pulse is scanned by a step of $\pi/20$, and the atom interference fringes are obtained by recording the population versus phase. We obtain interference fringes with a signal-to-noise ratio (SNR) of 100:1, and the fringe contrasts for IZ1 and IZ2 are 24% and 18%, respectively. The differential gravity, Δg, is given by the phase difference between IZ1 and IZ2; the phase difference, $\Delta\varphi$, is extracted by ellipse fitting[26]; and the gravity gradient is deduced from Δg and the baseline. In the ellipse fitting process, an ellipse is formed when the fringes from IZ1 are plotted versus those from IZ2, and the differential phase can be extracted from the ellipse fit parameters. Because the differential phase is too small to be exactly extracted by the ellipse fitting, we apply a pulsed magnetic field to the atoms in IZ2 to shift the differential phase by $\sim\pi/2$, whereupon the differential phase can be extracted by the ellipse fitting[27]. We estimate the ability of the gravity gradiometer to suppress the tidal effect by performing a 42-h-long measurement, which lasted from 20:00 of November 13 to 14:00 of November 15 in 2015, and from which the differential acceleration, Δg, can be obtained from the differential phase of the ellipses, $\Delta\varphi$. The measurements are shown in Fig. 5, plotting the local gravitational acceleration values measured by IZ1 (blue squares in Fig. 5) and IZ2 (green dots in Fig. 5), and the calculations using a tidal model (black lines in Fig. 5). It can be seen that the temporal variation of the local gravity measured by the two AIs agrees with the tidal model. Both of the AIs have sensitivities that are almost equivalent, and the standard deviations are better than $2\times10^{-8}g$ over the 42 h. The measured Δg (green circles in

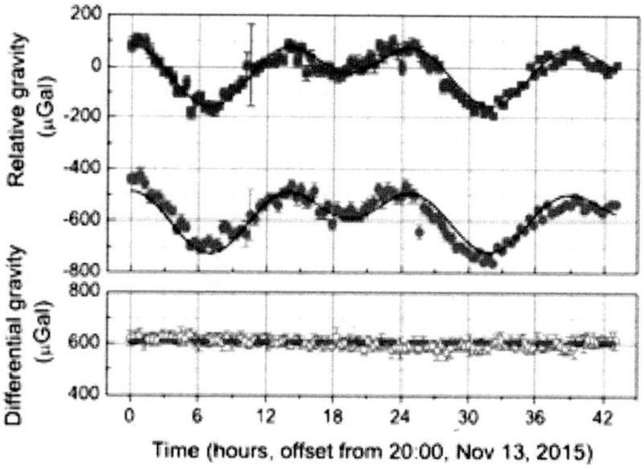

FIG. 5. Experimental data of the Earth's tides and differential gravity measured by the two AIs. The blue squares and the green dots are tidal signals monitored by interference zone 1 and interference zone 2, respectively. The black lines are calculations with tidal model. The green circles in the bottom panel are differential gravity measurements, and the blue dashed line is the fitted value.

Fig. 5)are scattered around the line where the tidal signal is canceled(blue dashed line in Fig. 5), and only the noncommon mode phase noise remains. The performance of the gravity—gradient measurement is evaluated by the Allan deviation, for which the two AIs are continuously run for 24 h with a sample rate of 0.68 Hz, and 56 000 data points are obtained. Each 40—point ellipse is used to extract a differential phase by ellipse fitting, and 1400 differential phases are obtained and converted to gravity—gradient values by dividing by a scale factor of $k_{\mathrm{eff}} T^2 L$, where, $k_{\mathrm{eff}} = 1.6 \times 10^5\ \mathrm{cm}^{-1}$ is the effective Raman wave vector, and L is the baseline. Meanwhile, both local gravity measurements are calculated by sine fitting from the individual fringes, and the Allan deviation is then calculated based on these measurements, as shown inFig. 6. The values of the differential gravity are plotted(left$y$axis in Fig. 6) with those of the corresponding gravity gradient (right side$y$axis in Fig. 6). Each data point in the Allan deviation curves(green squares for IZ1 and blue triangle blocks for IZ2 in Fig. 6)is obtained from a 40—point fringe. The resolutions of both AIs have the same Allan deviation curves and their resolutions reach $1.5 \times 10^{-7}\ \mathrm{m/s^2}$, which is mainly limited by vibrational noise. The gravity gradient is not affected by the common—mode phase noise and consequently its Allan deviation(red dots in Fig. 6) sharply decreases. The resolution of the differential gravity measurement is $4.9 \times 10^{-9}\ \mathrm{m/s^2}$@ 15 000 s, and the corresponding resolution of the gravity—gradient measurement according to the 66.35−cm baseline is 7.4 E @ 15 000s. We also demonstrate a static measurement of the gravity gradient using the mass−modulation method. The local gravity gradient is modulated by a testmass of 824 kg that is composed of 73 lead bricks. These lead bricks are stacked compactly and placed on a home-made wheeled table, where the centroid position of the bricks is as high as the geometric center of the two AIs. This test mass is placed alternately at two fixed positions with a constant spacing. The measured shift of the gravity gradient is found to be(221 ± 56)E, which is extracted by subtracting the averaged gravity gradients at the two positions. This measurement roughly agrees with the theoretical value of(251 ± 13)E.

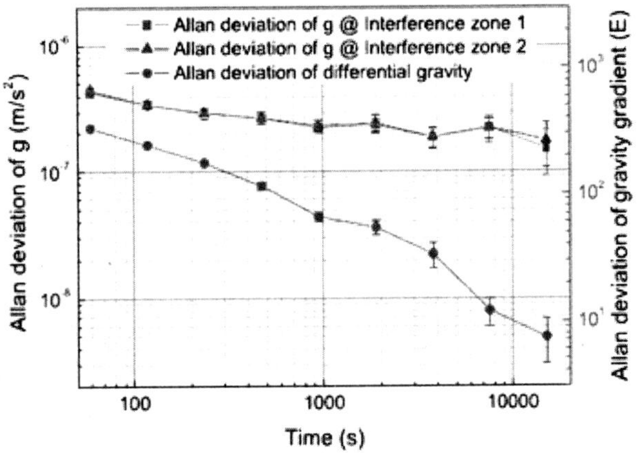

FIG. 6. Allan deviation of gravity and gravity-gradient measurements ($1\ E = 1 \times 10^{-9}\ s^{-2}$).

6. DISCUSSION

The resolution of the differential gravity acceleration reported herein is comparable to that of the dual−AI gravity gradiometers reported by McGuirk and co−workers[6,9]and Yu*et al*.[14]. Compared with the recent result of a single−AI gravity gradiometer reported by Sorrentino *et al*.[7], the advantage of our dual AI is the higher sampling rate, while there is still room for improvement in the resolution. The main limitations for the current resolution include a less−desirable SNR(100:1) and fringe contrast(18%−24%), amplitude noise, noncommon mode vibrational noise, and stray magnetic fields(1μT). To improve the SNR and fringe contrast, atom clouds with more than 5×10^8 atoms and colder than 1μK should be involved in the interferometers. A normalized detection is necessary for suppressing the amplitude noise. Also, the noncommon mode vibrational noise can be further suppressed by using a vibration isolation system, and a stray magnetic field less than 0.1μT is possible by designing new shielding. In addition, the current sampling rate(0.68 Hz) is expected to increase to 2 Hz, which can further improve the sensitivity. With these improvements, a better short—term sensitivity is expected in the near future. For some applications in navigation, geodesy, environmental survey, and resource exploration, the short—term sensitivity of the state—of—the—art AGG can meet the practical needs; however, the environmental adaptability, reliability, and sampling rate are necessary to be further improved.

7. CONCLUSION

We investigated the effect of the location—dependent Raman transition in gravity—gradient measurements with a dual AI, where the optimal location for the Raman transition was determined and then applied in gravity—gradient measurements. The gravity gradiometer was built based on a dual ^{85}Rb AI with a baseline of 66.35 cm. The saturated absorption spectra of ^{87}Rb atoms was used to control the detuning of the Raman beams for ^{85}Rb atoms. A gravity—gradient measurement using the differential AI with an optimized Raman transition was demonstrated, where the resolution of the gravity gradient is 7.4 E @ 15 000 s. The results are helpful for improving the design of compact and reliable gravity gradiometers. Further improvements will focus on improving the SNR, suppressing the noncommon mode vibration noise, and improving the magnetic field shielding.

8. ACKNOWLEDGEMENTS

This work was supported by the National Key Researchand Development Program of China under Grant No. 2016YFA0302002, and the National Natural Science Foundationof China under Grants No. 91536221, No. 11227803, andNo. 11504411.

REFERENCES

[1]L. Zhou, S. T. Long, B. Tang, X. Chen, F. Gao, W. C. Peng, W. T. Duan, J. Q. Zhong,

Z. Y. Xiong, J. Wang, Y. Z. Zhang, and M. S. Zhan, Test of Equivalence Principle at 10－8 Level by a Dual－Species Double－Diffraction Raman Atom Interferometer, Phys. Rev. Lett. 115,013004(2015).

[2]P. Berg, S. Abend, G. Tackmann, C. Schubert, E. Giese, W. P. Schleich, F. A. Narducci, W. Ertmer, and E. M. Rasel, Composite－Light－Pulse Technique for High－Precision Atom Interferometry, Phys. Rev. Lett. 114,063002(2015).

[3]D. Durfee, Y. Shaham, and M. A. Kasevich, Long－Term Stability of an Area－Reversible Atom－Interferometer Sagnac Gyroscope, Phys. Rev. Lett. 97,240801(2006).

[4]A. Peters, K. Y. Chung, and S. Chu, A measurement of gravitational acceleration by dropping atoms, Nature(London)400,849(1999).

[5]Z. K. Hu, B. L. Sun, X. C. Duan, M. K. Zhou, L. L. Chen, S. Zhan, Q. Z. Zhang, and J. Luo, Demonstration of an ultrahighsensitivity atom－interferometry absolute gravimeter, Phys. Rev. A88,043610(2013).

[6]J. M. McGuirk, G. T. Foster, J. B. Fixler, M. J. Snadden, and M. A. Kasevich, Sensitive absolute－gravity gradiometry using atom interferometry, Phys. Rev. A65,033608(2002).

[7]F. Sorrentino, Q. Bodart, L. Cacciapuoti, Y.－H. Lien, M. Prevedelli, G. Rosi, L. Salvi, and G. M. Tino, Sensitivity limits of a Raman atom interferometer as a gravity gradiometer, Phys. Rev. A89,023607(2014).

[8]X. C. Duan, M. K. Zhou, D. K. Mao, H. B. Yao, X. B. Deng, J. Luo, and Z. K. Hu, Operating an atom－interferometry－based gravity gradiometer by the dual－fringe－locking method, Phys. Rev. A90,023617(2014).

[9]J. B. Fixler, G. T. Foster, J. M. McGuirk, and M. A. Kasevich, Atom interferometer measurement of the Newtonian constant of gravity, Science315,74(2007).

[10]G. Lamporesi, A. Bertoldi, L. Cacciapuoti, M. Prevedelli, and G. M. Tino, Determination of the Newtonian GravitationalConstant Using Atom Interferometry, Phys. Rev. Lett. 100,050801(2008).

[11]G. Rosi, F. Sorrentino, L. Cacciapuoti, M. Prevedelli, and G. M. Tino, Precision measurement of the Newtonian gravitational constant using cold atoms, Nature(London)510,518(2014).

[12]M. J. Snadden, J. M. McGuirk, P. Bouyer, K. G. Haritos, and M. A. Kasevich, Measurement of the Earth's Gravity Gradient with an Atom Interferometer－Based Gravity Gradiometer, Phys. Rev. Lett. 81,971(1998).

[13]X. A. Wu, Gravity gradient survey with a moble atom interferometer, Ph. D. thesis, Stanford University, CA, 2009.

[14]N. Yu, J. M. Kohel, J. R. Kellogg, and L. Maleki, Development of an atom－interferometer gravity gradiometer for gravity measurement from space, Appl. Phys. B84,647(2006).

[15]M. Schmidt, M. Prevedelli, A. Giorgini, G. M. Tino, and A. Peters, A portable laser system for high—precision atom interferometry experiments, Appl. Phys. B102,11(2011).

[16]S. H. Yim, S. B. Lee, T. Y. Kwon, and S. E. Park, Optical phase locking of two extended—cavity diode lasers with ultra—lowphase noise for atom interferometry, Appl. Phys. B115,491(2014).

[17]O. Carraz, F. Lienhart, R. Charrière, M. Cadoret, N. Zahzam, Y. Bidel, and A. Bresson, Compact and robust laser system for onboard atom interferometry, Appl. Phys. B97,405 (2009).

[18]F. Theron, O. Carraz, G. Renon, N. Zahzam, Y. Bidel, M. Cadoret, and A. Bresson, Narrow linewidth single laser source system for onboard atom interferometry, Appl. Phys. B118,1(2015).

[19]J. K. Stockton, K. Takase, and M. A. Kasevich, Absolute Geodetic Rotation Measurement Using Atom Interferometry, Phys. Rev. Lett. 107,133001(2011).

[20]K. Takase, Precision rotation rate measurements with a mobile atom interferometer, Ph. D. thesis, Stanford University, CA, 2008.

[21]M. Kasevich and S. Chu, Measurement of the gravitational acceleration of an atom with a light—pulse atom interferometer, Appl. Phys. B54,321(1992).

[22]O. Carraz, R. Charrière, M. Cadoret, N. Zahzam, Y. Bidel, and A. Bresson, Phase shift in an atom interferometer induced by the additional laser lines of a Raman laser generated by modulation,

Phys. Rev. A86,033605(2012).

[23]G. W. Biedermann, X. Wu, L. Deslauriers, S. Roy, C. Mahadeswaraswamy, and M. A. Kasevich, Testing gravity with cold—atom interferometers, Phys. Rev. A91,033629(2015).

[24]M. Schmidt, A mobile high—precision gravimeter based on atom interferometry, Ph. D. thesis, Humboldt University, 2011.

[25]J. Ramirez—Serrano, N. Yu, J. M. Kohel, J. R. Kellogg, and L. Maleki, Multistage two—dimensional magneto—optical trap, Opt. Lett. 31,682(2006).

[26]G. T. Foster, J. B. Fixler, J. M. McGuirk, and M. A. Kasevich, Method of phase extraction between coupled atom interferometers using ellipse—specific fitting, Opt. Lett. 27, 951(2002).

[27]Y. P. Wang, J. Q. Zhong, X. Chen, R. B. Li, D. W. Li, L. Zhu, H. W. Song, J. Wang, and M. S. Zhan, Extracting the differential phase in dual atom interferometers by modulating magnetic fields, Opt. Commun. 375,34(2016).

(本文发表于 2017 年《PHYSICAL REVIEW A》95 期)

Codon—pair Usage and Genome Evolution

Fang—Ping Wang　Hong Li *

摘要：为了探索作用于密码对上下文的可能的进化约束，本文分析了110个物种的基因组中蛋白质编码序列中的密码对的模式数的分布和基因间序列中三联体对的模式数分布。我们发现这种分布和伽玛分布相符合。通过研究伽玛分布的形状参数α值，得到α值和基因组进化之间有显著的关系。对编码序列中密码对，α值按照古菌、细菌、真核生物的顺序增加；对基因间序列的三联体对，按α值110个物种分为两类，细菌为一类，古菌和真核生物为一类。这个发现推测密码对的上下文关系可能是种系发生的一个决定因素，表明在古菌、细菌和真核生物的编码序列和基因间序列的进化上存在着根本的区别。

The aim of this paper is to demonstrate possible evolutionary constraints that shape codon—pair context. The distributions of numbers of modes(DNM)of codon—pairs in protein coding sequences(CDSs)and the frequency of base triplet pairs in intergenic sequences(IGSs)are analyzed in 110 fully sequenced genomes. We propose that these distributions are in accordance with a gamma distribution. By studying the shape parameter α value of gamma distribution a distinct relation between the α value and the genome evolution is obtained. For codon—pairs in CDSs, the α value increases in the order Archaea, Bacteria, and Eukaryota, and divides the species into three evolutionary groups, Archaea, Bacteria and Eukaryota. For triplet pairs in IGSs, on the other hand, the α value classifies the species into two groups, one is Bacteria and the other is Archaea and Eukaryota. The findings suggest that the codon—pair context could be an important determinant

* 作者简介：王芳平(1977—)，女，甘肃省通渭县，博士，天水师范学院电子信息与电气工程学院副教授，主要从事理论生物物理研究。

for phylogeny of individual species, and indicate the existence of fundamental differences of evolutional constraints imposed on CDSs and IGSs among Archaea, Bacteria, and Eukaryota.

1. INTRODUCTION

The degeneracy of the genetic code has important implications for primary structure evolution ofgenes as it provides nature with a vast array of options for constructing coding sequences for any particular protein. Thebiased synonymous codon usage has been documented for many genes in various species and varies within and between organisms(Ikemura and Wada 1991; Andersson and Kurland 1990; Kurland 1993; Sharp and Matassi 1994). But the biological basis of codon choice has not been well understood yet. Nevertheless, it is becoming increasingly clear that codon usage bias reflects the action of two main evolutionary forces: selection and mutation. The first of which is supported by the evidence that highly expressed genes tend to use codons that are decoded by abundant cognate tRNAs (Moriyamaand Powell 1997; Ikemura 1985; Duret 2000). Support for the second force is arises from the fact that GC—rich synonymous codons are more frequentlyused in GC—rich organisms and vice versa(Bergand Silva 1977; Fedorovet al 2002; Mcveanand Hurst 2000; Jukesand Bhushan 1986; Sueoka 1992).

Codon—pairusage, like codon usage, has also been found to be biased. (Cheng and Goldman 2001; Yarus and Folley 1985). Initially, Gutman and Hatfield(1989)showed that codon—pair usage is non—randombased on a survey of 237 *E. coli* gene coding sequences. Later, several investigators expanded the analysis to a genome—wide survey, confirming the initial observation of codon—pair bias(Boycheva et al. 2003; Moura et al. 2005). It was alsofound that codon—pair bias seems to be different between highly and weaklyexpressed gene sets(Boycheva et al. 2003). Similarto codon bias, codon—pair context has effects on translational elongation rate *in vivo*(Irwin et al. 1995; Folley and Yarus 1989). Other studies suggest that codonpairinginfluences mRNA decoding accuracy and efficiency, indicating that the translational machinery imposes significant constraints on codon—pair context(Robinson et al. 1984; Smith and Yarus 1989; Percudani et al. 1997; Pooleet al. 1998; Curran et al. 1995; Shahet al. 2002; Murgolaet al. 1984; Torket al. 2004). Recently, a survey of codon—pair bias was undertaken on all

ORFs in 16 genomes(Buchan et al. 2006). The study revealed three distinct effects. Firstly, codon—pair preference is primarily determined by a tetranucleotide combination of the third nucleotide of the P—site codon, and all 3 nt of the A—site codon. Secondly, pairs of rare codons are generally under—used in eukaryotes, but over—used in prokaryotes. Thirdly, tRNA structural features have significant effects on codon pairing. More recently, Moura et al. (2007) showed that in *Bacterial* and *Archeal*codon—pair context is mainly dependent on constraints imposed by the translational machinery, while in *Eukaryotes* DNA methylation and tri—nucleotide repeats impose strong biases on codon—pair context. Despite these findings, due to the vast amount of information on the codon usage and codon—pair context, it remains a daunting task to decipher its meaning in many respects, especially for codon—pair bias. It is known that codon usage is associated with biological evolution(Goldman and Yang 1994; Nesti et al. 1995; Pouwels and Leunissen 1994). However, little studies have been carriedout to investigate whether codon—pair context can help unravel the evolutionary relationships between species to date(Moura et al. 2007).

Inthis paper, in order to discover possible evolutionary constraints that shape codon—pair context, a large scale survey of codon—pair context was undertaken in 110 genomes. The study revealed that the DNMs of codon—pairs in CDSs and the base triplet pairs in IGSs with their frequencies are in accordance with a gamma distribution. Our results show that the codon—pair context could be an important determinant on phylogeny of individual species, and indicate the existence of fundamental differences of evolutional constraints imposed on CDSs and IGSs among *Archaea*, *Bacteria*, and *Eukaryota*. Possible mechanismsof evolution for both codon—pairs in CDSs and the base triplet pairs in IGSsare discussed.

2. MATERIALS AND METHODS

2.1. Materials

The genomic sequences, the full CDSs components, and the annotation files of 40 *Archaea*, 60 *Bacteria*, and 10*Eukaryotes* were downloaded from GenBank (ftp://ftp. ncbi. nih. gov/genbank/genomes/). IGSs of *Drosophila melanogaster*were downloaded from(ftp://ftp. flybase. net/releases/FB2006_01/dpse_r20_20051018/fasta/). To reduce the effects on our results due to small genome size, we selected the genomes witha CDS number greater than 1800. In addition, to exclude the effect of complementary ingredients between the leading and lagging strands of replication, we used data only from the leading strand based on the annotation files. The procedures were as follows:

(1) A dataset of CDSs on the leading strand was obtained for each organism. CDSs with internal stop codons and those that do not start with start codons or end in stop codons were removed from the samples.

(2) Similarly, a dataset of IGSs on the leading strand was obtained for each organism. Intergenic random repeat ingredients, telomere sequences and intergenic complementary segments corresponding to the lagging strand existing in CDSs were excluded from our datasets.

2. 2. Methods

2. 2. 1. Calculation of absolute frequencies of codon—pairsin CDSsand triplet pairs in IGSs

The absolute frequenciesof codon—pairs in CDSs and triplet—pairs in IGSs were calculated by following methods. The row of codon—pairs for each CDS began with an initiator codon(A_1) and the second codon(A_2) to form the codon—pair(A_1A_2). It continued with the second and the third(A_2A_3), the third and the fourth(A_3A_4), and finished with the combination A_nA_{STOP} (where n is the last sense codon of the CDS, preceding the termination codon; see Fig. 1). We considered $61\times61=3721$ modes of codon—pairs(sense: sense codon—pairs; excluding sense: stop, stop: sense and stop: stoppairs combinations). For each mode of codon—pair, its occurrence frequency within a genome was computed.

Similarly, the row of triplet pairs for each IGS began with first triplet(B_1), movedalong the reading window three nucleotides at a time to obtain the second triplet(B_2) which formed the triplet pair(B_1B_2). It continued with the second and the third(B_2B_3), the third and the fourth(B_3B_4), and finishes with the combination($B_{n-1}B_n$), where n is the last triplet of the IGS. The total number of triplet pairs was 4096(64^2). The occurrence frequency of each mode of triplet pair in a genome was computed. We found that different methods used to identify triplet pairs had no effect on the results reported in this paper. Here, we began with the first nucleotide position of IGS to group the triplet.

2. 2. 2. *DNMs of codon — pairsin CDSsand triplet pairs in IGSs with*

A_1	A_2	A_3	A_4	A_5	A_6	…	A_x	A_{x+1}	…	A_n	A_{STOP}
AUG	CAC	CAA	GCG	ACC	*CCC*	…	CGG	CUA	…	CCG	UAG

Fig. 1. Grouping of codons into codon－pairs. Each codon from a CDS was assigned 'A_x'. The numeric identifier(x)shows the codon position and used only integers from 1 to n. We divide the CDS into the following codon－pairs:A_1A_2;A_2A_3;A_3A_4;A_4A_5;… A_xA_{x+1}…A_nA_{STOP}.

theirabsolute frequencies

Themodes of codon－pairs in CDSs and triplet pairs in IGSs were grouped by absolute frequency:1－10,11－20,21－40,etc for the *Archaeal* genomes;1－20,21－40,41－60,etc for the *Bacterial* genomes;and 1－50,51－100,101－150, etc for the *Eukaryotic* genomes. For each frequency group,we first computed C_k, the number of modes of codon－pairs,then obtained y_k,the ratio of the number of modes of codon－pairs to total number of modes3721,in the number k group. Similarly,we computed T_k,the number of modes of triplet pairs in the number k group,and obtained y'_k,the ratio of the number of modes of triplet pairs to total number of modes of triplet pairs 4096. The equations are:

$$y_k = C_k/3721(k = 1,2,\cdots,m) \tag{1}$$

$$y'_k = T_k/4096(k = 1,2,\cdots,n) \tag{2}$$

where k is the number of groups,m and n is the number of maximum group for codon－pairs and triplet pairs,respectively.

Then we obtained the distribution of the ratios versus the absolute frequencies of codon－pairs(or triplet pairs)and found that 99% of modes of codon－pairs were within 100 groups(see Fig. 2). We have also confirmed that the size of group has no impact on the shape of the distribution.

2.2.3. DNMs of codon－pairs and triplet pairs with their relativefrequencies

To exclude the effects of different genome sizesand to further validate that the DNMfollows a gamma distribution,all genomes included in the study were converted to a uniform size of one million codon－pairs,roughlyequivalent to the *Archaeoglobus_fulgidus* genome. Thus,the frequencies of codon－pairs and triplet pairs are called the "relative frequencies". If the DNM of codon－pairs with

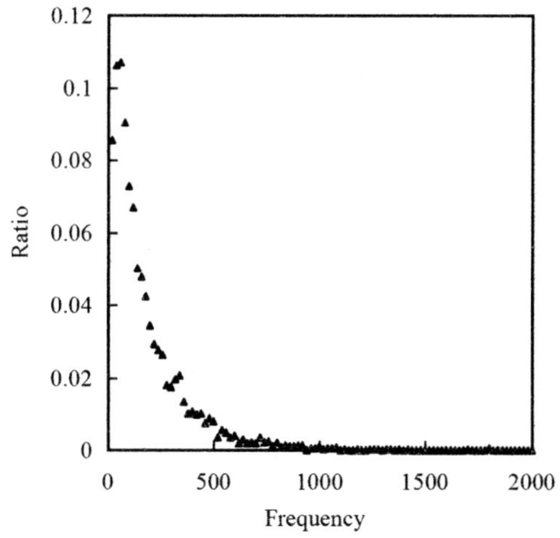

Fig. 2. The DNM of codon — pairs from *Escherichia _ coli _ CFT*073are plotted as absolute frequency vs. the ratio.

theirabsolutefrequencies are in accordance with gamma distribution, the DNM of-codon — pairs with their relativefrequenciesshould also be in accordance with gamma distribution. The values of the shape parameter α of the fitting results between the absolute and the relativefrequencies should be equal.

The relative frequencies $N_{Crelative}$ for codon—pair and $N_{Trelative}$ for triplet pair are defined as:

$$N_{Crelative} = (10^6 / N^C_{\mathrm{TOT}}) \times N_{\mathrm{Cabsolute}} \tag{3}$$

$$N_{Trelative} = (10^6 / N^6_{\mathrm{TOT}}) \times N_{\mathrm{Tabsolute}} \tag{4}$$

where $N_{\mathrm{Cabsolute}}$is the absolute frequency of a given codon—pair mode; N^C_{TOT} is the total number of codon—pairs in each organism; $N_{\mathrm{Tabsolute}}$ is the absolute frequency of a given triplet pair; and N^T_{TOT} is the total number of triplet pairs in each organism.

2. 2. 4. The fittingmethod andthe Chi—square goodness of fit test

We found that the regularities of the DNMsin 110analyzed genomes are similar. Therefore, we suggestthat the DNM might be combined with a common distribution. For this purpose, we usedthe software Origin 7. 0(http://www.originlab.com/)to carry out the fitting analysis. By fitting the DNMusing different distributions, we found that the gamma distributionis most accordant with the DNM. The gamma distribution function can be defined as follows:

$$f(x)=\begin{cases}1/\beta^{\alpha}\Gamma(\alpha)x^{\alpha-1}e^{-x/\beta}, & x\geqslant 0\\ 0, & else\quad where\end{cases} \tag{5}$$

Where α and β are the shape parameter and the scale parameter, respectively. Here $\alpha>0, \beta>0$.

The mean and the variance of gamma distribution are $\alpha\beta$ and $\alpha\beta^2$ respectively. The shape parameter α is closely correlated with the skewness($\gamma=2/\sqrt{\alpha}$) and the kurtosis($\gamma_2=6/\alpha$). Theoretically, a gamma distribution converges to the normal distribution when the skewness and the kurtosisapproach zero. Generally, when α approaches to 10, gamma distribution canbe considered as the normal distribution(Fang and Xu, 1987).

The goodness of fit is determined using Chi—square test, which answers the question of how well observed data fit expected data. Generally, for a good fit, the Chi—square value for every degree of freedom is less than 1(Du, 1999).

3. RESULTS AND DISCUSSION

3. 1. The DNMs of codon—pairsin CDSs and the base triplet pairs in IGSs with their absolute frequencies are in accordance with the gamma distribution

Some examples of the fitting results for the DNMs of codon—pairs and triplet pairsusing gamma distribution are listed in Tables 1 and 2. As a control, we also obtained fitting results of the DNM of triplet pairs for a random sequence produced by DAMBE(http://dambe. bio. uottawa. ca). The results from the data given in Appendix A and B are detailed below.

In all analyzed genomes: for codon—pairs, the R^2 value of the correlation coefficient is in the range of 0. 88 ~ 1. 00, and the Chi—square value for every degree of freedom is less than 10^{-4}; for triplet pairs, the R^2 value is in the range of 0. 89 ~ 1. 00, and the Chi—square value for every degree of freedom is less than 10^{-4}. As indicted by the lowChi—square values, our results show that the DNM is significantly coincident with a gamma distribution. Furthermore, the correlation coefficient$R$ is also an important indicator for the goodness of fit. The higher R^2 value also suggests that these distributions are in accordance with a gamma distribution(see Tables 1, 2, Fig. 3).

Table1. Some examples of the fitting results for the DNM of codon—pairswith their absolute frequencies using gamma distribution[a]

Species	No. of CDS	$\alpha \pm \sigma$	$\beta \pm \sigma$	Chi^2/DoF	R^2
Archaea					
Halobacterium_sp	1008	0.68±0.001	79±0.241	2.54E−6	0.98
Archaeoglobus_fulgidus	1178	0.74±0.006	130±4.184	7.15E−7	0.99
Pyrococcus_abyssi	929	0.79±0.002	80±1.303	1.43E−6	0.98
Pyrococcus_furiosus	1054	0.73±0.001	105±0.794	4.32E−7	0.99
Methanocorpusculum_labreanum_Z	866	2.65±0.022	45±0.397	2.00E−5	0.90
Bacteria					
Anabaena_variabilis_ATCC_29413	2668	2.09±0.001	96±0.528	1.35E−6	0.99
Arthrobacter_aurescens_TC1	1968	1.01±0.000	185±1.342	4.03E−6	0.98
Bacillus_subtilis	1941	1.08±0.000	192±3.367	6.67E−6	0.96
Bacillus_thuringiensis_Al_Hakam	2320	1.10±0.000	172±0.983	1.54E−6	0.99
Escherichia_coli_K12	2070	1.11±0.000	178±2.687	5.31E−6	1.00
Eukaryota					
Schizosaccharomyces_pombe	2429	2.13±0.001	138±0.853	1.59E−6	0.97
Saccharomyces erevisiae	2943	2.05±0.001	296±1.750	5.31E−6	0.99
Caenorhabditis elegans	11569	2.02±0.001	754±14.251	2.00E−5	0.96
Arabidopsis thaliana	15310	2.02±0.000	884±5.200	1.01E−6	0.98
Drosophila_melanogaster	9871	2.05±0.002	800±15.206	9.39E−6	0.91
Random sequence	—	9.99±0.002	45±1.368	1.58E−6	0.99

[a] α and β are the parametersof gamma distribution, $Chi^2/DoF$is the Chi—square value for every degree of freedom, R^2 is the square of correlation coefficient, and the σ is the standard deviation.

Table 2. Some examples of the fitting results for the DNM of triplet pairswith their absolute frequenciesusing gamma distribution[b]

Species	No. of IGS	$\alpha \pm \sigma$	$\beta \pm \sigma$	Chi^2/DoF	R^2
Archaea					
Halobacterium_sp	867	2. 10±0. 001	12±0. 193	5. 00E−5	0. 99
Archaeoglobus_fulgidus	850	2. 10±0. 004	25±0. 989	2. 60E−4	0. 93
Pyrococcus_abyssi	800	2. 10±0. 002	17±0. 175	1. 00E−4	0. 98
Pyrococcus_furiosus	807	2. 11±0. 001	20±0. 224	2. 00E−5	0. 99
Methanocorpusculum_labreanum_Z	634	2. 10±0. 002	18±0. 342	8. 00E−5	0. 94
Bacteria					
Anabaena_variabilis_ATCC_29413	2535	2. 47±0. 004	30±0. 290	2. 11E−6	0. 98
Arthrobacter_aurescens_TC1	1650	3. 24±0. 002	38±0. 330	3. 03E−6	0. 97
Bacillus_subtilis	1680	3. 14±0. 005	18±0. 037	2. 51E−6	0. 98
Bacillus_thuringiensis_Al_Hakam	2026	3. 97±0. 110	16±0. 110	9. 11E−6	0. 90
Escherichia_coli_K12	1756	3. 16±0. 005	27±0. 059	1. 99E−6	0. 98
Eukaryota					
Schizosaccharomyces_pombe	2020	2. 12±0. 001	272±12. 281	7. 94E−7	0. 97
Saccharomyces erevisiae	2859	2. 11±0. 005	129±14. 779	4. 00E−5	0. 90
Caenorhabditis elegans	10143	2. 02±0. 001	667±21. 236	9. 59E−6	0. 90
Arabidopsis thaliana	11459	2. 03±0. 003	1456±28. 221	2. 52E−6	0. 89
Drosophila_melanogaster	6886	3. 37±0. 010	558±15. 830	1. 60E−6	0. 93

[b] α and β are the parametersof gamma distribution, $Chi^2/DoF$is the Chi−square value for every degree of freedom, R^2 is the square of correlation coefficient, and the σ is the standard deviation.

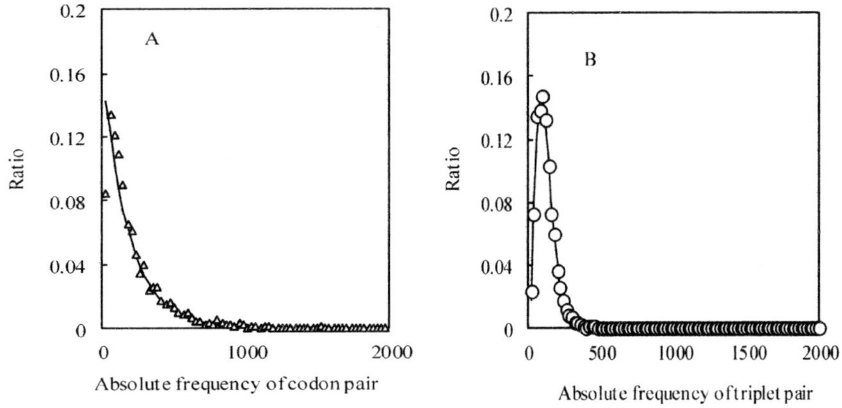

Fig. 3. The fitting curves using gamma distribution for the DNMs of codon—pairs in CDSs(A)and triplet pairs in IGSs(B)from *Escherichia_coli_CFT073*. The *x*—axis denotes the absolute frequencies of codon—pairs and triplet pairs, and the *y*—axis denotes the ratios of numbers of modes of codon—pairs(triangles)and triplet pairs(circles). Black lines represent the gamma distribution.

3.1.1. The usages of codon—pairs and triplet pairs are significantly non—random

The parameters α and β of gamma distribution are showed in Fig. 4 for all analyzed organisms. From these numbers it can be seen that: *i.* for CDSs, the α value of gamma distribution from *Archaea* is in the range of 0.60 ~ 0.86 except for *Methanocorpusculum_labreanum_Z*(α=2.65±0.022), while the α value from *Bacteria* in the range of 1.00 ~ 1.33, except for the genome of *Anabaena_variabilis_ATCC_29413*(α=2.09±0.001), and from *Eukaryotes*, it is in the range of 2.02 ~ 2.21. *ii.* for IGSs, the α values from *Archaea*, *Bacteria*, and *Eukaryotes* are in the range of 2.09 ~ 2.19, 2.45 ~ 3.97, and 2.02 ~ 3.37, respectively. *iii.* the value of parameter α for a random sequence, however, is close to 10(see Table 1). Evidently, the α values for CDSs and IGSs are smaller than 4(see Fig. 4), suggesting a non—random distribution, which means that the usages of codon—pairs and triplet pairs are non—random.

3.1.2. General codon—pair context rules

The mean values of α and β of gamma distribution from *Archaea*, *Bacteria*, and *Eukaryota* are tabulated in table 3, from which we can arrive at the following conclusions:

i. from *Archaea*, the mean α value is smaller than 1, while the mean α value from *Bacteria* is larger than 1. *ii.* the mean α value from *Eukaryota*, however, is larger than 2 and is 2.69 times and 1.94 times larger than those from *Archaea* and *Bacteria*, respectively. Apparently, the α value increases in the order *Archaea*, *Bacteria* and *Eukaryota*. Interestingly, the shape parameter α classifies the coding sequences into three categories, *Archaea*, *Bacteria*, and *Eukaryota* (see Fig. 4, and Table 3). The results demonstrate that the DNMs of codon—pairs in CDSs of the three domains of life are significantly different from each other, indicating that the DNM of codon—pairs is a good indicator of phylogeny.

According to the gamma distribution, the randomness of gamma distributions increases with increase in α value. The α values are smaller than 1 for *Archaea* with the exception of *Methanocorpusculum_labreanum_Z*. Hence, the distribution is similar to the exponential distribution, indicating that the ratio with the

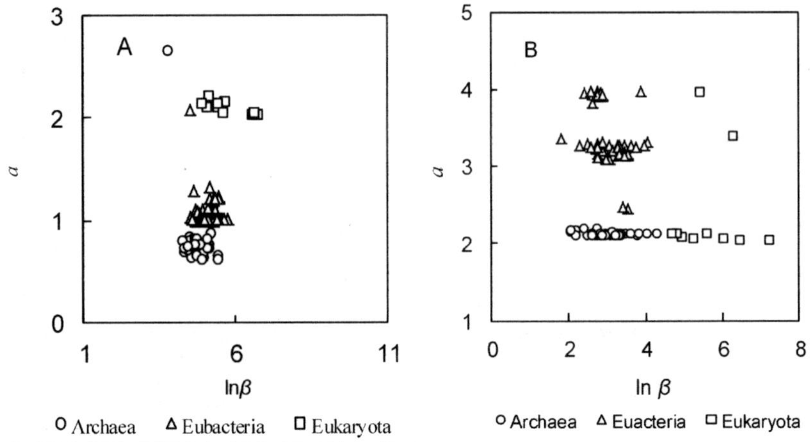

Fig. 4. Two parameters α and ln β of the fitting results using gamma distribution for the DNMs of codon—pairs in CDSs(A)and triplet pairs in IGSs(B)from 110 genomes analyzed. Circles, triangles and squares denote *Archaea*, *Bacteria* and *Eukaryota* respectively.

lowest frequenciesof codon—pairs is the largest(see Fig. 3). However, for the *Bacteria* and *Eukaryotes* genomes analyzed, the α values are larger than 1. The corresponding distribution curve has an apex. The peculiarity is that the ratios(see Fig. 3)withthe lowestand highest frequenciesof codon—pairs are the smallest. The α value increasesin the order of *Archaea*, *Bacteria*, and *Eukaryota*, indicating that the randomness of DNM of codon—pairs increases with evolution. This reveals that there are general rules governing the evolution of the DNM of codon—pairs in the 3 domains of life. Thereby, we infer that the existence of fundamental differences on constraints imposed by the translational machinery between *Archaea*, *Bacteria*, and *Eukaryota*. According to therecent results by Moura et al. (2007), in *prokaryotes* codon—pair context is mainly determined by constraints imposed by the translational machinery, while in *Eukaryotes* the emergence of DNA methylation and tri—nucleotide repeats influence codon—pair context.

One of the complexities of living organism is owing to its specificity. Interestingly, an exception in *Bacteria* is the genome of *Anabaena_variabilis_ATCC_29413*, of which α value is 2. 09±0. 001; and an exception in *Archaea* is the genome of *Methanocorpusculum_labreanum_Z*, of which α value is 2. 65±0. 022 (Fig. 4). Both are located within the region of *Eukaryote*, suggesting that codon—pairing in the two genomestakes on *eukaryotic* characteristics incertainaspects. This suggests that the non—random usage of codon—pairs also reflects the action of species—specific evolutionary forces besides its own commonness.

3. 1. 3. General triple—pair context rules in IGSs

For triplet pairs, the characteristics of α valuesof gamma distributions aretotally different from that for codon—pairs, as shown in Tables2 and 3. From *Bacteria*, the mean α value is the largest and is 1. 59 times and 1. 40 times larger than those from *Archaea* and *Eukaryota*, respectively. The mean α from *Archaea*, however, is close tothat from *Eukaryote*. The higher α value suggests that the DNMs of triplet pairs from *Bacteria* have a higher degree of randomness.

Although both *Archaea* and *Bacteria* are *prokaryote*, their IGSs followed different evolutionary paths. From this perspective, *Archaea* is close to *Eukaryotes*, and is significantly different from Bacteria. This result agrees with the theory of two domains of life, indicating that triplet—pair context in IGSs is influenced by evolutionary forces and also is an indicator of phylogeny.

Table 3. The mean α and β values of the gamma distribution for the DNMs of codon—pairs in CDSs and triplet pairs in IGSs with their absolute frequencies from 110 genomes analyzed. The subscript 1 denotes codon—pairs and the subscript 2 denotes triplet pairs

Domain	$\bar{\alpha}_1$	$\bar{\beta}_1$	$\bar{\alpha}_2$	$\bar{\beta}_2$
Archaea	$0.78\pm0.31(0.73\pm0.06)$[c]	120 ± 42	2.11 ± 0.02	23 ± 14
Bacteria	$1.08\pm0.15(1.06\pm0.08)$[d]	180 ± 54	3.37 ± 0.36	24 ± 10
Eukaryota	2.10 ± 0.06	400 ± 291	2.39 ± 0.68	418 ± 411

[c] the result in bracket is the mean value of the parameter α of gamma distributionfor 39 of 40 organisms from *Archaea* except for the genome of *Methanocorpusculum_labreanum_Z*, its α value is 2.65 ± 0.022; [d] the result in bracket is the mean value of the parameter α of gamma distributionfor 59 of 60 organisms from *Bacteria* except for *Anabaena_variabilis_ATCC_29413*, the genome of its α value is 2.07 ± 0.001.

3.1.4. Comparison of the results between codon—pairs and triplet pairs

A comparative analysis showed that theDNMs betweencodon—pairs in CDSs and triplet pairs in IGSs have significant differenceswithinand between genomes (Fig. 3). In *Archaea*, the mean α value of gamma distributionfor triplet pairs is 2.67times larger than that for codon—pairs; in *Bacteria*, that fortriplet pairs is 3.12times larger than that for codon—pairs; and in *Eukaryotes*, that fortriplet pairs is only 1.14 times larger than that for codon—pairs. This shows that the DNM of triplet pairs is more random than that of codon—pairs, implying that the correlation of codon—pair context is stronger than that of triplet pairs in *prokaryotic* genomes. In other words, the structure of CDSs is more regular than that of IGSs. Thisresulthighlights the fundamental differences between coding and non—codingsequences, suggesting that the evolution paths are significantly different between CDSs and IGSs. Therefore, evolution studies should use CDSs and ICGs separately instead of the entire genome DNA sequences.

3.2. The DNMs of codon—pairs and triplet pairs with their relative frequencies are in accordance with gamma distribution

To exclude the effect of different genome sizes and to validate further that the DNM follows a gamma distribution, all genomes included in the study were converted into a uniform size(see method 3.3). The fitting results using the gamma distribution for the DNMs of codon—pairs and triplet pairs with relative frequencies are obtained in 110 organisms. The data are presented in Appendix C and D, while some examples are shown in Tables4—6 and Figs. 5,6.

Table4. Some examples of the fitting results forthe DNM of codon—pairs with their relative frequencies using gamma distribution[e]

Species	$\alpha\pm\sigma$	$\beta\pm\sigma$	Chi^2/DoF	R^2
Archaea				
Halobacterium_sp	0.66±0.008	378±13.753	2.54E−6	0.98
Archaeoglobus_fulgidus	0.74±0.022	337±16.743	3.34E−6	0.96
Pyrococcus_abyssi	0.79±0.006	316±13.768	7.09E−6	0.92
Pyrococcus_furiosus	0.72±0.006	347±13.058	2.75E−6	0.97
Methanocorpusculum_labreanum_Z	2.59±0.009	108±1.860	2.43E−6	0.90
Bacteria				
Anabaena_variabilis_ATCC_29413	2.07±0.001	130±4.602	1.43E−6	0.98
Arthrobacter_aurescens_TC1	1.00±0.000	274±13.367	2.95E−6	0.97
Bacillus_subtilis	1.07±0.001	233±10.041	6.10E−6	0.93
Bacillus_thuringiensis_Al_Hakam	1.12±0.000	245±11.578	1.32E−6	0.99
Escherichia_coli_K12	1.12±0.000	243±13.874	2.58E−6	0.97
Eukaryota				
Schizosaccharomyces_pombe	2.13±0.001	127±0.783	1.92E−6	0.97
Saccharomyces erevisiae	2.05±0.001	120±1.055	5.34E−6	0.99
Caenorhabditis elegans	2.02±0.000	142±1.155	8.62E−6	0.98
Arabidopsis thaliana	2.02±0.000	157±1.373	9.99E−6	0.97
Drosophila_melanogaster	2.05±0.002	125±3.862	7.00E−5	0.93

[e] α and β are the parametersof gamma distribution, Chi^2/DoF is the Chi—square value for every degree of freedom, R^2 is the square of correlation coefficient, and the σ is the standard deviation.

Table5. Some examples of the fitting results for the DNM of triplet pairswith their relative frequencies using gamma distribution[f]

Species	$\alpha\pm\sigma$	$\beta\pm\sigma$	Chi^2/DoF	R^2
Archaea				
Halobacterium_sp	2.14±0.005	116±2.122	4.00E−5	0.93
Archaeoglobus_fulgidus	2.11±0.004	119±3.611	3.00E−5	0.94
Pyrococcus_abyssi	2.11±0.005	115±4.371	5.00E−5	0.91
Pyrococcus_furiosus	2.12±0.003	104±2.313	2.00E−5	0.96

续表

Species	$\alpha \pm \sigma$	$\beta \pm \sigma$	Chi^2/DoF	R^2
Archaea				
Methanocorpusculum_labreanum_Z	2.10±0.002	134±2.727	1.00E−5	0.91
Bacteria				
Anabaena_variabilis_ATCC_29413	2.47±0.002	110±1.062	3.08E−6	0.91
Arthrobacter_aurescens_TC1	3.24±0.004	85±1.431	2.65E−6	0.91
Bacillus_subtilis	3.10±0.010	90±0.257	2.69E−6	0.93
Bacillus_thuringiensis_Al_Hakam	3.87±0.016	68±0.301	5.13E−6	0.89
Escherichia_coli_K12	3.14±0.008	79±0.262	1.94E−6	0.95
Eukaryota				
Schizosaccharomyces_pombe	2.12±0.001	118±0.684	1.60E−6	0.98
Saccharomyces erevisiae	2.11±0.003	105±2.271	2.00E−5	0.93
Caenorhabditis elegans	2.06±0.002	106±3.300	3.00E−5	0.89
Arabidopsis thaliana	2.06±0.001	121±1.872	2.00E−5	0.96
Drosophila_melanogaster	3.41±0.003	76±0.079	2.35E−6	0.99

[f] α and β are the parametersof gamma distribution, $Chi^2/DoF$is the Chi−square value for every degree of freedom, R^2 is the square of correlation coefficient, and the σ is the standard deviation.

Table6. The mean α and β values of the gamma distribution for the DNMs of codon−pairs in CDSs and triplet pairs in IGSs with their relative frequencies from 110 genomes analyzed. The subscript 1 represents codon−pairs and the subscript 2 denotes triplet pairs.

Domain	$\bar{\alpha}_1$	$\bar{\beta}_1$	$\bar{\alpha}_2$	$\bar{\beta}_2$
Archaea	0.78±0.30(0.73± 0.07)[g]	349±52	2.12±0.02	118±9
Bacteria	1.08±0.15(1.06±0.08)[h]	244±26	3.35±0.36	78±11
Eukaryota	2.10±0.06	127±13	2.40±0.68	113±26

[g]the result in bracket is the mean value of the parameter α of gamma distributionfor 39 of 40 organisms from *Archaea* except for*Methanocorpusculum_labreanum_Z*, its α value is 2.59 ±0.009; [h]the result in bracket is the mean value of the parameter α of gamma distributionfor 59 of 60 organisms from *Bacteria* except for *Anabaena_variabilis_ATCC_29413*, the genome of its α value is 2.07 ±0.001.

For codon−pairs, theR^2 value of the correlation coefficient is in the range of 0.89 ~ 1.00, and the Chi−square value for every degree of freedom is less than 6.00E−5. Fortriplet pairs, the R^2 value is in the range of 0.89 ~ 0.99, and the Chi−square value for every degree of freedom is less than 7.00E−5. The small Chi−square value and the higher R^2 values show that the DNM significantly

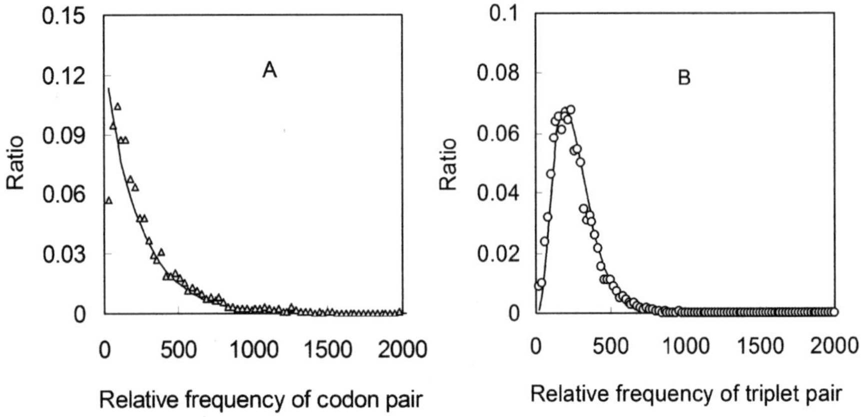

Fig. 5. The fitting curves using a gamma distribution for the DNMs of codon－pairs in CDSs(A) and triplet pairs in IGSs(B) with their relative frequencies from*Escherichia_coli_CFT*073. The *x*－axis denotes the relative frequencies of codon－pairs and triplet pairs, and the *y*－axis denotes the ratios of numbers of modes of codon－pairs(triangles) and triplet pairs(circles) respectively. Black lines represent the gamma distribution.

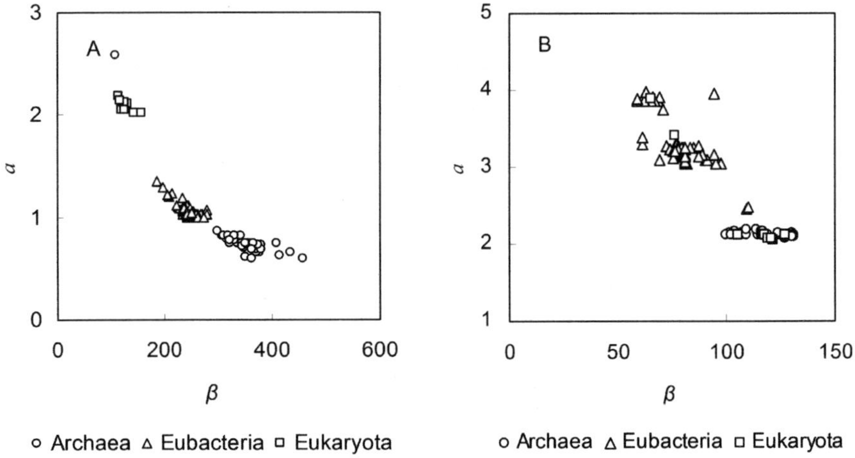

Fig. 6. Two parameters $\alpha$and β of the fitting resultsusing gamma distribution for the DNMs of codon－pairs(A) and triplet pairs(B) in IGSs with their relative frequencies from 110 genomes analyzed. Circles, triangles and squares denote *Archaea*, *Bacteria*an-d*Eukaryota* respectively.

agreeswith the gamma distribution.

Importantly, the data in Tables 3 and 4 show that the differences of the α

values of gamma distribution between absolute and relative frequenciesare smaller than 3% for all analyzed organisms. For example, in *Bacillus subtilis*, the α valuesare1.07 ±0.001 and 1.08 ±0.000, the difference between absolute and relative frequencies of codon-pairs is only 0.9%. This indicates the shape factor α is almost invariable compared with the previous result(see Tables 1-6). Thus, the results confirm that the parameter α does not depend on genome size and only reflects the shape of distribution, implying that the DNMsof codon-pairs and triplet pairs do follow gamma distribution. The resultssuggest that despite the species specificity of codon-pair context, at least some of the evolutionary constraints that shaped codon-pair context are conserved across species in the three domains of life.

4. CONCLUSION

Gamma distribution has recently been used extensively in phylogenetics (Hsieh et al. 2003; Wang et al. 2006; Wang and Li 2007; Feng and Li 2004). In previous work, we have proved that the length distribution of CDS in genomes agrees with gamma distribution(Feng and Li 2004). The results from this study show that the DNMsof codon-pairs in CDSs and triplet pairs in IGSs are in accordance with gamma distribution. The data also demonstrates that the parameter α of gamma distribution can be used as an index of codon-pair usage(or triplet pair usage) with genome evolution. For codon-pairs, this index divides the species into three evolutionary groups, *Archaea*, *Bacteria*, and *Eukaryota*, *as* shown inFig. 7. While for triplet pairs, the index divides the species into two groups, one is *Bacteria and* the other is *Archaea* and *Eukaryota*. Furthermore, the data demonstrates that the evolutionpath of CDSs is clearly different from that of IGSs, and the DNM of codon-pairs is more non-random than that of triplet pairs in three domains of life(i. e. the parameter α from codon-pairs is smaller than that from triplet pairs). The findings establish that codon-pair(or triplet pair)context follows phylogeny, strongly suggesting the existence of fundamental differences between *Eukaryota*, *Bacteria* and *Archaea* in the evolutionary forces that shape CDSs(or IGSs)primary structure, and consequently, bringing out different codon-pair(or triplet pair)context.

In addition, it is well known that the most disputed species is *Archaea* with

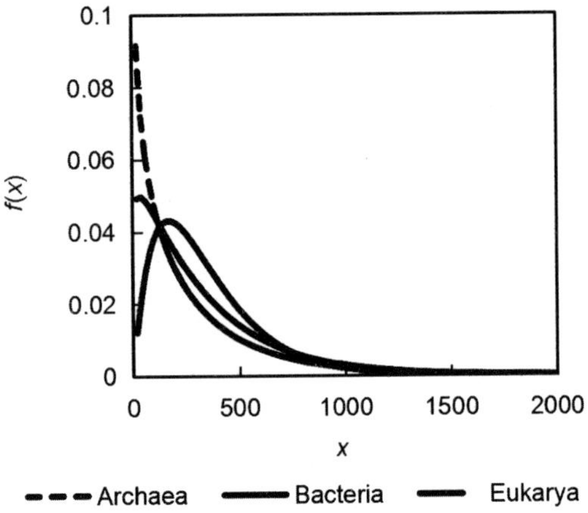

Fig. 7. The gamma distribution curves of DNM of codon — pairs with their frequencies. For the purpose of showingthe differences of DNMs of codon — pairs in the three domains of life, here, the gamma distribution curves corresponding to codon — pairs are showed using the meanαandβvalues in 40 *Archaea*, 60 *Bacteria* and 10 *Eukaryotes* genomes respectively.

regard toevolution due to its particular features that are different from *Bacteria* and *Eukaryote* as opposed to those features that are similar to that of *Bacterial* and *Eukaryotic*(Saito and Tomita, 1999). Here, on the rule of DNM of codon — pairs, our results indicate that *Archaea* is close to *Bacteria* and significantly different from*Eukaryota*. However, for the DNM of triplet pairs, *Archaea* is close to *Eukaryote*. This impliesthat different evolution paths between CDSs and IGSs have formed the unique way of entire genome evolution of *Archaea*.

Meanwhile, our conclusion is that the randomness of DNM of codon — pairs increases with evolution as suggested by the observation that the αvalue increases with evolution. This agrees with the randomness of nucleotides in coding sequences, where the randomness is high in *Eukaryotes*and low in *prokaryotes* (Takeuchi et al., 2003), indicating that degree of randomness increases as the living organismsbecome more complex. One explanation for this observation is that higher organisms have accumulated random mutations (Takeuchi et al., 2003). Another possible contributing factor to consider is the environment. For*Archaea*, as a result of extremeliving environment, codon — pair context has been affected by severe selective forces, thus, there must be stricter restriction on amino acid composition, and leading to the non－randomness combinations of codons in tandem. Therefore, the non－random DNM of codon－pairs is very significant. Complex organisms with many gene repertoire can move on to better conditions or keep their own optimum inner environment and thus might have less restrictions and more freedom in tolerating randomness.

In summary, codon－pair bias has been shown to be a universalphenomenon that results from a balance of competing selectiveforces. Our results show that the codon－pair context could be an important determinant on phylogeny of individual species, and indicate the existence of fundamental differences of evolutional constraints imposed on CDSs and IGSs among *Archaea*, *Bacteria*, and *Eukaryota*. The findings in this study may shed new light on our understanding of the rules that govern codon－pairing and genome evolution.

Acknowledgements

We thankDr. Chi－Hao Luan and Sara Fernandez Dunne in U. S. Northwestern University for assistance in revising the English expression and improving the manuscript. This work was supported by National Natural Science Foundation of China (30660044) and The Ph. D. Programs Foundation of Ministry of Education of China (20050126003).

References

[1]Andersson, S. G. E., Kurland, C. G., 1990. Codon preferences in free livingmicroorganisms. Microbiol. Rev. 54, 198－210.

[2]Berg, O. G., Silva, P. J., 1977. Codon bias in Escherichia coli: the influence of codon context on mutation and selection. Nucl. Acids Res. 25, 397－404.

[3]Boycheva, S., Chkodrov, G., Ivanov, L., 2003. Codon pairs in the genome of *Escherichia coli*. J. Bioinformatics. 19, 987－998.

[4]Buchan, J. R., Aucott, L. S., Stansfield, I., 2006. tRNA properties help shape codon pair preferences in open reading frames. Nucl. Acids Res. 34, 1015－1027.

[5]Cheng, L., Goldman, E., 2001. Absence of effect of varying Thr－Leu codon pairs on protein synthesis in a T7 system. Biochemistry40, 6102－6106.

[6]Curran, J. F., Poole, E. S., Tate, W. P., Gross, B. L., 1995. Selection of aminoacyl－tRNAs at sense codons: the size of the tRNA variable loop determines whether the immediate 39 nucleotide to the codon has a context effect. Nucl. Acids Res. 23, 4104－4108.

[7]Du R. Q. ,1999. Biostatistics,Second ed. Liu,L. ,Beijing.

[8]Duret,L. ,2000. tRNA gene number and codon usage in the C. elegans genome are co—adapted for optimal translation of highly expressed genes. Trends Genet. 16,287—289.

[9]Fang,K. T. ,Xu,J. L. ,1987. StatisticalDistributions,First ed. Xiao,Y. ,Beijing.

[10]Fedorov,A. ,Saxonov,S. ,Gilbert,W. ,2002. Regularities of context—dependent codon bias in eukaryotic genes. Nucl. Acids Res. 30,1192—1197.

[11]Feng,L. Q. ,Li,H. ,2004. The distribution mode of open reading frame length in different genomes and the genome evolution. Acta Biophysica Sinca. 20,375—381.

[12]Folley,L. S. ,Yarus,M. ,1989. Codon contexts from weakly expressed genes reduce expression in vivo. J. Mol. Biol. 209,359—378.

[13]Goldman,N. ,Yang,Z. H. ,1994. Codon based model of nucleotide substitution for protein codingDNA sequences. Mol. Biol. Evol. 11,725—736.

[14]Gutman,G. A. ,Hatfield,G. W. ,1989. Non—random utilization of codon pairs in *Escherichia coil*. Proc. Natl. Acad. Sci. USA,86,3699—3703.

[15]Hsieh,L. C. ,Luo,L. F. ,Ji F. M. ,2003. Minimal model for genome evolution and growth. Physical Review Letters90,1—4.

[16]Ikemura,T. ,1985. Codon usage and tRNA content in unicellular and multicellular organisms. Mol. Biol. Evol. 2,12—34.

[17]Ikemura,T. ,Wada,K. N. ,1991. Evident diversity of codon usage patterns of human genes with respect to chromosome banding patterns and chromosome numbers; relation between nucleotide sequence data and cytogenetic data. Nucl. Acids Res. 19,4333—4339.

[18]Irwin,B. ,Heck,J. D. ,Hatfield,G. W. ,1995. Codon pair utilization biases influence translational elongation step times. J. Biol. Chem. 270,22801—22806.

[19]Jukes,T. H. ,Bhushan,V. ,1986. Silent nucleotide substitutions and G+C content of some mitochondrial and Bacterial genes. J. Mol. Evol. 24,864—875.

[20]Kurland,C. G. ,1993. Major codon preference: theme and variations. Biochem. Soc. Trans. 21,841—845.

[21]Mcvcan,G. A. T. ,Hurst,G. D. D. ,2000. Evolutionary lability of context—dependent codon bias in bacteria. J. Mol. Evol. 50,264—275.

[22]Moriyama,E. N. ,Powell,J. R. ,1997. Codon usage bias and tRNA abundance in *Drosophila*. J. Mol. Evol. 45,514—523.

[23]Moura,G. ,Pinheiro,M. ,Silva,R. ,Miranda,I. ,Afreixo,V. ,Dias,G. ,Freitas,A. ,Oliveira,J. L. ,Santos,M. A. ,2005. Comparative context analysis of codon pairs on an ORFeome scale. Genome Biol. 6,R28.

[24]Moura,G. ,Pinheiro,M. ,Arrais,J. ,Gomes,A. C. ,Carreto,L. ,Freitas,A. ,et al. ,2007. Large Scale Comparative Codon—Pair Context Analysis Unveils General Rules that

Fine—Tune Evolution of mRNA Primary Structure. Plos. one. 2,e847.

[25]Murgola,E. J. ,Pagel,F. T. ,Hijazi,K. A. ,1984. Codon context effects in missense-suppression. J. Mol. Biol. 175,19—27.

[26] Nesti, C. , Poli, G. , Chicca, M. , Ambrosino, P. , Scapoli, C. , 1995. Phylogeny inferred from codonusage pattern in 31 organisms. Computer Applications for the Biosciences 2, 167—171.

[27]Percudani,R. ,Pavesi,A. ,Ottonello,S. ,1997. Transfer RNA gene redundancy and translational selection in Saccharomyces cerevisiae. J. Mol. Biol. 268,322—330.

[28] Poole, E. S. , Major, L. L. , Mannering, S. A. , 1998. Translational termination in Escherichia coli: three bases following the stop codon crosslink to release factor 2 and affect the decoding efficiency of UGA—containing signals. Nucl. Acid Res. 26,954—960.

[29]Pouwels,P. H. ,Leunissen,J. A. M. ,1994. Divergence in codon usage of *Lactobacillus*species. Nucl. Acids Res. 22,929—936.

[30]Robinson,M. ,Lilley,R. ,Little,S. ,Emtage,J. S. ,Yarranton,G. ,Stephens,P. ,Millican,A. ,Eaton,M. ,Humphreys,G. ,1984. Codon usage can affect efficiency of translation of genes in Escherichia coli. Nucl. Acids Res. 12,6663—6671.

[31]Saito,R. ,Tomita,M. ,1999. Computer analyses of complete genomes suggest that some archaebacteria employ both eukaryotic and bacterial mechanisms in translation initiation. Gene 238,79—83.

[32]Shah,A. A. ,Giddings,M. C. ,Gesteland,R. F. ,Atkins,J. F. ,Ivanov,I. P. ,2002, Computational identification of putative programmed translational frameshift sites. Bioinformatics 18,1046—1053.

[33]Sharp,P. M. ,Matassi,G. ,1994. Codon usage and genome evolution. Curr. Opin. Genet. Dev. 4,851—860.

[34]Smith,D. ,Yarus,M. ,1989. tRNA—tRNA interactions within cellular ribosomes. Proc. Natl. Acad. Sci. USA. 86,4397—4401.

[35] Sueoka, N. , 1992. Directional mutation pressure, selective constraints, and genetic equilibria. J. Mol. Evol. 34,95—114.

[36]Takeuchi,F. ,Futamura,Y. ,Yoshikur,H. ,Yamamoto,K. ,2003. Statistics of trinucleotidesin coding sequences and evolution. J. Theor. Biol. 222,139—149.

[37]Tork,S. ,Hatin,I. ,Rousset,J. P. ,Fabret,C. ,2004. The major 59 determinant in stopcodon read—through involves two adjacent adenines. Nucl. Acids Res. 32,415—421.

[38]Wang,F. P. ,Li,H. ,2007. Codon pairs usage and genome evolution. Acta Biophysica Sinca. 23,176—184.

[39]Wang,S. L. ,Wang,J. ,Chen,H. W. ,Zhang D. X. ,2006. The research of the occurrence frequency distribution of k—mer in whole DNA sequence. Acta Biophysica Sinca. 22,

178－195.

[40]Yarus,M.,Folley,L. S.,1985. Sense codons are found in specific contexts. J. Mol. Biol. 182,529－540.

（本文发表于 2009 年《Gene》433 卷 1－2 期）

Polarization of M2 Line Emitted Following Electron—Impact Excitation of Beryllium—Like Ions

Ying—Long Shi *

摘要:利用全相对论扭曲波程序(RDW)计算得到了类铍离子 1s→2p 内壳层电子碰撞激发到激发态 $1s2s^2 2p_{3/2} J=2$ 的总截面和磁子能级截面,由此得到了激发态的磁子能级布局信息。系统地分析了电子—电子间相互作用的相对论修正(即 Breit 相互作用)对高电荷态类铍离子(42≤Z≤92)退激发过程 $1s2s^2 2p_{3/2} J=2$ → $1s^2 2s^2 J=0$ 发出谱线极化特性的影响。研究结果表明,Breit 相互作用对该磁四极谱线的线性极化度有重要的影响,随着入射电子能量的增加或原子序数的增大,该去极化影响将进一步变大。

Detailed calculations have been carried out for the electron—impact excitation cross sections from the ground state to the individual magnetic sublevels of the $1s2s^2 2p_{3/2} J=2$ excited state of highly—charged beryllium—like ions by using a fully relativistic distorted—wave(RDW)method. The contributions of the Breit interaction to the linear polarization of the $1s2s^2 2p_{3/2} J=2$ → $1s^2 2s^2 J=0$ magnetic quadrupole(*M*2)line have been investigated systematically for the beryllium isoelectronic sequence with $42 \leqslant Z \leqslant 92$. It is found that the Breit interaction depolarizes significantly the linear polarization of the *M*2 fluorescence radiation and that these depolarization effects increase as the incident electron energy and/or the atomic number is enlarged.

---

* 作者简介:师应龙(1982—),男,甘肃天水人,天水师范学院电子信息与电气工程学院副教授、博士,主要从事原子结构、光谱及其碰撞动力学过程的理论研究。

When the highly charged ions are excited by a(directed)beam of electron or, more generally, by electrons with a cylindrical symmetric but otherwise anisotropic velocity distribution, the excitedionic state can be aligned with respect to the direction of the incident beam after the collision, *i. e.* the ionic sublevels with different(modulus of the)magnetic quantum numbers are unequally populated. In the subsequent radiative stabilization of the excited ion, this alignment is then partially transferred to the lower—lying levels of the final ion and also affects the emitted radiation, leading in many cases to an anisotropic angular distribution and linear polarization of the characteristic radiation [1]. From the analysis of these photons, therefore, valuable information can be obtained for both the dynamical process and the magnetic sublevel population of the excited states, and complementary to other conventional observables. Based on these features, anisotropic electron beams provide an important tool for the diagnostics of laboratory and astrophysical plasmas, and they have been applied successfully to the study of solar plasmas [2], laser — produced plasmas [3], and Z — pinch[4]. Furthermore, the knowledge of polarization is necessary, for example, when one wishes to extract the excitation cross section from measurement of emission following the excitation process in an electron—beam ion trap(EBIT)[5].

During the last two decades, therefore, a number of experimental and theoretical studies have been carried out on the behaviour and degree of linear polarization of x—ray emitted from highly charged ions and if they collide with beam of electrons. Henderson*et al.* [6] reported the first polarization measurement of the x—ray emission of the highly charged helium—like Sc ion. Later, also several other polarization measurements were reported [7-9]. On theoretical side, detailed calculations of electron—impact excitation have been made mainly for *K*—and *L*—shell transitions in few—electron ions[10-21]. For instance, Reed *et al.* [18] investigated the relativistic effects on the angular distribution and the degree of linear polarization of x — ray emission following the electron — impact excitation for highly charge helium—like ions. They found that the polarization of the resulting radiation is independent of the atomic number Z in the nonrelativistic limit but that is becomes strongly Z—dependent due to the relativistic effects. Moreover, Fontes *et al.* [19] and Bostock *et al.* [20] studied the contribution of the Breit interaction to the collision strength and the degree of linear polarization of the radiation emission following the electron－impact excitation of helium－like ions. They found that the contribution of the Breit interaction is very important for high－Z ions and is even non－negligible for Fe^{24+} ions. Wu *et al*.[21] analyzed the influence of the Breit interaction on the degree of linear polarization of radiation following the inner－shell electron impact excitation processes. When compared with the photon emission following dielectronic recombination process[22], a quite different behavior was found for the polarization of the same x－ray line. In practice, both the angular distribution and the polarization of radiation following the electron－ion collisions can provide us with an important route to explore the electron－electron ($e-e$) interaction in the presence of strong electromagnetic fields for middle－and high－Z ions. To our knowledge, only little[19-21] was known how the Breit interaction affects the population of the ionic substates and, hence, the angular distribution and polarization of the subsequent photon emission following the inner－shell electron－impact excitation. With the recent advancements in the design of position－sensitive solid－state detectors[23], these studies have become feasible experimentally, which requires detailed theoretical predictions in order to select the candidate system and advance the field of x－ray based polarization diagnostics[24].

In this letter, we apply the fully relativistic distorted－wave (RDW) method and the density matrix theory to explore the magnetic sublevels population of the $1s2s^2 2p_{3/2} J=2$ excited state and the polarization properties of the subsequent x－ray emission following electron－impact excitation for beryllium－like ions, where the $1s2s^2 2p_{3/2} J=2$ excited state is well separated in energy from other inner－shell excitations for all medium and heavy elements. The formation as well as the $M2$ decay of the $1s2s^2 2p_{3/2} J=2$ can be well resolved with present－day (position－sensitive) x－ray detectors with a resolution of $\leqslant$ 50 eV. Here, our main attention is paid to the $1s2s^2 2p_{3/2} J=2 \rightarrow 1s^2 2s^2 J=0$ magnetic quadrupole ($M2$) transitions. Emphasis is placed on the question of how the population of the excited state and polarization properties of subsequent radiation is affected by the relativistic terms (Breit) in the $e-e$ interaction. In the following, we first introduce the basic theory in brief and later present and discuss the results of our computations.

To describe the excitation and subsequent photon emission, it is convenient to choose the z axis to be in the direction of the motion of the incident electron. For this choice of the quantization axis, the electron—impact excitation cross section of the target ion from the initial state $\beta_i J_i M_i$ to the final state $\beta_f J_f M_f$ is written as [25]

$$\sigma_{\varepsilon i}(\beta_i J_i M_i - \beta_f J_f M_f) = \frac{2\pi a_0^2}{k_i^2} \sum_{l_i l'_i j_i j'_i m_{l_i} l_f j_f m_f} \sum_{JJ'M} (i)^{l_i - l'_i}[(2l_i+1)(2l'_i+1)]^{1/2}$$

$$\times \exp[i(\delta_{\kappa_i} - \delta_{\kappa'_i})] C\left(l_i \frac{1}{2} m_{l_i} m_{s_i}; j_i m_i\right) C\left(l'_i \frac{1}{2} m_{l'_i} m_{s_i}; j'_i m_i\right) \quad (1)$$

$$\times C(J_i j_i M_i m_i; JM) C(J_i j'_i M_i m_i; J'm) C(J_f j_f M_f m_f; JM)$$

$$\times C(J_f j_f M_f m_f; J'M) R(\gamma_i, \gamma_f) R(\gamma'_i, \gamma'_f),$$

in which the subscripts i and f refer to the initial and final states, respectively; ε_i is the incident electron energy in Rydberg; a_0 is the Bohr radius; C represents the Clebsch—Gordan coefficients; R is the collision matrix elements; $\gamma_i = \varepsilon_i l_i j_i \beta_i J_i JM$ and $\gamma_f = \varepsilon_f l_f j_f \beta_f J_f JM$, J and M are the quantum numbers corresponding to the total angular momentum of the impact system (target ion plus free electron) and its z component, respectively; m_s, l_i, j_i, m_{l_i}, and m_i are the spin, orbital angular momentum, total angular momentum, and its z component quantum numbers, respectively, for the incident electron e_i; β represents the all additional quantum numbers required to specify the initial and final states of the target ion in addition to its J and M; δ_{k_i} is the phase shift for the continuum electron; κ is the relativistic quantum number, which is related to the orbital and total angular momentum l and j; k_i is the relativistic wave number of the incident electron, which is given by $k_i^2 = \varepsilon_i(1+\alpha^2\varepsilon_i/4)$, where α is the fine—structure constant. It turns out that the electron— electron interaction matrix elements $R(\gamma_i, \gamma_f)$ are independent of M[21,25,26],

$$R(\gamma_i, \gamma_f) = \langle \Psi_{\gamma_f} \mid V \mid \Psi_{\gamma_i} \rangle, \quad (2)$$

where Ψ_{γ_i} and Ψ_{γ_f} are the antisymmetric N+1 electron wave functions for the initial and final states of the collision system "bound ion+free electron", respectively. In this notation, V is the $e-e$ interaction operator. In the relativistic atomic theory, as appropriate for the medium and heavy elements, the (frequency—dependent) $e-e$ interaction [22]

$$V = V^C + V^B = \sum_{p<q} \left(\frac{1}{r_{pq}} - (\alpha_p \cdot \alpha_q) \frac{\cos(\omega r_{pq})}{r_{pq}}\right.$$

$$+ (\alpha_p \cdot \nabla_p)(\alpha_q \cdot \nabla_q) \frac{\cos(\omega r_{pq}) - 1}{\omega_{pq}^2 r_{pq}} \Bigg), \tag{3}$$

consists of both the instantaneous Coulomb repulsion(first term)and the Breit interaction, that is, the magnetic and retardation contributions(second and third terms). In this expression ω_{pq} denotes the frequency of the virtual photon and α_p the vector of Dirac matrices associated with the p' th particle. The $e-e$ interaction operator has been derived rigorously within the framework of quantum electrondynamics and applied to a large number of calculations of electronic structure and dynamical processes in recent years[see Ref. [27]and references therein].

If we assume that electron—impact excitation is the dominant mechanism for populating the upper magnetic sublevel and that the scattered electron remains undetected, we can related the degree of linear polarization for the$J=2$ to the $J=0$ $M2$ line emitted at angle 90° with respect to the electron beam to the excitation cross sections as [10,17]

$$P_L = (J = 2 \rightarrow 0; M2) = \frac{\sigma_2 - \sigma_1}{\sigma_2 + \sigma_1}, \tag{4}$$

where σ_2 and σ_1 are the cross sections for electron—impact excitation from the ground state to the $M_f=2$ and $M_f=1$ magnetic sublevels of the excited state, respectively. P_L can be easily determined experimentally by measuring the intensities of the light that is linearly polarized in parallel and perpendicular to the reaction plane: $P_L=(I_{\parallel}-I_{\perp})/(I_{\parallel}+I_{\perp})$.

For the computation of the partial excitation cross sections and the degree of linear polarization, we made use of a fully relativistic distorted—wave(RDW) REIE06 program[18,22,28] which was developed recently. The initial—and final—state wave functions of target ions are generated with the use of the atomic structure package GRASP92 based on the multiconfiguration Dirac—Fock(MCDF) method [27]. Here, we considered the configurations $1s^2 2s^2$, $1s2s^2 2p$, and $1s^2 2p^2$ which give rise to a total of 10 levels. The continuum electron wave functions are produced by the component COWF of the RATIP package [29,30] by solving the coupled Dirac equation in which the exchange effect between the bound and continuum electrons are considered. In the calculations of cross sections, partial waves with maximal angular momentum k=50 have been taken into account in order to ensure the convergence of the final results. To analyze the contribution of the Breit interaction to the excitation cross section, we have carried out two kinds of calculations. For the Coulomb calculations (labeled with "C"), we just use the Coulomb excitation energies and the Coulomb operator in the electron−impact excitation matrix elements. For the Coulomb−plus−Breit calculations (labeled with "C+B"), the Breit interaction is included in addition in the computation of both the excitation energies and the electron−impact excitation matrix elements. The calculated excitation cross sections were then used to compute the polarization parameter P_L of the subsequent $M2$ line according to Eq. (4).

In order to illustrate the accuracy of the present calculations, Table 1 compares the cross sections from this work with previous relativistic results[31]; data are shown especially for the excitation cross section from the ground state to the $1s2s^2 2p_{3/2} J=2$ excited state for the beryllium−like Mo^{38+} ion at 20 keV energy of the incident electron. Excellent agreement is found for the two computations for the total and magnetic sublevel cross sections if the $e-e$ interaction (Coulomb + Breit) in Eq. (3) is taken fully into account. As seen from this table, the influences of the Breit interaction on the excitation cross sections to the $1s2s^2 2p_{3/2} J=2$ state are relatively small at present specific low energy (nearly 1.1 times the threshold energy). For example, the total cross section increases by about 6% when the contribution of the Breit interaction isconsidered.

Table 1. The total and magnetic sublevels cross sections (in barn; 1 barn = 10^{-24} cm^2) for excitation from the ground stateto the individual magnetic sublevels M_f of the excited state $1s2s^2 2p_{3/2} J=2$ for the beryllium−like Mo^{38+} ion. The incident electron energy is 20 keV.

	Methods	M_f			Total
		0	±1	±2	
Previous	C+B[31]	6.31	5.15	1.68	19.97
Present	C	6.17	5.05	1.65	19.57
	C+B	6.98	5.17	1.67	20.67

Figure 1 displays the total cross sections from the ground state $1s^2 2s^2 J=0$ to the excited state $1s2s^2 2p_{3/2} J=2$ for the beryllium−like Mo^{38+}, Ho^{63+}, and U^{88+} ions as function of incident electron energy in threshold units. For an increasing energy of the incident electrons, the total cross sections decrease monotonically in a similar way for both case, namely the complete Coulomb－plus－Breit interaction as well as if only the Coulomb repulsion is taken into account. While the Breit interaction generally increases the cross section for this transition, the differences between these two computational models become more apparent with increasing nuclear charge Z. For a given beryllium－like ion, moreover, the influence of Breit interaction becomes more evident if the incident electron energy increases.

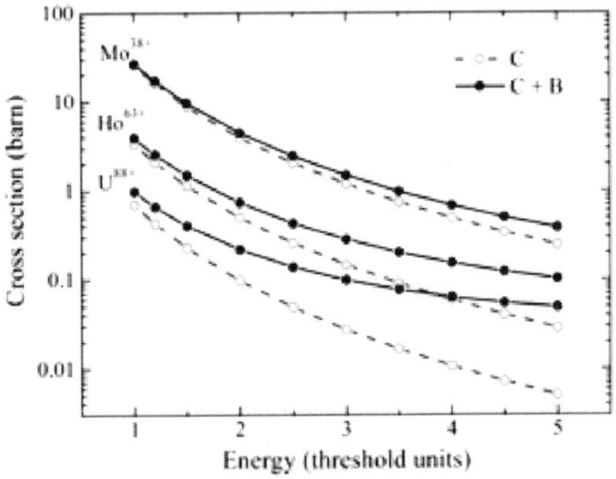

Fig. 1. Total electron－impact excitation cross sections (in barn) from the ground state to the $1s2s^2 2p_{3/2}$ $J=2$ excited state for the beryllium－like Mo^{38+}, Ho^{63+} and U^{88+} ions as function of the incident electron energy in threshold units.

In Figs. 2(a)－2(c), we show the influence of the Breit interaction on the cross sections for excitation into individual magnetic sublevels of the $1s2s^2 2p_{3/2}$ $J=2$ excited state of the beryllium－like Mo^{38+}, Ho^{63+}, and U^{88+} ions. For both computational models, with and without the Breit interaction cases in the excitation amplitudes, the cross sections for the three magnetic sublevels decrease monotonically with the increasing of incident electron energy. If only the Coulomb repulsion is included, for instance, the $M_f=0$ sublevel is preferentially populated at given incident electron energies for the beryllium－like Mo^{38+}, Ho^{63+}, and U^{88+} ions. However, this dominance becomes weaker for increasing nuclear charge Z owing the excitation of the $M_f=\pm 1$ sublevels. This is similar to the case of the helium－like ions where the magnetic sublevel with the smaller magnetic quantum number also is preferentially populated [10,18]. When the Breit interaction is taken into account, it is found that the Breit interaction increases the cross section for electron—impact excitation into the sublevels with $M_f=0, \pm1, \pm2$ for any given energy of the incident electrons, and that its effects are enhanced as the electron energy is increased. It is also found that the influences of Breit interaction on cross sections for each sublevel become more evident as nuclear charge Z increases. Especially for the beryllium—like U^{88+} ion, the population of the magnetic sublevel changes substantially at higher energies of the incident electrons. For example, the two curves of cross sections for the sublevels $M_f=0$ and $M_f=1$ intersect each other at about 3 times the threshold energy, while they cross at about 4.5 times the threshold energy for $M_f=1$ and $M_f=2$, respectively.

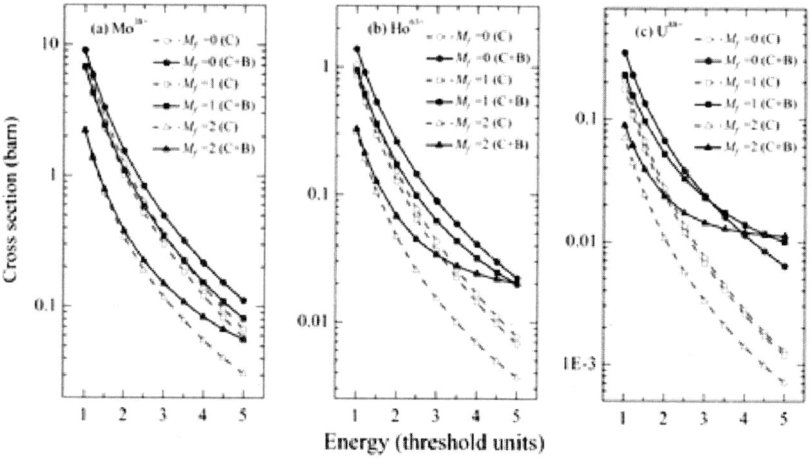

Fig. 2. The same as figure 1 but for the individual magnetic sublevels $M_f=0$, $M_f=\pm1$ and $M_f=\pm2$ of the $1s2s^2 2p_{3/2} J=2$ excited state: (a) Mo^{38+}, (b) Ho^{63+} and (c) U^{88+}.

Since the degree of linear polarization is determined uniquely by the cross sections σ_2 and σ_1 according to Eq. (4), it is sensitive to the population of the magnetic sublevels. A remarkable effect of the Breit interaction is predicted for the linear polarization of the $1s2s^2 2p_{3/2} J=2 \rightarrow 1s^2 2s^2 J=0$ $M2$ line. For the photon emission perpendicular to the quantization axis z, Fig. 3 displays the degree of linear polarization of this $M2$ line in the rest frame of the target as function of incident electron energy for the different beryllium—like ions. While the degree of linear polarization increases in its absolute value for both computational models, with and without the Breit interaction, near to the threshold, it later starts to decrease again at higher energies of the incident electrons for beryllium—like Mo^{38+} and Ho^{63+} ions. As seen from this figure, the Breit interaction decreases the degree of linear polarization for each ion and at any given energy of the incident electrons. Therefore, a strong depolarization effect due to the Breit interaction becomes evident especially if the electron energy is enhanced. For beryllium—like U^{88+} ion, the Breit interaction even causes a change of the sign of the linear polarization at about 4.5 times the threshold energy. The reason can be seen clearly from the combination of Fig. 2(c) and Eq. (4): The crossing of the $M_f=1$ and $M_f=2$ cross sections at about 4.5 times threshold energies for the U^{88+} ions results in a sign change of the linear polarization.

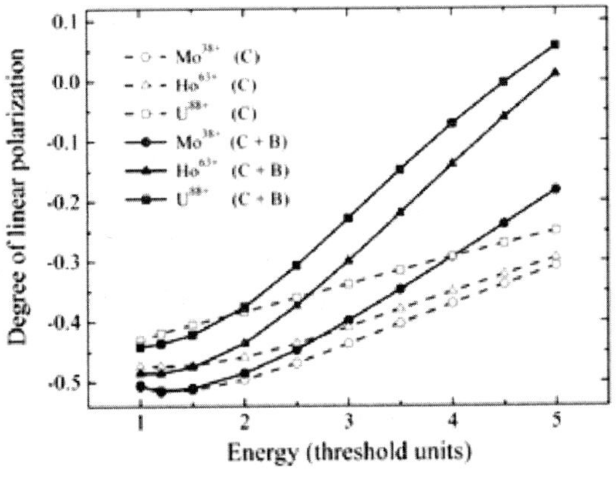

Fig. 3. Degree of linear polarization of the $1s2s^2 2p_{3/2} J=2 \rightarrow 1s^2 2s^2$ $J=0$ *M*2 line for the beryllium—like Mo^{38+}, Ho^{63+} and U^{88+} ions as function of incident electron energy in threshold units.

Fig. 3 displays how the Breit interaction affects the degree of linear polarization of the $1s2s^2 2p_{3/2} J=2$ *M*2 line for three different nuclear charges and as the energy of the incident electron is increased. To provide a more detailed account on these relativistic effects, Fig. 4 shows the degree of linear polarization as function of the nuclear charge $Z(42 \leqslant Z \leqslant 92)$ for incident electron energies of just 5 times the corresponding threshold energy. While the linear polarization of the *M*2 line decreases slowly (in magnitude) with increasing nuclear charge of the ions if only the Coulomb repulsion is included, a much stronger depolarization is found for this transition for the full account of the $e-e$ interaction (C+B). For Z ≈ 66, moreover, the linear polarization of the $M2$ line becomes *zero* and changes its behaviour from a dominantly perpendicular to an *in-plane* polarization with regard to the scattering plane if the nuclear charge Z is further increased.

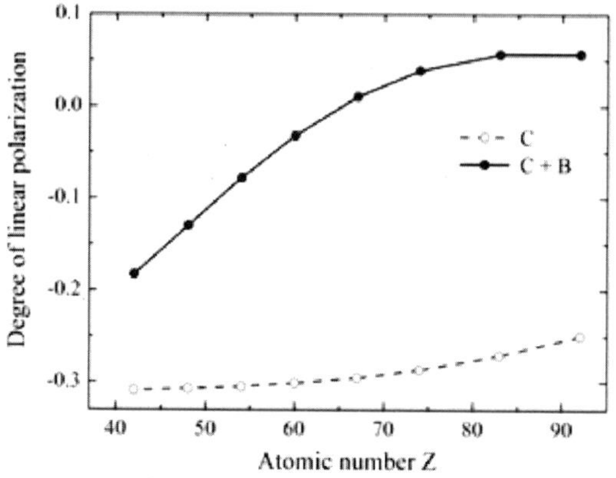

Fig. 4. Degree of linear polarization of the $1s2s^2 2p_{3/2} J=2 \rightarrow 1s^2 2s^2 J=0$ $M2$ line for the beryllium-like ions as function of the atomic number at 5 times the threshold energy. Calculations that only included the Coulomb repulsion in the electron-impact excitation amplitudes (C, dashed line) are compared with the full account of the Coulomb-plus-Breit interaction (C+B, solid line).

In conclusion, the influence of the Breit interaction on the $1s \rightarrow 2p$ electron-impact excitation cross sections and the degree of linear polarization of the subsequent $1s2s^2 2p_{3/2} J=2$ $M2$ x-ray emission has been analyzed for highly-charged beryllium-like ions. Using MCDF wave functions in the framework of the fully relativistic distorted-wave (RDW) theory, computations are performed for the excitation cross sections to the magnetic sublevels $M_f=0$, $M_f=\pm 1$, and $M_f=\pm 2$ of the $1s2s^2 2p_{3/2} J=2$ excited state for beryllium-like ions. These cross sections provide the valuable information about the dynamical excitation process and the magnetic sublevel population of the excited states, which are then employed in calculating the degree of linear polarization for the $1s2s^2 2p_{3/2} J=2 \rightarrow 1s^2 2s^2 J=0$ $M2$ emission. It is found that the Breit interaction considerably alters the degree of linear polarization, and that the changes resulting from the Breit interaction become progressively more significant with increasing of incident electron energy and/or atomic number Z. In particular, a qualitative change in the polarization at electron energies of five times the threshold is predicted for $Z \geqslant 66$ if the complete "Coulomb+Breit" interaction is taken into account in the computation of the excitation amplitudes. With the recent advancements in the design of position—sensitive solid—state detectors, such a strong relativistic effect will become measurable in experiment with the present—day facilities, both at heavy—ion storage rings and EBIT devices. This may provide a promising and alternative route for studying the relativistic corrections to the $e-e$ interaction in electron—ions collisions.

Acknowledgements: This word has been supported by the National Natural Science Foundation of China under Grant Nos. 11274254, 91126007. Dr. Ying—Long Shi would like to thanks Professor Fumihiro Koike and Dr. Andrey Surzhykov for their warmly help about the manuscript and density matrix theory.

References

[1]Berezhko E G and Kabachnik N M 1977*J. Phys. B: At. Mol. Phys.* 10 2467

[2]Haug E 1981*Sol. Phys.* 71 77

[3]Kieffer J C, Matte J P, Chaker M et al 1993*Phys. Rev. E*48 4648

[4]Shlyaptseva A S, Hansen S B, Kantsyrev V L et al 2001*Rev. Sci. Instrum.* 72 1241

[5]Beiersdorfer P, Phillips T W, Wong K L et al 1992*Phys. Rev. A*46 3812

[6]Henderson J R, Beiersdorfer P, Bennett C L et al 1990*Phys. Rev. Lett.* 65 705

[7]Takács E, Meyer E S, Gillaspy J D et al. 1996*Phys. Rev. A*54 1342

[8]Nakamura N, Kato D, Miura N et al 2001*Phys. Rev. A*63 024501

[9]Robbins D L, Beiersdorfer P, Faenov A Ya et al 2006*Phys. Rev. A*74 022713

[10]Inal M K and Dubau J 1987*J. Phys. B: At. Mol. Phys.* 20 4221

[11]Itikawa Y, Srivastave R and Sakimoto K 1991*Phys. Rev. A*44 7195

[12]Chen C Y, Qi J B, Wang Y S, Xu X J and Sun Y S 2000*Chin. Phys. Lett.* 17 403

[13]Zeng J L, Zhao G and Yuan J M 2005*Chin. Phys. Lett.* 22 1972

[14]Kai T, Nakazaki S, Kawamura T, Nishimura H and Mima K 2007*Phys. Rev. A*75 062710

[15]Wu Z Q, Li Y M, Duan B, Zhang H, Yan J 2009*Chin. Phys. Lett.* 26 123202

[16]Chen C Y, Wang K, Huang M et al 2010*J. Quant. Spectrosc. Radiat. Transfer*111

843

[17]Jiang J,Dong C Z,Xie L Y and Wang J G 2008*Phys. Rev.* A 78 022709

[18]Reed K J and Chen M H 1993*Phys. Rev. A*48 3644

[19]Fontes C J,Zhang H L and Sampson 1999*Phys. Rev.* A 59 295

[20]Bostock C J,Fursa D V and Bray I 2009*Phys. Rev.* A 80 052708

[21]Wu Z W,Jiang J and Dong C Z 2011*Phys. Rev.* A 84 032713

[22]Fritzsche S,Surzhykov A and Stöhlker Th 2009*Phys. Rev. Lett.* 103 113001

[23]Weber G,Bräuning H,Surzhykov A et al 2010 *Phys. Rev. Lett.* 105 243002

[24]Nakano Y 2012 Private communication

[25]Zhang H L,Sampson D H and Clark R E H 1990*Phys. Rev.* A 41 198

[26]Zhang H L,Sampson D H and Mohanty A K 1989*Phys. Rev.* A 40616

[27]Grant I P 2007 *Relativistic Quantum Theory of Atoms and Molecules: Theory and Computation*(New York:Springer)

[28]Jiang J,Dong C Z,Xie L Y et al 2007 *Chin. Phys. Lett.* 24 691

[29]Fritzsche S 2001 *J. Electron Spectrosc. Relat. Phenom.* 114—116 1155

[30]Fritzsche S 2012 *Comput. Phys. Commun.* 183 1525

[31]Chen M H and Reed K J 1994 *Phys. Rev.* A 50 2279

(本文发表于 2013 年《Chinese Physics Letter》30 卷 6 期)

Experimental Observation of Topological Edge States at the Surface Step Edge of the Topological Insulator $ZrTe_5$

Xiang—Bing Li　Wen—Kai Huang　Yang—Yang Lv
Kai—Wen Zhang　Chao—Long Yang　Bin—Bin Zhang
Y. B. Chen　Shu—Hua Yao　Jian Zhou　Ming—Hui Lu
Li Sheng　Shao—Chun Li　Jin—Feng Jia　Qi—Kun Xue
Yan—Feng Chen　Ding—Yu Xing*

摘要:我们利用扫描隧道显微镜报到了 $ZrTe_5$ 原子级的特征。我们观测到约 80 meV 的体带隙和台阶边缘处位于能隙中的拓扑边缘态,因此,证明了 $ZrTe_5$ 是二维拓扑绝缘体。我们还发现,施加的磁场引起拓扑边缘态的劈裂,归因于拓扑边缘态与体能态之间强关联效应。$ZrTe_5$ 相对较大的带隙,使得其在未来的基础研究和器件应用中具有潜在的应用价值。

We report an atomic—scale characterization of $ZrTe_5$ by using scanning tunneling microscopy. We observe a bulk band gap of ~80 meV with topological edge states at the step edge and, thus, demonstrate that $ZrTe_5$ is a two—dimensional topological insulator. We also find that an applied magnetic field induces an energetic splitting of the topological edge states, which can be attributed to a strong link between thetopological edge states and bulk topology. The relatively large band gap makes $ZrTe_5$ a potential candidate for future fundamental studies and device applications.

---

* 作者简介:李向兵(1986—),男,甘肃张家川人,天水师范学院电子信息与电气工程学院讲师,博士,主要从事新型量子拓扑材料新奇物理效应的研究。

A two-dimensional topological insulator(2D TI)is a new type of quantum matter and hosts the quantumspinHall effect(QSHE)[1,2]. It features an insulating bulk band gap and time-reversal-invariant topological edge states(TESs) protected against localization and backscattering. Since the discovery of the QSHE in the inverted HgTe/CdTe quantum wells[3,4],progress has been made in the predictions and characterizations of 2D TI materials[5-21]. Most of the currently confirmed 2D TIs are of limited practical use,due to either small bulk band gaps or difficulty in achieving thin sheets down to a single layer. Searching for a 2D TI material with practically ideal properties is still of great importance, which,however,is rather challenging. Furthermore,the response of TESs to an applied magnetic field has not been fully understood. Theoretical models have been established to address the spatial distribution of TESs under magnetic fields[22,23],but the energetic evolution has never been studied. Tuning the external magnetic fields or gating voltages,the transition from a 2D TI to a normal insulator or a quantum Hall state could be realized[4,21].

Recently,Weng et al.[24]predicted that single layer $ZrTe_5$ is a 2D TI with a large band gap,and the 3D crystal of $ZrTe_5$ is located near the phase boundary between weak and strong topological insulators sensitive to lattice parameters. In contrast,several experimental studies suggested that $ZrTe_5$ might be a Dirac semimetal[25-27]. Therefore,a direct characterization is rather important to clarify the bulk band topology of $ZrTe_5$. In fact,studies of $ZrTe_5$ can be tracked back two decades due to its anomalous magnetoresistance and low-temperature thermoelectric power[28-30].

In this study,we use scanning tunneling microscopy(STM)to characterize the surface of cleaved single crystal $ZrTe_5$. We observe not only a gapped bulk band structure at the surface terrace of the a-c plane but also TESs located at the step edge. It then follows that the $ZrTe_5$ surface is a 2D TI,suggesting that the $ZrTe_5$ crystal is a weak 3D TI,in good agreement with the theoretical prediction[24]. In the presence of magnetic field,the edge states undergo a large energy splitting,which could be explained by time-reversal-symmetry-broken edge states closely linked to the bulk band topology[22,23,31].

Single crystal $ZrTe_5$ was grown by the chemical vapor transport method with iodine(I2)as the transport agent. Polycrystalline $ZrTe_5$ was first synthesized by a solid—state reaction(at about 500 °C for seven days)in a fused silica tube sealed under vacuum (~ 4 × 10^{-6} mbar). A ratio of 1: 5 for the high purity Zr (99.999%) and Te powder (99.999%) was adopted. The mixture of prepared $ZrTe_5$ polycrystal—line and I2(~5 mg/L)powder were loaded into a sealed evacuated quartz tube and then put into a two—zone furnace. After growing in a temperature profile of 520 °C—450 °C for over ten days, the single crystal was successfully grown, with a typical size of ~35 × 1 × 0.5 mm^3.

All STM characterizations were carried out in ultrahigh vacuum(UHV)with a Unisoku LT—STM at~4 K. The base pressure was 5×10^{-11} mbar. The $ZrTe_5$ single crystal was cleaved in situ in UHV and quickly cooled down to 4 K prior to the STM and STS measurements. The constantcurrent mode was adopted. dI/dV spectra were taken using a lock—in amplifier. A modulation of 5 mV at 1000 Hz was applied. A mechanically polished PtIr tip was used.

The $ZrTe_5$ crystal has an orthorhombic layered structure[32] and contains 2D sheets of $ZrTe_5$ in the a—c plane, which stack along the b axis via interlayer van der Waals interactions, as shown in Fig. 1(a). Each 2D sheet consists of alternating prismatic $ZrTe_3$ chains along the a axis that are linked by parallel zigzag Te chains. The prism of $ZrTe_3$ is formed by a dimer of Te and an apical Te atom surrounding a Zr atom.

$ZrTe_5$ is easily cleavable along the a—c plane, while it also exhibits a quasi—1D preference along the a axis. Figures 1(b)and 1(c)show the topography of such a cleaved $ZrTe_5$ surface, in which Te dimer and Te apical atoms can be well identified. The zigzag chain of Te atoms is right intercalated in between. Note, no charge density wave phase is observed at either 4 or 80 K. The steps are dominantly along the a axis due to the quasi—1D preference, and they are superstraight in mesoscopic scale(approx—imately a few μm; see Ref. [33] Fig. S1). Such long and straight surface steps make the contact to other quantum material, such as superconductor or ferromagnetic metal, easily controllable. The measured height of a single step is~0.8 nm. The present STM results are in good agreement with the crystal structure of $ZrTe_5$[24]. The steps along thec axis are seldom observed, but they are not straight at all(see Ref. [33] Fig. S2).

The local density of states(LDOS, from spectroscopic measurement)obtained at the surface terrace is plotted in Fig. 1(d)and Ref. [33] Fig. S3. Its high homogeneity indicates the high quality of the $ZrTe_5$ single crystal. Anenergy gap as large as ~80 meV is identified, with the topof the valence band and the bottom of the conductance band located at ~35 and ~45 meV, respectively. This result will be further proven below by the quasiparticle interference analysis at the step edges. The Fermi level is located inside the band gap, but slightly towards thevalence band, indicating that the surface terrace is a nearly intrinsic semiconductor. The spectroscopic feature at the $ZrTe_5$ terrace agrees well with the density-functional theory calculation for single layer $ZrTe_5$[24]. Furthermore, no observable residue states were detected in the band gap region, supporting the model of weak TI rather than strong TI. If it were a strong TI, the residual intensity due to topological surface states would be detectable within the energy gap[24].

To ascertain the electronic structure at the step edge, the LDOS along a line perpendicular to the step edge is plotted in Fig. 2(a). Clearly, the quasiparticle interference due to confinement by the step is observed in the valence and conductance band regions, indicating a 2D-like bulk band structure. In Fig. 2(b), the corresponding 1D fast Fourier transform shows the scattering interference dispersion. The dispersion below the Fermi energy resembles the recent ARUPS results[25]. An energy gap can be also identified between the dispersions for the valence band and conductance band. When approaching the step edge, a couple of new discontinuous peaks appear in the gap region (see, also, the in-gap feature in the fast Fourier transform). In contrast to the spectra at the terrace, the intensity is pronouncedly enhanced near the step edge and spans thewhole gap region, indicating the existence of TESs[1,2]. The penetration depth, as measured in the profile perpendicular to the step edge in Fig. 2(c), is ~7.7 and ~6.5 nm for energy near and away from the Fermi level, respectively. This subtle difference can be understood bythe argument that the wave function of the edge state penetrates further into the terrace as its energy is closer to the valence or conductance band. The STS measurements at the step along the c axis are analyzed in the Supplemental Material[33] Fig. S4.

Figure 2(d) shows two STM spectra obtained in thesurface terrace region [black triangle in Fig. 2(a)] and at the step edge [red triangle in Fig. 2(a)], respectively. The broad bumps located in the valence and conductance band regions arise from the interference at the step edge. The peaks in the gap region are

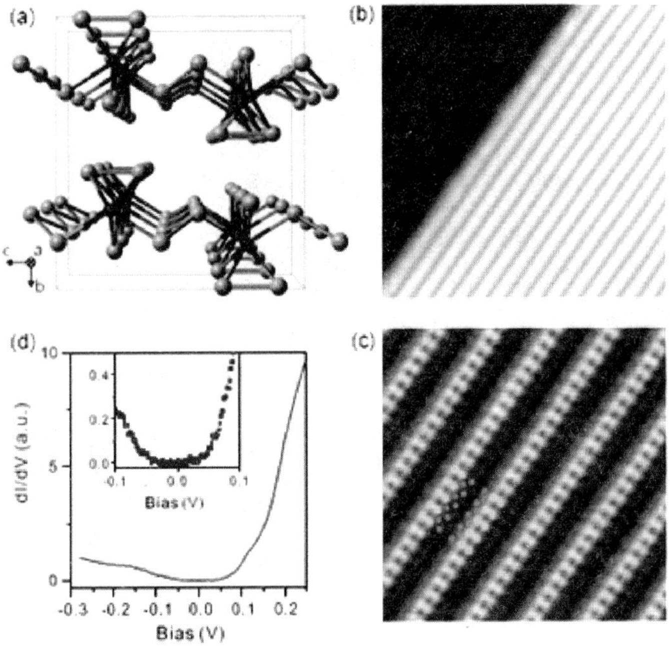

FIG. 1. (a) Crystal structure of $ZrTe_5$. (b) STM image of 25×25 nm^2 obtained at 4 K (bias voltage U=1.0 V, tunneling current It=100 pA). The height of the monolayer step is ~ 0.8 nm. (c) Atomic-resolution image of the $ZrTe_5$ surface (8×8 nm^2, U=350 mV, It=130 pA). The yellow balls mark the Te dimers, the red apical Te atoms, and the green zigzag Te atoms. (d) Differential conductance (dI=dV spectra, U=250 mV, It=200 pA, Vosc=5 mV) taken over the terrace and far away from the step edge.

asymmetric, and the 1D character can be further confirmed by fitting the peaks with a 1D density of states exhibiting an inverse-square-root singularity. This characteristic of the edge states recalls the theoreticalcalculations for two different types of cuttings of an isolated $ZrTe_5$ single layer[24]. According to the calculated results, the observation of inverse-square-root singularities is expected for the step edge terminated by the zigzag chain rather than by the prismatic chain[24]. Our spectra are more consistent with the zigzag chain terminated steps. Because of the tip curvature effect, it isdifficult to identify the exact atomic geometry at the step edge.

The 1D character of the edge states can be further revealed by the spatially resolved spectroscopic mapping along another step. The edge states stay exactly

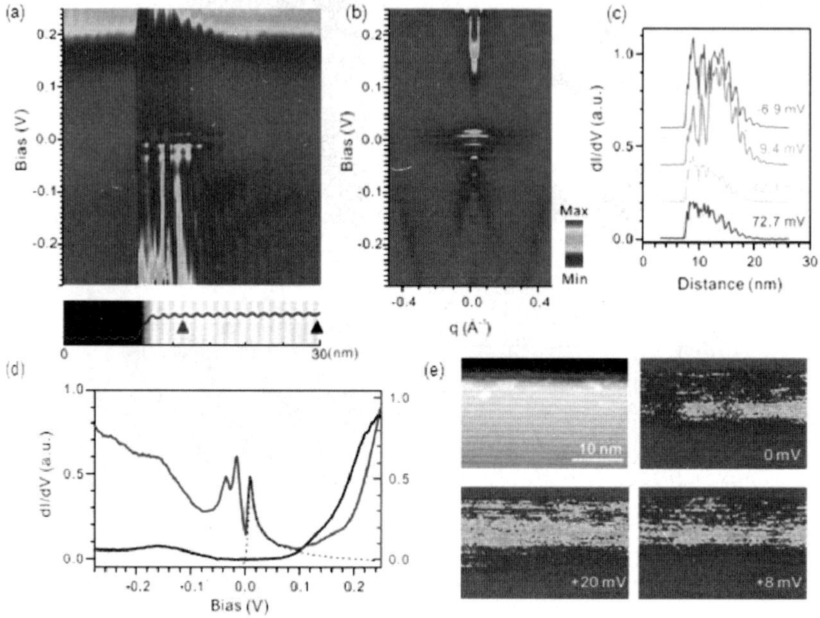

FIG. 2. (a) Differential conductance (dI/dV spectra, U = 250 mV, It = 200 pA, Vosc = 5 mV) taken across the step. The corresponding step topography is shown below, and the red line profile illustrates the step geometry (30 nm in length). (b) 1D fast Fourier transform of (a) showing the quasiparticle interference. (c) Line cuts at various energies extracted from (a) showing the variation of penetration depth over energy. (d) Two spectra extracted from (a) at the positions marked by black and red triangles. The dotted blue line is the fitting results with a 1D density of states exhibiting the inverse—square—root singularity. The broadening used is ~3 meV. (e) STM topography (34×23 nm^2, U=1.05 V, It=100 pA) of another step and the corresponding dI/dV mapping taken atvarious bias voltages.

along the step edge, as shown in Fig. 2(e). Since the penetration depth of the edge states is relatively large, the edge channels can circumvent a large—sized perturbation of a few nm, as shown in Fig. 2(e) and Ref.[33] Fig. S5, which is just comparable to the spatial resolution limits for microfabrication techniques. Moreover, the edge states can be observed at all kinds of steps (see Ref.[33] Fig. S5), unlike in the Bi bilayer where the edge states are sensitively coupled to the step geometry[19]. Therefore, the edge states can, in principle, form looped electron channels along the periphery of a $ZrTe_5$ sheet, which can be easily made by microfabrication technology and are rather robust against external perturbations.

An applied magnetic field normal to the surface will break the time—reversal symmetry and drive a complicated evolution of the edge state, as shown in Fig. 3(a)(see, also, Ref. [33] Fig. S6). Three characteristic peaks marked as A, B, and C are tracked for increased magnetic fields. Each peak at energy E is split into two branches with energy difference ΔE, with the corresponding energies shifting up and down, respectively. The intensity for the higher—energy branch is lower than that for the lower—energy one, under relatively high fields. We rule out the possibility of observed bulk Landau quantization based on the following arguments. First, the peak splitting emerges mainly at the step edge, not observable away from the step edge[see Fig. 3(c)]. Second, the splitting is prominent in the gap region where there is no bulk band state. Finally, the evolutions of E [Fig. 3(b)and Ref. [33] Fig. S7]and ΔE[Fig. 3(d)and Ref. [33] Fig. S7]with applied field show neither linear nor parabolic dependence. If the evolutions are the bulk effect, both E vs B and ΔE vs B will exhibit either a linear or parabolic behavior, respectively, corresponding to the bulk Dirac or normal electrons. As a result, the STS spectra in Figs. 3(b)and 3(d) and Ref. [33] Fig. S7 do not come from the bulk band structure but from the TESs. For edge state A_1, the above argument seemssomewhat lacking of convincing evidence, since its energy is close to the top of the valence band, and there may be a coupling between edge state A_1 and the bulk state. However, states B and C are right inside the band gap and well isolated from the bulk. It was theoretically suggested that an applied magnetic field may open a tiny gap of the edge states at the Dirac point. However, such a tiny gap cannot be distinguished in the present experiment due to the complication of the STS spectra.

Owing to the time—reversal symmetry, the two helical edge states have the symmetric dispersion for opposite momentums, k and k, and the identical spatial distribution. Thus, the two edge states are not distinguishable by STM in the absence of a magnetic field. When the time—reversal symmetry is broken by an external magnetic field, it was theoretically suggested that the two helical edge states persist but are delocalized from the step edge. The favored edge state is pushed further toward the stepedge and the unfavored one gradually merges into the bulk[22,23,31]. This scenario can qualitatively explain the decrease in intensity

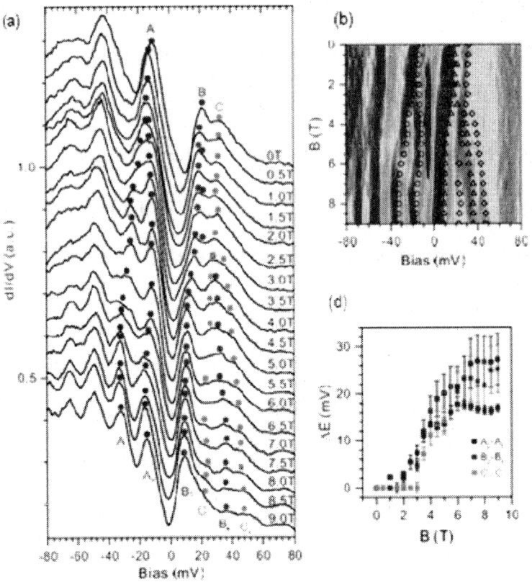

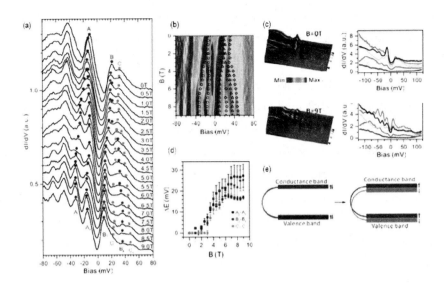

FIG. 3. (a) Differential conductance(*dI/dV* spectra) obtained at a certain position upon magnetic field, varying from 0 to 9 T, with an increment of 0. 5 T. The three peaks in the band gap region are marked as A, B, and C and traced via colored dots, where A splits into A_1, A_2, the same for B and C. (b) The same spectra after differentiation is plotted in false colored mode. The peaks A, B, and C are marked by open circles, triangles, and diamonds. (c) *dI/dV* spectra(*U* =250 mV, *It* =200 pA, *Vosc* =5 mV) taken across the step edge under the magnetic field of 0(up) and 9 T(down), respectively. Five spectra are extracted and plotted in the right. The corresponding positions for taking the spectra are marked with colored triangles in the left. (d) Energy gap defined as ΔE = E2 − E1, vs B. (e) Sketch illustrating the close link between the topological edge states splitting and the magnetic—field—induced bulk quantization. Left and right are sketches without and with an applied magnetic field, respectively.

for the higher—energy branch of two splitting peaks. Regardless of the nonlinear dependence ofΔE on B[see Fig. 3(d)], a fitting with $\Delta E = g\mu_B B$ roughly gives an effective g factor of 40—60, suggesting that the splitting arises from not only the Zeeman effect but also theorbital effect due to vector potential[27,34]. Furthermore, the effective g factor is comparable to the value for bulk Landau quantization obtained in a magnetoinfrared spectroscopy study[27]. Therefore, we conclude that the edge state evolution is highly correlated with the bulk band topology. The schematic shown in Fig. 3(e) illustrates the magnetic—field—induced Landau quantization and Zeeman splitting of the bulk band, which results in an effective splitting of the corresponding edge states. This simple model can qualitatively explain the experiment, but, in fact, the evolution will be much more complicated if the bulk band and edge state symmetries are taken into account, as seen from the unusual dependence of ΔE on B. Further theoretical and experimental investigations, especially under higher magnetic field, are required to fully under—stand the physical mechanism.

In summary, we have shown that the $ZrTe_5$ surface is a 2D TI with bulk band gap and topological edge stateslocated at the surface step edge. This suggests that the$ZrTe_5$ is a weak 3D TI rather than a Dirac semimetal. This material is easily cleavable, the band gap of the surface terrace is relatively large, and the surface step edges obtained are ultralong and ultrastraight. The topological edge states are found rather robust against nonmagnetic perturbations. In addition, we have also studied the evo—lution of the edge states in the presence of an increased magnetic field, which can be understood by a theoretical model of the time—reversal—symmetry—broken topological edge states.

This work was supported by the State Key Program for Basic Research of China (Grants No. 2014CB921103, No. 2015CB921203, and No. 2013CB922103), NationalNatural Science Foundation of China (Grants No. 11374140, No. 11374149, No. 51032003, No. 50632030, No. 10974083, No. 51002074, No. 10904092, and No. 51472112), and the New Century Excellent Talents in University (Grant No. NCET—09—0451). X. —B. L., W. —K. H., and Y. —Y. L contributed equally to this work.

Note added.—Recently, we found that similar conclusions were reached independently in Ref.[35].

References

[1]M. Z. Hasan and C. L. Kane, Rev. Mod. Phys. 82, 3045(2010).

[2]X. —L. Qi and S. —C. Zhang, Rev. Mod. Phys. 83, 1057(2011).

[3]B. A. Bernevig, T. L. Hughes, and S. —C. Zhang, Science314, 1757(2006).

[4]M. König, S. Wiedmann, C. Brüne, A. Roth, H. Buhmann, L. W. Molenkamp, X. —L. Qi, and S. —C. Zhang, Science 318, 766(2007).

[5]X. Qian, J. Liu, L. Fu, and J. Li, Science 346, 1344(2014).

[6]S. M. Nie, Z. D. Song, H. M. Weng, and Z. Fang, Phys. Rev. B 91, 235434(2015).

[7]M. A. Cazalilla, H. Ochoa, and F. Guinea, Phys. Rev. Lett. 113, 077201(2014).

[8]R. Roy, Phys. Rev. B 79, 195322(2009).

[9]C. C. Liu, W. X. Feng, and Y. G. Yao, Phys. Rev. Lett. 107, 076802(2011).

[10]Y. Xu, B. H. Yan, H. J. Zhang, J. Wang, G. Xu, P. Z. Tang, W. H. Duan, and S. C. Zhang, Phys. Rev. Lett. 111, 136804(2013).

[11]C. L. Kane and E. J. Mele, Phys. Rev. Lett. 95, 226801(2005).

[12]A. Shitade, H. Katsura, J. Kunes, X. L. Qi, S. C. Zhang, and N. Nagaosa, Phys. Rev. Lett. 102, 256403(2009).

[13]C. Liu, T. L. Hughes, X. －L. Qi, K. Wang, and S. －C. Zhang, Phys. Rev. Lett. 100, 236601(2008).

[14]S. Murakami, Phys. Rev. Lett. 97, 236805(2006).

[15]B. Rasche, A. Isaeva, M. Ruck, S. Borisenko, V. Zabolotnyy, B. Buchner, K. Koepernik, C. Ortix, M. Richter, and J. van den Brink, Nat. Mater. 12, 422(2013).

[16]A. Roth, C. Brune, H. Buhmann, L. W. Molenkamp, J. Maciejko, X. L. Qi, and S. C. Zhang, Science 325, 294(2009).

[17]F. F. Zhu, W. J. Chen, Y. Xu, C. L. Gao, D. D. Guan, C. H. Liu, D. Qian, S. C. Zhang, and J. F. Jia, Nat. Mater. 14, 1020(2015).

[18]C. Pauly, B. Rasche, K. Koepernik, M. Liebmann, M. Pratzer, M. Richter, J. Kellner, M. Eschbach, B. Kaufmann, L. Plucinski, C. M. Schneider, M. Ruck, J. van den Brink, and M. Morgenstern, Nat. Phys. 11, 338(2015).

[19]I. K. Drozdov, A. Alexandradinata, S. Jeon, S. Nadj－Perge, H. Ji, R. J. Cava, B. Andrei Bernevig, and A. Yazdani, Nat. Phys. 10, 664(2014).

[20]F. Yang, L. Miao, Z. F. Wang, M. Y. Yao, F. Zhu, Y. R. Song, M. X. Wang, J. P. Xu, A. V. Fedorov, Z. Sun, G. B. Zhang, C. Liu, F. Liu, D. Qian, C. L. Gao, and J. F. Jia, Phys. Rev. Lett. 109, 016801(2012).

[21]I. Knez, R. －R. Du, and G. Sullivan, Phys. Rev. Lett. 107, 136603(2011).

[22]H. Li, L. Sheng, R. Shen, L. B. Shao, B. Wang, D. N. Sheng, and D. Y. Xing, Phys. Rev. Lett. 110, 266802(2013).

[23]G. Tkachov and E. M. Hankiewicz, Phys. Rev. Lett. 104, 166803(2010).

[24]H. Weng, X. Dai, and Z. Fang, Phys. Rev. X 4, 011002(2014).

[25]Q. Li, D. E. Kharzeev, C. Zhang, Y. Huang, I. Pletikosic, A. V. Fedorov, R. D. Zhong, J. A. Schneeloch, G. D. Gu, and T. Valla, arXiv:1412. 6543v1.

[26]R. Y. Chen, S. J. Zhang, J. A. Schneeloch, C. Zhang, Q. Li, G. D. Gu, and N. L. Wang, Phys. Rev. B 92, 075107(2015).

[27]R. Y. Chen, Z. G. Chen, X. Y. Song, J. A. Schneeloch, G. D. Gu, F. Wang, and N. L.

Wang, Phys. Rev. Lett. 115, 176404(2015).

[28]T. M. Tritt, N. D. Lowhorn, R. T. Littleton, A. Pope, C. R. Feger, and J. W. Kolis, Phys. Rev. B 60, 7816(1999).

[29]R. T. Littleton, T. M. Tritt, J. W. Kolis, and D. R. Ketchum, Phys. Rev. B 60, 13453 (1999).

[30]T. E. Jones, W. W. Fuller, T. J. Wieting, and F. Levy, Solid State Commun. 42, 793 (1982).

[31]H. Li, L. Sheng, and D. Y. Xing, Phys. Rev. Lett. 108, 196806(2012).

[32]H. Fjellvag and A. Kjekshus, Solid State Commun. 60, 91(1986).

[33]See Supplemental Material at http://link. aps. org/ supplemental/10. 1103/PhysRevLett. 116. 176803 for supple-mental STM and STS data taken on $ZrTe_5$ surface terrace and steps.

[34]S. Jeon, B. B. Zhou, A. Gyenis, B. E. Feldman, I. Kimchi, A. C. Potter, Q. D. Gibson, R. J. Cava, A. Vishwanath, and

A. Yazdani, Nat. Mater. 13, 851(2014).

[35]R. Wu, J. -Z. Ma, L. -X. Zhao, S. -M. Nie, X. Huang, J. -X. Yin, B. -B. Fu, P. Richard, G. -F. Chen, Z. Fang, X. Dai, H. -M. Weng, T. Qian, H. Ding, and S. H. Pan, arXiv:1601. 07056[Phys. Rev. X(to be published)].

（本文发表于 2016 年《Physical Review Letters》第 116 卷 17 期）

后 记

六十年风雨历程，六十年求索奋进。编辑出版《天水师范学院60周年校庆文库》（以下简称《文库》），是校庆系列活动之“学术华章”的精彩之笔。《文库》的出版，对传承大学之道，弘扬学术精神，展示学校学科建设和科学研究取得的成就，彰显学术传统，砥砺后学奋进等都具有重要意义。

春风化雨育桃李，弦歌不辍谱华章。天水师范学院在60年办学历程中，涌现出了一大批默默无闻、淡泊名利、潜心教学科研的教师，他们奋战在教学科研一线，为社会培养了近10万计的人才，公开发表学术论文10000多篇（其中，SCI、EI、CSSCI源刊论文1000多篇），出版专著600多部，其中不乏经得起历史检验和学术史考量的成果。为此，搭乘60周年校庆的东风，科研管理处根据学校校庆的总体规划，策划出版了这套校庆《文库》。

最初，我们打算策划出版校庆《文库》，主要是面向校内学术成果丰硕、在甘肃省内外乃至国内外有较大影响的学者，将其代表性学术成果以专著的形式呈现。经讨论，我们也初步拟选了10位教师，请其撰写书稿。后因时间紧迫，入选学者也感到在短时期内很难拿出文稿。因此，我们调整了《文库》的编纂思路，由原来出版知名学者论著，改为征集校内教师具有学科代表性和学术影响力的论文分卷结集出版。《文库》之所以仅选定教授或具有博士学位副教授且已发表在SCI、EI或CSSCI源刊的论文（已退休教授入选论文未作发表期刊级别的限制），主要是基于出版篇幅的考虑。如果征集全校教师的论文，可能卷帙浩繁，短时间内难以出版。在此，请论文未被《文库》收录的老师谅解。

原定《文库》的分卷书名为“文学卷”“史地卷”“政法卷”“商学卷”“教育卷”“体艺卷”“生物卷”“化学卷”“数理卷”“工程卷”，后出版社建议，总名称用“天水师范学院60周年校庆文库”，各分卷用反映收录论文内容的卷名。经编委会会议协商论证，分卷分别定为《现代性视域下的中国语言文学研究》《“一带一路”视域下的西北史地研究》《“一带一路”视域下的政治经济研究》《“一带一路”视域下的教师教育研究》《“一带一路”视域下的体育艺术研究》《生态文明视域下的生物学研究》《分子科学视域下的化学前沿问题研究》《现代科学思维视域下的数理问题研究》《新工科视域下的工程基础与应用研究》。由于收录论文来自不同学科领域、不同研究方向、不同作者，这些卷名不一定能准确反映所有论文的核心要义。但为出版策略计，还请相关论文作者体谅。

鉴于作者提交的论文质量较高，我们没有对内容做任何改动。但由于每本文集都有既定篇幅限制，我们对没有以学校为第一署名单位的论文和同一作者提交的多篇论文，在收录数量上做了限制。希望这些论文作者理解。

这套《文库》的出版得到了论文作者的积极响应，得到了学校领导的极大关怀，同时也得到了光明日报出版社的大力支持。在此，我们表示深切的感谢。《文库》论文征集、编校过程中，王弋博、王军、焦成瑾、贾来生、丁恒飞、杨红平、袁焜、刘晓斌、贾迎亮、付乔等老师做了大量的审校工作，以及刘勍、汪玉峰、赵玉祥、施海燕、杨婷、包文娟、吕婉灵等老师付出了大量心血，对他们的辛勤劳动和默默无闻的奉献致以崇高的敬意。

《天水师范学院60周年校庆文库》编委会

2019年8月